AF882025

Logic and Algebraic Structures in Quantum Computing

Arising from a special session held at the 2010 North American Annual Meeting of the ASL, this volume is an international cross-disciplinary collaboration with contributions from leading experts exploring connections across their respective fields. Themes range from philosophical examination of the foundations of physics and quantum logic, to exploitations of the methods and structures of operator theory, category theory, and knot theory in an effort to gain insight into the fundamental questions in quantum theory and logic.

The book will appeal to researchers and students working in related fields, including logicians, mathematicians, computer scientists, and physicists. A brief introduction provides essential background on quantum mechanics and category theory, which, together with a thematic selection of articles, may also serve as the basic material for a graduate course or seminar.

JENNIFER CHUBB is Assistant Professor of Mathematics at the University of San Francisco, where she teaches a wide range of courses, including quantum computing, to students in physics, computer science, and mathematics. She has a background in physics, dynamical systems, and pure and applied math. Her current research focuses on computable structure theory and algorithmic mathematics.

ALI ESKANDARIAN holds the positions of Dean and Professor at The George Washington University. He is a theoretical physicist and a founding member of the groups in astrophysics and quantum computing/information. He serves as co-director of the Center for Quantum Computing, Information, Logic, and Topology.

VALENTINA HARIZANOV is a Professor of Mathematics at The George Washington University, where she also serves as co-director of the Center for Quantum Computing, Information, Logic, and Topology. She is internationally recognized for her research in mathematical logic, particularly in computability theory and computable model theory.

LECTURE NOTES IN LOGIC

A Publication of The Association for Symbolic Logic

This series serves researchers, teachers, and students in the field of symbolic logic, broadly interpreted. The aim of the series is to bring publications to the logic community with the least possible delay and to provide rapid dissemination of the latest research. Scientific quality is the overriding criterion by which submissions are evaluated.

Editorial Board

Jeremy Avigad
Department of Philosophy, Carnegie Mellon University

Zoe Chatzidakis
DMA, Ecole Normale Supérieure, Paris

Peter Cholak, Managing Editor
Department of Mathematics, University of Notre Dame, Indiana

Volker Halbach
New College, University of Oxford

H. Dugald Macpherson
School of Mathematics, University of Leeds

Slawomir Solecki
Department of Mathematics, University of Illinois at Urbana-Champaign

Thomas Wilke
Institut für Informatik, Christian-Albrechts-Universität zu Kiel

More information, including a list of the books in the series, can be found at http://www.aslonline.org/books-lnl.html

LECTURE NOTES IN LOGIC 45

Logic and Algebraic Structures in Quantum Computing

Edited by

JENNIFER CHUBB

University of San Francisco

ALI ESKANDARIAN

George Washington University, Washington DC

VALENTINA HARIZANOV

George Washington University, Washington DC

ASSOCIATION FOR SYMBOLIC LOGIC

CAMBRIDGE
UNIVERSITY PRESS

CAMBRIDGE
UNIVERSITY PRESS

University Printing House, Cambridge CB2 8BS, United Kingdom

Cambridge University Press is part of the University of Cambridge.

It furthers the University's mission by disseminating knowledge in the pursuit of education, learning, and research at the highest international levels of excellence.

www.cambridge.org
Information on this title: www.cambridge.org/9781107033399

Association for Symbolic Logic
Richard Shore, Publisher
Department of Mathematics, Cornell University, Ithaca, NY 14853
http://www.aslonline.org

© Association for Symbolic Logic 2016

This publication is in copyright. Subject to statutory exception and to the provisions of relevant collective licensing agreements, no reproduction of any part may take place without the written permission of Cambridge University Press.

First published 2016

A catalogue record for this publication is available from the British Library.

Library of Congress Cataloguing in Publication Data
Names: Chubb, Jennifer. | Eskandarian, Ali. | Harizanov, Valentina S.
Title: Logic and algebraic structures in quantum computing / edited by Jennifer Chubb, University of San Francisco, Ali Eskandarian, George Washington University, Washington DC, Valentina Harizanov, George Washington University, Washington DC.
Description: Cambridge : Cambridge University Press, 2016. | Series: Lecture notes in logic | Includes bibliographical references and index.
Identifiers: LCCN 2015042942 | ISBN 9781107033399 (hardback : alk. paper)
Subjects: LCSH: Quantum computing–Mathematics. | Logic, Symbolic and mathematical. | Algebra, Abstract.
Classification: LCC QA76.889 .L655 2016 | DDC 006.3/843–dc23 LC record available at http://lccn.loc.gov/2015042942

ISBN 978-1-107-03339-9 Hardback

Cambridge University Press has no responsibility for the persistence or accuracy of URLs for external or third-party Internet Web sites referred to in this publication and does not guarantee that any content on such Web sites is, or will remain, accurate or appropriate.

CONTENTS

PREFACE

This project grew out of a Special Session on *Logic and the Foundations of Physics* at the 2010 North American Annual Meeting of the Association for Symbolic Logic[1]. Many of the session's lecturers investigated the role of algebraic structures in the context of the foundations of quantum physics, especially in quantum information and computation. In addition to this session, attendees heard tutorial lectures on quantum computing (given by Bob Coecke, University of Oxford) and an invited lecture on intuitionistic quantum logic (by Klaas Landsman, Radboud University, Nijmegen). The talks were so well-received by conference participants that we felt a volume of collected works on this subject would be a valuable addition to the literature.

The articles in this volume by mathematicians, philosophers, and scientists address foundational issues and fundamental abstract structures arising in highly active areas of theoretical, mathematical, and even experimental physics relevant to quantum information and quantum computation. We hope that the present collection advances this worthwhile program of scientific and mathematical progress.

We would like to thank the authors that contributed to this volume, and the ASL and Cambridge University Press for publishing it. This project was partially supported by the George Washington University Centers & Institutes Facilitating Fund Grant and by the University of San Francisco Faculty Development Fund. Many thanks also to Bryan Fregoso (a University of San Francisco student) for his invaluable assistance in assembling this volume.

Jennifer Chubb
Ali Eskandarian
Valentina Harizanov
Summer, 2015, Washington, D.C.

[1] The full program is available in the ***Bulletin of Symbolic Logic***, vol. 17 (2011), no. 1, pp. 135–137, available online at https://www.math.ucla.edu/~asl/bsl/1701-toc.htm.

INTRODUCTION

JENNIFER CHUBB, ALI ESKANDARIAN, AND VALENTINA HARIZANOV

In the last two decades, the scientific community has witnessed a surge in activity, interesting results, and notable progress in our conceptual understanding of computing and information based on the laws of quantum theory. One of the significant aspects of these developments has been an integration of several fields of inquiry that not long ago appeared to be evolving, more or less, along narrow disciplinary paths without any major overlap with each other. In the resulting body of work, investigators have revealed a deeper connection among the ideas and techniques of (apparently) disparate fields. As is evident from the title of this volume, logic, mathematics, physics, computer science and information theory are intricately involved in this fascinating story. The inquisitive reader might focus, perhaps, on the marriage of the most unlikely and intriguing fields of quantum theory and logic and ask: *Why quantum logic?*

By many, "logic" is deemed to be panacea for faulty intuition. It is often associated with the rules of correct thinking and decision-making, but not necessarily in its most sublime role as a deep intellectual subject underlying the validity of mathematical structures and worthy of investigation and discovery in its own right. Indeed, within the realm of the classical theories of nature, one may encounter situations that defy comprehension, should one hold to the intuition developed through experiencing familiar macroscopic scenarios in our routine impressions of natural phenomena.

One such example is a statement within the special theory of relativity that the speed of light is the same in all inertial frames. It certainly defies the common intuition regarding the observation of velocities of familiar objects in relative motion. One might be tempted to dismiss it as contrary to observation. However, while analyzing natural phenomena for objects moving close to the speed of light and, therefore, unfamiliar in the range of velocities we are normally accustomed to, logical deductions based on the postulates of the special relativity theory lead to the correct predictions of experimental observations.

There exists an undeniable interconnection between the deepest theories of nature and mathematical reasoning, famously stated by Eugene Wigner as the unreasonable efficacy of mathematics in physical theories. The sciences,

Logic and Algebraic Structures in Quantum Computing
Edited by J. Chubb, A. Eskandarian and V. Harizanov
Lecture Notes in Logic, 45
© 2016, Association for Symbolic Logic

and in particular physics, have relied on, and benefited from, the economy of mathematical expressions and the efficacy and rigor of mathematical reasoning with its underlying logical structure to make definite statements and predictions about nature. Mathematics has become the *de facto* language of the quantitative sciences, particularly scientific theories, and the major discoveries and predictive statements of these theories (whenever possible) are cast in the language of mathematics, as it affords them elegance as well as economy of expression. *What happens if the syntax and grammar of such a language become inadequate?*

This seems to have been the case when some of the more esoteric predictions of the then new theory of quantum mechanics began to challenge the scientific intuition of the times around the turn of the 20th century. This violation of intuition was so severe that even the most prominent of scientists were not able to reconcile the dictates of their intuition with the experimentally confirmed predictions of the theory. The discomfort with some of the features and predictions of quantum theory were, perhaps, most prominently brought out in the celebrated work of Einstein, Podolsky, and Rosen (EPR) in the mid 1930s. EPR fueled several decades of investigations on the foundations of quantum theory that continue to this day. The main assertion of the EPR work was that quantum theory had to be, by necessity, incomplete. Otherwise, long held understanding of what should be taken for granted as "elements of reality" had to be abandoned. Here, according to EPR, logical deductions based on primitives that were the very essence of reality and logical consistency forced the conclusion of the incompleteness of quantum theory; as if considering quantum theory as complete would question one's logical fitness and one's understanding of reality! Yet, in the decades since, with increasing sophistication in experimentation, and multiple ways of testing the theory, quantum theory has consistently outshined the alternatives. In particular, many predictions relying on the sensibilities of classical theories, where concepts such as separability, locality, and causality are the seemingly indispensable factors in our *understanding* of reality, are found to be entirely inconsistent with the *actual* reality around us. Quantum theory has not (as yet) suffered any such blow.

Confronted with the stark inability to reconcile the predictions of a theory, which are shown to be correct every time subjected to experimental verification, and a logical structure that seems to fall short in facilitating correct thinking and correct decision making (at least, in so far as the behavior of natural phenomena at the quantum level is concerned), one is forced to consider and question the validity of the premises on which that logical structure is built, or to discover alternative structures. Furthermore, the striking applications of quantum theory in the theory of computation, development of new algorithms, and the promising prospects for the building of a computing machine operating on the basis of the laws of quantum theory, necessitate a deeper investigation of alternative logical structures that encompass the elements of this new quantum reality. One must then give credence to the argument that, perhaps, the fault is not with the revolutionary quantum theory; rather, it is with the inadequacies of logical structures that were insufficient to be expanded and applied to a world that does not comply with the notions embodied in our understanding of the macroscopic classical physical theories of nature.

The utility of logical rules is most pronounced when applied to the building and operation of computing machines. With the advent of computing that takes advantage of the laws of quantum theory, i.e., quantum computing, it is only natural to search for those logical and algebraic structures that underlie the scaffolding of the quantum rules in computations. As obvious as it is that Boolean logic underlies classical computing and much of classical reasoning, it is equally obvious that it is not sufficient to express the logic underlying quantum mechanics or quantum computing. Birkhoff and von Neumann were among the first to propose a generalization of Boolean logic in which propositions about quantum systems could be formulated. While their endeavor was revolutionary, the Birkhoff-von Neumann quantum logic was not to be the final word on the subject of a logic for quantum mechanics, and indeed the investigation continues with increasing urgency.

In this volume, we present the work of a select group of scholars with an abiding interest in tackling some of the fundamental issues facing quantum computing and information theory, as investigated from the perspective of logical and algebraic structures. This selection, no doubt, reflects the intellectual proclivities and curiosities of the editors, within the reasonable limitations of space and coverage of topics for a volume of this size, and for the purpose of generating ideas that would fuel further investigation and research in these and related fields.

The first two articles, by Stairs and Parke, address philosophical and historical issues. Brandenburger and Keisler use ideas from continuous model theory to explore determinism and locality in quantum mechanical systems. Abramsky and Heunen, and Jacobs and Mandemaker describe the relationship between the category-theoretic and operator-theoretic approaches to the foundations of quantum physics. Döring gives a topos-based distributive form of quantum logic as an alternative to the quantum logic of Birkhoff and von Neumann. The papers by Coecke and Kartsaklis et al. use a diagrammatic calculus in analyzing quantum mechanical systems and, very recently, in computational linguistics. Kauffman's article presents an extensive treatment of the prominent role of algebraic structures arising from topological considerations in quantum information and computing; the pictorial approach used in knot theory is closely related to the quantum categorical logic presented in other articles in this volume.

Could logic be empirical? The Putnam-Kripke debate, by Allen Stairs. In his article in the present volume, Stairs outlines Hilary Putnam's position that quantum mechanics provides an empirical basis for a re-evaluation of our idea of logic and Saul Kripke's response, in which he takes issue with the very idea of a logic that is based on anything empirical. Stairs carefully interprets their positions, and in the end offers the beginnings of a compromise, which includes "disjunctive facts," which can be true even if their disjuncts are not, and the notion of "l-complementarity," to describe the relationship between statements having non-commuting associated projectors. The article wrestles with the idea of whether and how quantum mechanics should inform our logic and reasoning processes.

The essence of quantum theory for computers, by William C. Parke. In this article, Parke provides a thorough yet succinct introduction to the elements of physical theories, classical and quantum, which are relevant to a deeper understanding of the mathematical and logical structures underlying (or derived) from such theories, and important in the appreciation of the more subtle quandaries of quantum theory, leading to its utilization in computation. The emphasis has been placed on the physical content of information and elements of computation from a physicist's point of view. This includes a treatment of the role of space-time in the development of physical theories from an advanced point of view, and the limitations that our current understanding of space-time imposes on building and utilizing computing machines based on the rules of quantum theory. The treatment of the principles of quantum theory is also developed from an advanced point of view, without too much focus on unnecessary details, but covering the essential conceptual ingredients, in order to set the stage properly and provide motivation for the work of the others on logical and algebraic structures.

Fiber products of measures and quantum foundations, by Adam Brandenburger and H. Jerome Keisler. In this model-theoretic article, the authors use fiber products of (probability) measures within a framework they construct for empirical and hidden-variable models to prove determinization theorems. These objects (fiber products) were conceived by Rae Shortt in a 1984 paper, and were used recently by Itaï Ben Yaacov and Jerome Keisler in their work on continuous model theory (2009). Techniques in continuous model theory are relevant to the notion of models of quantum structures as in that context the "truth value" of a statement may take on a continuum of values, and can be thought of as probabilistic. In this case, a technique employed in continuous model theory is used in the construction of models in proofs of theorems that assert that every empirical model can be realized by an extension that is a deterministic hidden-variable model, and for every hidden-variable model satisfying locality and λ-independence, there is a realization-equivalent (both models extend a common empirical submodel) hidden-variable model satisfying determinism and λ-independence. The latter statement, together with Bell's theorem, precludes the existence of a hidden-variable model in which both determinism and λ-independence hold. The notion of λ-independence was first formulated by W. Michael Dickson (2005). It says that the choices made by an entity as to which observable to measure in a system are not influenced by the process of the determination of the value of a relevant hidden-variable.

Operational theories and categorical quantum mechanics, by Samson Abramsky and Chris Heunen. There are two complementary research programs in the foundations of quantum mechanics, one based on operational theories (also called general probabilistic theories) and the other on category-theoretic foundation of quantum theory. Samson Abramsky and Chris Heunen establish strong and important connections between these two formalisms. Operational theories focus on empirical and observational content, and quantum mechanics occupies one point in a space of possible theories. The authors define a symmetric monoidal categorical structure of an operational theory, which they call process category, and exploit the ideas of categorical quantum mechanics to obtain an operational theory as a certain representation of this process category. They lift the notion of non-locality to the general level of operational category. They further propose to apply a similar analysis to contextuality, which can be viewed as a broader phenomenon than non-locality.

Relating operator spaces via adjunctions, by Bart Jacobs and Jorik Mandemaker. By exploiting techniques of category theory, Jacobs and Mandemaker clarify and present in a unified framework various, seemingly different results in the foundation of quantum theory found in the literature. They use category-theoretic tools to describe relations between various spaces of operators on a finite-dimensional Hilbert space, which arise in quantum theory, including bounded, self-adjoint, positive, effect, projection, and density operators. They describe the algebraic structure of these sets of operators in terms of modules over various semirings, such as the complex numbers, the real numbers, the non-negative real numbers. The authors give a uniform description of such modules via the notion of an algebra of the multiset monad. They show how some spaces of operators are related by free constructions between categories of modules, while the other spaces of operators are related by a dual adjunction between convex sets (conveniently described via a monad) and effect modules.

Topos-based logic for quantum systems and bi-Heyting algebras, by Andreas Döring. Döring replaces the standard quantum logic, introduced by Birkhoff and von Neumann, which comes with a host of conceptual and interpretational problems, by the topos-based distributive form of quantum logic. Instead of having a non-distributive orthomodular lattice of projections, he considers a complete bi-Heyting algebra of propositions. More specifically, Döring considers clopen subobjects of the presheaf attaching the Gelfand spectrum to each abelian von Neumann algebra, and shows that these clopen subojects form a bi-Heyting algebra. He gives various physical interpretations of the objects in this algebra and of the operations on them. For example, he introduces two kinds of negation associated with the Heyting and co-Heyting algebras, and gives physical interpretation of the two kinds of negation. Döring considers the map called outer daseinisation of projections, which provides a link between the usual Hilbert space formalism and his topos-based quantum logic.

The logic of quantum mechanics – Take II, by Bob Coecke. Schrödinger maintained that composition of systems is the heart of quantum computing, and Coecke agrees. He suggests that the Birkhoff-von Neumann formulation of quantum logic fails to adequately and elegantly capture composition of quantum systems. The author puts forth a model of quantum logic that is based on composition rather superposition. He axiomatizes composition without reference to underlying systems using strict monoidal categories as the basic structures and explains a graphical language that exactly captures these structures. Imposing minimal additional structure on these categories (to obtain dagger compact categories) allows for the almost trivial derivation of a number of quantum phenomena, including quantum teleportation and entanglement swapping. This (now widely adopted) formalism has been used not only to solve open problems in quantum information theory, but has also provided new insight into non-locality.

Coecke's framework has been applied both to logic concerned with natural language interpretations, and to more formal automated reasoning processes. In this article, the focus is on the former. Coecke applies the graphical language of dagger compact categories to natural language processing—"from word meaning to sentence meaning"—implementing Lambek's theory of grammar and the notion of words as "meaning vectors." He argues that sentence meaning amounts to more than the meanings of the constituent words, but also the way in which they *compose*.

In the end, Coecke confesses that dagger compact categories do not capture all we might want them to, in particular, measurement, observables, and complementarity are left by the wayside. The model can be expanded (using spiders!) in such a way that all these are captured. Coecke closes with speculation about an important question: Where is the traditional logic hiding in all this?

Reasoning about meaning in natural language with compact closed categories and Frobenius algebras, by Dimitri Kartsaklis, Mehrnoosh Sadrzadeh, Stephen Pulman, and Bob Coecke. The authors apply category-theoretic methods to computational lingustics by mapping the derivations of the grammar logic to the distributional interpretation *via* a strongly monoidal functor. Such functors are structure preserving morphims. Grammatical structure is modeled through the derivations of pregroup grammars. A pregroup is a partially ordered monoid with left and right adjoints for every element in the partial order. The authors build tensors for linguistic constructs with complex types by using a Frobenius algebra. The Frobenius operations allow them to assign and compare the meanings of different language constructs such as words, phrases, and sentences in a single space. The authors present their experimental results for the evaluation of their model in a number of natural languages.

Knot logic and topological quantum computing with Majorana fermions, by Louis H. Kauffman. Kauffman presents several topics exploring the relationship between low-dimensional topology and quantum computing. These topics have been introduced and developed by Kauffman and Samuel J. Lomonaco over the last ten years. Kauffman uses the diagrammatic approach, and is particularly interested in models based upon the Temperley-Lieb categories. He discusses from several different perspectives the Fibonacci model related to the Temperley-Lieb algebra at fifth roots of unity. Kauffman shows how knots are related to braiding and quantum operators, as well as to quantum set-theoretic foundations. For example, the negation can generate the fusion algebra for a Majorana fermion, which is a particle that interacts with itself and can even annihilate itself. Thus, Kauffman calls the negation the mark.

He investigates the relationship between knot-theoretic recoupling theory and topological quantum field theory. Kauffman works with braid groups and their representations, and produces unitary representations of the braid groups that are dense in the unitary groups. He describes the Jones polynomial in terms of his bracket polynomial and applies his approach to design a quantum algorithm for computing the colored Jones polynomials for knots and links. Kauffman also gives a quantum algorithm for computing the Witten-Reshetikhin-Turaev invariant of three manifolds.

DEPARTMENT OF MATHEMATICS
UNIVERSITY OF SAN FRANCISCO
SAN FRANCISCO, CA 94117
E-mail: jcchubb@usfca.edu

DEPARTMENT OF PHYSICS
VIRGINIA SCIENCE & TECHNOLOGY CAMPUS
GEORGE WASHINGTON UNIVERSITY
ASHBURN, VIRGINIA 20147
E-mail: ea1102@gwu.edu

DEPARTMENT OF MATHEMATICS
GEORGE WASHINGTON UNIVERSITY
WASHINGTON, D.C. 20052
E-mail: harizanv@gwu.edu

A (VERY) BRIEF TOUR OF QUANTUM MECHANICS, COMPUTATION, AND CATEGORY THEORY

JENNIFER CHUBB AND VALENTINA HARIZANOV

This chapter is intended to be a brief treatment of the basic mechanics, framework, and concepts relevant to the study of quantum computing and information for review and reference. Part 1 (sections 1–4) surveys quantum mechanics and computation, with sections organized according to the commonly known postulates of quantum theory. The second part (sections 5–7) provides a survey of category theory. Additional references to works in this volume are included throughout, and general references appear at the end.

PART 1: QUANTUM MECHANICS & COMPUTATION

§1. Qubits & quantum states.

Postulate of quantum mechanics: Representing states of systems. The state of a quantum system is represented by a unit-length vector in a complex Hilbert space[1], $\mathcal{H}$, that corresponds to that system. The state space of a composite system is the tensor product of the state spaces of the subsystems.

The Dirac bra-ket notation for states of quantum systems is ubiquitous in the literature, and we adopt it here. A vector in a complex Hilbert space representing a quantum state is written as a *ket*, $|\psi\rangle$, and its conjugate-transpose (adjoint, or sometimes Hermitian conjugate) is written as a *bra*, $\langle\psi|$. In this notation, a bra-ket denotes an inner product, $\langle\varphi|\psi\rangle$, and a ket-bra denotes an outer product, $|\varphi\rangle\langle\psi|$.

Each one-dimensional subspace of $\mathcal{H}$ corresponds to a possible state of the system, and a state is usually described as a linear combination in a relevant orthonormal basis. The basis elements are often thought of as *basic states*. Quantum systems can exist in a *superposition* of more than one basic state: If a quantum system has access to two basic states, say $|\alpha\rangle$ and $|\beta\rangle$, then, in general, the system's "current state" can be represented by a linear combination of these states in complex Hilbert space:

$$|\psi\rangle = c_1|\alpha\rangle + c_2|\beta\rangle, \text{ where } ||\psi\rangle| = 1.$$

[1]A Hilbert space is a complete, normed metric space, where the norm and distance function are induced by an inner product defined on the space.

Logic and Algebraic Structures in Quantum Computing
Edited by J. Chubb, A. Eskandarian and V. Harizanov
Lecture Notes in Logic, 45
© 2016, ASSOCIATION FOR SYMBOLIC LOGIC

The complex coefficients, c_1 and c_2, of $|\alpha\rangle$ and $|\beta\rangle$ give classical probabilistic information about the state. For example, the value $|c_1|^2$ is the probability that the system would be found to be in state $|\alpha\rangle$ upon measurement. The coefficient itself, c_1, is called the *probability amplitude*. Two vectors in $\mathcal{H}$ represent the same state if they differ only by a global phase factor: If $|\psi\rangle = e^{i\theta}|\varphi\rangle$, then $|\psi\rangle$ and $|\varphi\rangle$ represent the same state, and the (real) probabilities described by the coefficients are the same.

The squared norm of the state vector $|\psi\rangle$ is the inner product of $|\psi\rangle$ with itself, i.e., the bra-ket $\langle\psi|\psi\rangle$. The quantity $|\langle\varphi|\psi\rangle|^2$ is the probability that upon measurement, $|\psi\rangle$ will be found to be in state $|\varphi\rangle$, and $\langle\varphi|\psi\rangle$ is the corresponding probability amplitude. (More about measurement of quantum systems can be found in Section 3 below.)

1.1. Qubits. A classical bit can be in only one of two states at a given time, $|0\rangle$ or $|1\rangle$. A quantum bit or *qubit* may exist in a *superposition* of these basic (orthogonal) states, $|\psi\rangle = c_1|0\rangle + c_2|1\rangle$, where c_1 and c_2 are complex probability amplitudes. More precisely, a qubit is a 2-dimensional quantum system, the state of which is a unit-length vector in $\mathcal{H} = \mathbb{C}^2$. The basic states for this space are usually thought of as $|0\rangle$ and $|1\rangle$, but at times other bases are used (for example, $\{|+\rangle, |-\rangle\}$ or $\{|\uparrow\rangle, |\downarrow\rangle\}$). Basic states are typically the eigenstates (eigenvectors) of an observable of interest (see discussion of measurement below).

Any unit vector that is a (complex) linear combination of the basic states is a *pure* state and non-trivial linear combinations are *superpositions*. So-called *mixed* states are not proper state vectors, they are classical probabilistic combinations of pure states and are best represented by *density matrices*.

The state space of a qubit is often visualized as a point on the *Bloch sphere*. The norm of a state vector is always one, and states that differ only by a global phase factor are identified, so two real numbers, θ and ϕ, suffice to specify a distinct state via the decomposition

$$|\psi\rangle = \cos\left(\frac{\theta}{2}\right)|0\rangle + e^{i\phi}\sin\left(\frac{\theta}{2}\right)|1\rangle.$$

Respectively, the range of values taken on by θ and ϕ may be restricted to the intervals $[0, \pi]$ and $[0, 2\pi)$ without any loss of generality, and so the corresponding distinct states may be mapped uniquely onto the unit sphere in $\mathbb{R}^3$. In this visualization, the basic vector $|0\rangle$ points up and $|1\rangle$ points down, θ describes the latitudinal angle, and φ the longitudinal angle. Orthogonal states are antipodal on the Bloch sphere. Note that states that differ by a global phase factor will (by design) coincide in this visualization.

1.2. Composite quantum systems. As described above, a single quantum system (for example, a single qubit) exists in a pure state that may be a superposition of basic states. A composition of systems may exist either in a *separable* or an *entangled* state. Separable states are states that can be written as tensor products of pure states of the constituent subsystems. Entangled states cannot be so written; they are non-trivial (complex) linear combinations of separable states. In the case of an entangled state, the subsystems cannot be thought of as existing in states independent of the composed system.

Example 1.1. Suppose we have a system of two qubits, the first in state $|\psi\rangle = (|0\rangle + |1\rangle)/\sqrt{2}$ and the second in state $|\varphi\rangle = (|0\rangle - |1\rangle)/\sqrt{2}$. The state of the combined system is

$$|\psi\rangle \otimes |\varphi\rangle = |\psi\rangle|\varphi\rangle = \frac{1}{2}(|00\rangle - |01\rangle + |10\rangle - |11\rangle).$$

Such a state of the composite system that can be written as a tensor product of pure states is called *separable*.

Example 1.2. The *Bell states* of a 2-qubit system are not separable; they are important and canonical examples of *entangled* states:

$$\frac{|00\rangle + |11\rangle}{\sqrt{2}} \qquad \frac{|00\rangle - |11\rangle}{\sqrt{2}}$$

$$\frac{|01\rangle + |10\rangle}{\sqrt{2}} \qquad \frac{|01\rangle - |10\rangle}{\sqrt{2}}$$

Example 1.3. The GHZ *states* (for Greenberger-Horne-Zeilinger) are examples of entangled states in composite systems that have three or more subsystems. The GHZ state for a system with n subsystems is

$$\frac{|0\rangle^{\otimes n} + |1\rangle^{\otimes n}}{\sqrt{2}}.$$

For more on entangled states, see Parke's article in this volume, or Section 6 of Kauffman's article.

§2. Transformations and quantum gates.

Postulate of quantum mechanics: Evolution of systems. The time evolution of a closed quantum system is described by a unitary transformation.

A transformation is *unitary* if its inverse is equal to its adjoint. Such transformations preserve inner products and are reversible, deterministic, and continuous. In quantum computing, algorithms are often described as circuits in which information (and time) flows from left to right. Quantum gates represent unitary transformations applied to qubits in such a circuit.

Example 2.1. *The Hadamard gate.* The 1-qubit Hadamard gate has as input and output one qubit, as shown in the simple circuit diagram below:

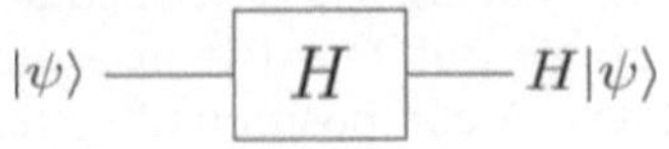

Its matrix representation (with respect to the basis $\{|0\rangle = [1\ 0]^T, |1\rangle = [0\ 1]^T\}$) is:

$$H = \frac{1}{\sqrt{2}}\begin{bmatrix} 1 & 1 \\ 1 & -1 \end{bmatrix}.$$

This transformation applied to the basic state $|0\rangle = \begin{bmatrix} 1 \\ 0 \end{bmatrix}$ results in the superposition $H|0\rangle = \frac{1}{\sqrt{2}}(|0\rangle + |1\rangle) = \frac{1}{\sqrt{2}}\begin{bmatrix} 1 \\ 1 \end{bmatrix}$.

Example 2.2. *The controlled-not gate.* Another important quantum gate is the *controlled-not* or CNOT gate. The gate requires two inputs, one designated as the *control* input (passing through the solid dot) and the other as the *target* input:

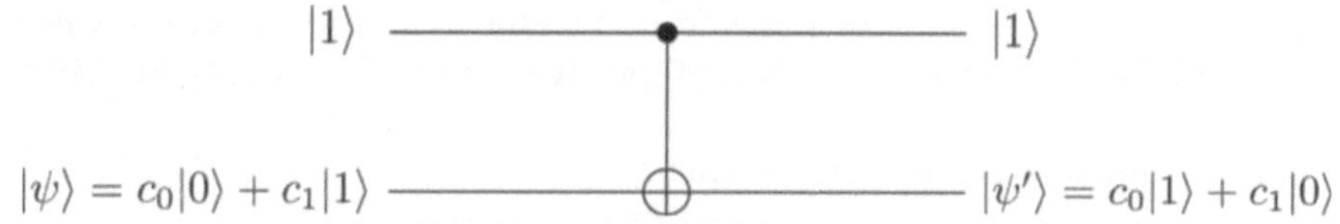

When the control input is in state $|0\rangle$, the gate does nothing. If the control is in state $|1\rangle$ (as it is in the diagram above), the gate acts by "flipping" the non-control (target) input as follows: If the target input is in state $|\psi\rangle = c_0|0\rangle + c_1|1\rangle$, then flipping transforms the state to $|\psi'\rangle = c_0|1\rangle + c_1|0\rangle$. The gate does not alter the control bit. The matrix representation of CNOT is the following (given with respect to the basis $\{|00\rangle = [1\ 0\ 0\ 0]^T, |01\rangle = [0\ 1\ 0\ 0]^T, |10\rangle = [0\ 0\ 1\ 0]^T, |11\rangle = [0\ 0\ 0\ 1]^T\}$):

$$\text{CNOT} = \frac{1}{2}\begin{bmatrix} 1 & 0 & 0 & 0 \\ 0 & 1 & 0 & 0 \\ 0 & 0 & 0 & 1 \\ 0 & 0 & 1 & 0 \end{bmatrix}.$$

For more on quantum gates and unitary transformations of quantum systems, see Parke's and Kauffman's articles in this volume.

§3. Measurement.

Postulate of quantum mechanics: Measurement. The notion of measurement is described in terms of *observables* represented by Hermitian (self-adjoint) matrices. (It should be noted that not all such matrices describe physically meaningful measurements.)

A Hermitian matrix has all real eigenvalues, and these represent the possible values obtained upon measurement of the observable. Moreover, distinct eigenvalues yield orthogonal eigenvectors. These matrices are often described in terms of their spectral decompositions. Upon measurement, a system's state (or wave function) experiences a "collapse" and is not preserved. After measurement, the state of the system is the eigenvector corresponding to the eigenvalue that was the result of the measurement.

Example 3.1. If the matrix A corresponding to an observable $\mathcal{A}$ has (real) eigenvalue a and corresponding unit-length eigenvector $|v_a\rangle$, then the probability that measuring $\mathcal{A}$ on state $|\varphi\rangle$ will yield the value a is given by $|\langle v_a|\varphi\rangle|^2$. If a is the result of the measurement of $\mathcal{A}$ on $|\varphi\rangle$, the system is left in state $|v_a\rangle$. If we consider the result of such a measurement as a random variable, the expected value (expectation value) of that quantity is given by $\langle\varphi|A|\varphi\rangle$.

Very briefly, if the matrices representing two different observables are non-commuting, then the observables are often referred to as *complementary* and measurements of these observables are subject to uncertainty limits. Complementary observables suffer from necessarily limited precision when measured simultaneously as a result of the Heisenberg Uncertainty Principle.

§4. No-go theorems and teleportation.

4.1. No cloning. In classical computation, it is possible to implement error correction by simply duplicating the classical data as needed. This is not the case in quantum computations.

Let $|\psi\rangle$ be an arbitrary state in state space $\mathcal{H}$, and $|e\rangle$ be an ancillary state (independent of $|\psi\rangle$) in an identical state space. To "clone" the state $|\psi\rangle$, we would need to have a unitary transformation that when applied to $|\psi\rangle|e\rangle$ replaces the ancillary state with a copy of $|\psi\rangle$, yielding $|\psi\rangle|\psi\rangle$.

THEOREM 4.1 (No-cloning theorem). *There is no unitary operator U so that for all states $|\psi\rangle$ and ancillary states $|e\rangle$,*

$$U|\psi\rangle|e\rangle = |\psi\rangle|\psi\rangle.$$

To see why, consider the possibility that there does exist such an operator U. As U must be unitary, it must preserve inner products, hence for any ψ and φ, we must have the following:

$$\langle\varphi|\psi\rangle = \langle e|\langle\varphi|\psi\rangle|e\rangle = \langle e|\langle\varphi|U^\dagger U|\psi\rangle|e\rangle = \langle\varphi|\langle\varphi|\psi\rangle|\psi\rangle = (\langle\varphi|\psi\rangle)^2.$$

We see that $\langle\varphi|\psi\rangle$ must be either 0 or 1 in order for this equality to hold, and so such a U preserves inner product only selectively—the states $|\varphi\rangle$ and $|\psi\rangle$ must be identical or orthogonal.

4.2. The EPR paradox, hidden variables, and Bell's Theorem. In 1935, Einstein, Podolsky, and Rosen (EPR) questioned the completeness of quantum mechanics in the form of a thought experiment involving the measurement of one part of a 2-particle entangled system. According to EPR, two mutually exclusive conclusions may be reached regarding quantum mechanics: either quantum mechanics is incomplete, or the physical quantities associated with two non-commuting operators cannot have simultaneous reality. Subsequently, building on the behavior of a two-component system under the laws of quantum theory, EPR argue for the incompleteness of quantum theory.

The following scenario captures the idea of the quandary they posed. Imagine that two particles, A and B, interact and then part ways. If one measures the momentum of particle A, he may *compute* the momentum of particle B exactly due to entanglement. If he subsequently *measures* the momentum of particle B, the result will be exactly that computed value. Similarly, the particles' positions may be observed, computed, and checked. However, the measurement operators corresponding to these observables (position and momentum) do not commute, and hence an exact knowledge of position entails some uncertainty in the value of momentum. The EPR argument makes a case for being able to assign two different wave functions (or states) to the same reality (particle B), by judicious choice of measurements on particle A, which leads to the conclusion that quantum mechanics must be incomplete.

A related question is this: How does particle B "know" to have a precisely defined momentum and an uncertain position when particle A's momentum is measured? According to the principle of locality, a physical process occurring in one place should not be able to affect a physical process in another location (outside the light cone of the first process). This scenario seems to entail either superluminal transmission of information between the particles (violating locality), or some "hidden variable" or "element of reality" encoding the information as yet unaccounted for by quantum mechanics (assuming determinism or realism). This is the idea underlying the famous EPR paradox.

In 1964, John Stewart Bell formalized (mathematically) the notions of locality and realism, and gave a set of inequalities that would provide a test of quantum mechanics against a local hidden variable theory. In the 1970s and 1980s, physical experiments (carried out most famously by Alain Aspect) demonstrated in favor of the former. What is known as *Bell's Theorem* is the summary of all this, asserting that no locally realistic theory can make the predictions of quantum mechanics.

Another related theorem is the *Kochen-Specker Theorem*, which says that a non-contextual hidden variable theory (one in which the value of an observable in a system is independent of the apparatus used to measure it) is unable to make the predictions of quantum mechanics.

4.3. Quantum teleportation. It would be difficult to overstate the importance of entanglement in quantum computing and the difficulty in representing and interpreting this phenomenon in possible quantum logics. A basic illustration of the power of entanglement is in the *quantum teleportation* protocol: An EPR *pair*, that is, a pair of qubits in a (entangled) Bell state, are prepared. One qubit is in the possession of entity A (Alice) and the other is in the possession of entity B (Bob). Alice also has a qubit, $|\psi\rangle$, which she would like to send to Bob. To do this, Alice applies a CNOT transformation to her two qubits, using $|\psi\rangle$ as the control, followed by an application of the Hadamard transformation to $|\psi\rangle$.

She then measures both of her qubits[2] (they are destroyed in the process), and (classically) communicates to Bob the (classical) information that results of her measurements. Upon receiving this information, Bob preforms one of four corresponding transformations, T, resulting in the transformation of his qubit into the state $|\psi\rangle$, which Alice wished to transmit to him.

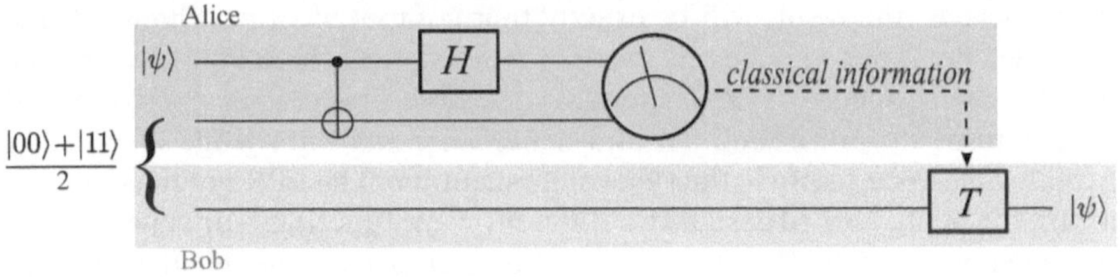

Note that this protocol does not violate the no-cloning theorem (Alice's copy is destroyed), nor Bell's Theorem (classical information must be transmitted subluminally).

For alternative formulations of the quantum teleportation protocol in a graphical language and another (similar) formulation in quantum topology, see Coecke's and Kauffman's (respectively) articles in this volume.

For more detailed exposition on all these ideas and topics, the following texts may be useful:

Textbooks at the undergraduate level

- *Quantum Computing for Computer Scientists*, by Noson S. Yanofsky and Mirco A. Mannucci, Cambridge University Press, 2008.
- *An Introduction to Quantum Computing*, by Phillip Kaye, Raymond Laflamme, and Michele Mosca, Oxford University Press, 2007.
- *Quantum Computing: A Gentle Introduction*, by Eleanor Rieffel and Wolfgang Polak, MIT Press, 2011.
- *Quantum Computer Science*, by N. David Mermin, Cambridge University Press, 2007.

At the graduate or research level

- *Quantum Computation and Quantum Information*, by Michael A. Nielsen and Isaac L. Chuang, Cambridge University Press, 2011.

PART 2: CATEGORY THEORY FOR QUANTUM COMPUTING

In physics, in the 1970s, Penrose used graphical language to represent linear operators, their products, and tensor products: boxes for operators, incoming wires for superscripts, and outgoing wires for subscripts. These diagrams represented various categories, which are of importance in physics

[2]This entire process is sometimes called a *Bell measurement*.

and quantum computing. Of particular importance are *tensor categories*, also called *monoidal categories*, which have been used by S. Abramsky and B. Coecke as a framework for quantum theory. Their categorical quantum mechanics can be also viewed as a suitable quantum logic. We will give a brief survey of monoidal categories. For more details see [3] and [1].

§5. **Basic category theory.** A *category* $\mathcal{C}$ consists of a class of *objects*, $ob(\mathcal{C})$, and a class of *morphisms*, $\hom(\mathcal{C})$, also called maps or arrows with specific abstract properties. For every pair of objects, A and B, there is a class of *morphisms* denoted by $\hom_{\mathcal{C}}(A, B)$, or simply $\hom(A, B)$ when the category is clear from the context. A morphism f has a domain $\mathrm{dom}(f)$ (also called source) and a codomain $\mathrm{cod}(f)$ (also called target), which we write $f : \mathrm{dom}(f) \to \mathrm{cod}(f)$. The morphisms are equipped with composition $\circ$, which is an associative operation that respects domain and codomain information. That is,

$$\text{(i) } (f \circ g) \circ h = f \circ (g \circ h),$$

where $f : A \to B$, $g : D \to A$, and $h : C \to D$. For every object A, the set $\hom(A, A)$ contains the identity morphism id_A such that for every $f : A \to B$, we have

$$\text{(ii) } f \circ id_A = f$$

and

$$\text{(iii) } id_B \circ f = f.$$

The equations (i)–(iii) can be viewed as the axioms for the categories. The *opposite category* (also called *dual category*) of $\mathcal{C}$ is formed by reversing the morphisms, that is, by interchanging the domain and the codomain of each morphism. It is denoted by $\mathcal{C}^{op}$. A category $\mathcal{C}$ is called *small* if both $ob(\mathcal{C})$ and $\hom(\mathcal{C})$ are sets, and it is called *locally small* is for every pair of objects A, B, the class $\hom(A, B)$ is a set.

A morphism $f : A \to B$ is called a *monomorphism* or *monic* if $f \circ g_1 = f \circ g_2$ implies $g_1 = g_2$ for all morphisms $g_1, g_2 : C \to A$. A morphism $f : A \to B$ has a *left inverse*, also called a *retraction* of f, if there is a morphism $g : B \to A$ such that $g \circ f = id_A$. Clearly, a morphism with a left inverse is a monomorphism. The converse may not be true. A morphism $f : A \to B$ is called an *epimorphism* or *epic* if $g_1 \circ f = g_2 \circ f$ implies $g_1 = g_2$ for all morphisms $g_1, g_2 : B \to C$. A morphism $f : A \to B$ has a *right inverse*, also called a *section* of f, if there is a morphism $g : B \to A$ such that $f \circ g = id_B$. A morphism with a right inverse is an epimorphism, but the converse may not be true. If a morphism has both a left inverse and a right inverse, then the two inverses are equal. Hence we have the following definition. A morphism $f : A \to B$ is called an *isomorphism* if there exists a morphism $g : B \to A$ such that $f \circ g = id_B$ and $g \circ f = id_A$. If it exists, g is unique and is called the *inverse* of f, and hence f is the inverse of g.

Examples of well-known categories include the category of sets as objects with functions as morphisms, the category of vector spaces as objects with linear maps as morphisms, and the category of Hilbert spaces as objects with unitary transformations as morphisms. In the graphical representation, object variables label edges ("wires") and morphism variables label nodes ("boxes"). The composition is represented by connecting the outgoing edge of one diagram to the incoming edge of another, while the identity morphism is represented as a continuing edge.

Functors capture the notion of a homomorphism between two categories. They preserve identity morphisms and composition of morphisms. More precisely, a *functor* Φ from a category $\mathcal{C}$ to a category $\mathcal{D}$ is a function that maps every object A of $\mathcal{C}$ to an object $\Phi(A)$ of $\mathcal{D}$, as well as every morphism of $\mathcal{C}$ to a corresponding morphism of $\mathcal{D}$ such that the following is satisfied. For every pair A, B of objects from $\mathcal{C}$, each morphism $f \in \hom(A, B)$ in $\mathcal{C}$ is mapped to a morphism $\Phi(f) \in \hom(\Phi(A), \Phi(B))$ in $\mathcal{D}$ such that

$$\Phi(g \circ h) = \Phi(g) \circ \Phi(h) \wedge \Phi(id_A) = id_{\Phi(A)}.$$

A functor from $\mathcal{C}$ to $\mathcal{D}$ is also called a *covariant functor*, in order to distinguish it from a contravariant functor, which reverses the order of composition. A *contravariant functor* Ψ from $\mathcal{C}$ to $\mathcal{D}$ is a map that associates to each object A in $\mathcal{C}$ an object $\Psi(A)$ in $\mathcal{D}$, and associates to each morphism $f \in \hom(A, B)$ in $\mathcal{C}$ a morphism $\Psi(f) \in \hom(\Psi(B), \Psi(A))$ in $\mathcal{D}$ such that

$$\Psi(g \circ h) = \Psi(h) \circ \Psi(g) \wedge \Psi(id_A) = id_{\Psi(A)}.$$

A functor Φ between locally small categories $\mathcal{C}$ and $\mathcal{D}$ is called *faithful* if it is injective when restricted to each set of morphisms that have a given domain and codomain. That is, for every pair A, B of objects in $\mathcal{C}$, the induced function

$$\Phi_{A,B} : \hom_{\mathcal{C}}(A, B) \to \hom_{\mathcal{D}}(\Phi(A), \Phi(B))$$

is injective. On the other hand, a faithful functor may not be injective on objects or morphisms. A functor Φ is called *full* if the induced functions $\Phi_{A,B}$ are surjective.

Natural transformations capture the notion of a homomorphism between two functors. That is, given two categories, $\mathcal{C}$ and $\mathcal{D}$, and two functors from $\mathcal{C}$ to $\mathcal{D}$, Φ and Ψ, a *natural transformation* $\mathcal{N} : \Phi \to \Psi$ consists of the family of morphisms for every object A of $\mathcal{C}$, $\mu_A : \Phi(A) \to \Psi(A)$, such that for every $f \in \hom_{\mathcal{C}}(A, B)$, we have

$$\Psi(f) \circ \mu_A = \mu_B \circ \Phi(f).$$

The content of the equation is captured by the following diagram.

$$\begin{array}{ccc} \Phi(A) & \xrightarrow{\mu_A} & \Psi(A) \\ {\scriptstyle \Phi(f)}\big\downarrow & & \big\downarrow{\scriptstyle \Psi(f)} \\ \Phi(B) & \xrightarrow{\mu_B} & \Psi(B) \end{array}$$

§6. **Monoidal categories.** A *monoidal category* captures the notion of a tensor product as a binary operation of objects, $A \otimes B$, and of morphisms, $f \otimes g$. The domain of $f \otimes g$ is the tensor product of the domains of f and g, and the codomain of $f \otimes g$ is the tensor product of the codomains of f and g. The tensor product of objects is associative in the sense that for every triple (A, B, C) of objects, there is an isomorphism

$$\alpha_{A,B,C} : (A \otimes B) \otimes C \to A \otimes (B \otimes C).$$

The tensor product is a *bifunctor*, which means that it satisfies the following equations for morphisms:

$$(f_1 \otimes f_2) \circ (f_3 \otimes f_4) = (f_1 \circ f_3) \otimes (f_2 \circ f_4)$$

and

$$id_{A \otimes B} = id_A \otimes id_B.$$

(See Coecke's article in this volume for a wire diagram representation of this equation.)

A monoidal category also has a constant unit object denoted by I. For every object A, there is an isomorphism (left)

$$\lambda_A : I \otimes A \to A$$

and an isomorphism (right)

$$\rho_A : A \otimes I \to A.$$

For morphisms $f : A \to A'$, $g : B \to B'$, $h : C \to C'$, we have

$$(f \otimes (g \otimes h)) \circ \alpha_{A,B,C} = \alpha_{A',B',C'} \circ ((f \otimes g) \otimes h),$$
$$f \circ \lambda_A = \lambda_{A'} \circ (id_I \otimes f),$$
$$f \circ \rho_A = \rho_{A'} \circ (f \otimes id_I).$$

In addition, the following *triangle axiom* is satisfied for every pair of objects A, B:

$$\rho_A \otimes id_B = (id_A \otimes \lambda_B) \circ \alpha_{A,I,B}.$$

Both sides map $(A \otimes I) \otimes B$ to $A \otimes B$. This equation is captured in the following diagram.

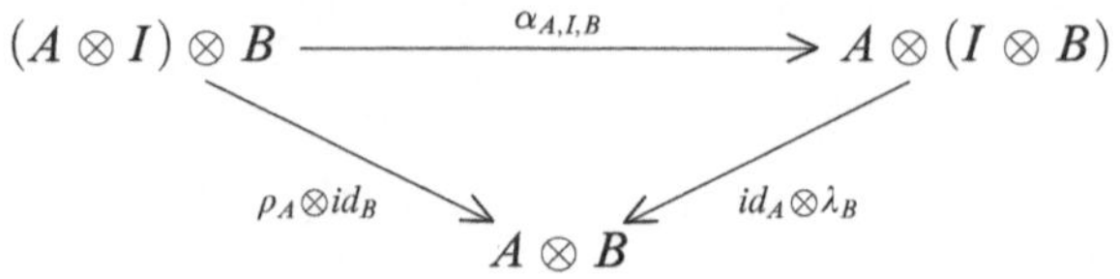

Also, the following *pentagon axiom* is satisfied for every quadruple of objects A, B, C, D:

$$(id_A \otimes \alpha_{B,C,D}) \circ (\alpha_{A,B \otimes C,D} \circ (\alpha_{A,B,C} \otimes id_D)) = \alpha_{A,B,C \otimes D} \circ \alpha_{A \otimes B,C,D}.$$

Both sides map $((A \otimes B) \otimes C) \otimes D$ to $A \otimes (B \otimes (C \otimes D))$. This relationship is visualized in the following diagram.

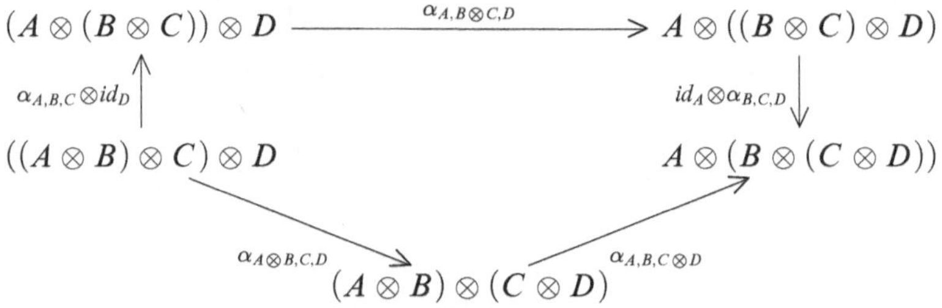

In the graphical language, the tensor product of objects is represented by parallel wires (input or output) from the bottom to the top, and the unit object is represented by no wire. Tensor product of morphisms is represented by stacking their diagrams. Examples of monoidal categories are vector spaces, or Hilbert spaces, with either direct sum or tensor product, as well as sets with direct products or disjoint unions. When no additional properties are assumed for a monoidal category, we often call it *planar monoidal category*.

Joyal and Street [2] established a coherence theorem for planar monoidal categories, which captures the correspondence between the formal language and the graphical language we described. The formal language of categories uses object variables and morphism variables, and object constants (such as I) and morphism constants (such as id_A), and operation symbols (such as $\circ$ and $\otimes$). These are used to form terms and equations (formulas). The *coherence theorem* of Joyal and Street states that an equation in the language of monoidal categories follows from the axioms of monoidal categories if and only if it holds in the graphical language, up to planar equivalence. Roughly speaking, here, a diagram D_1 is planar equivalent to a diagram D_2 if it is possible to transform D_1 to D_2 by continuously moving the boxes and wires of D_1 (without crossing or cutting). Other coherence theorem for special categories are of the similar nature. The part of a coherence theorem that states that an equation following from the axioms holds in the graphical language is called a *soundness theorem*, and its converse is called a *completness theorem*. Soundness is guaranteed by assuring that the axioms hold in the graphical language.

A *braided monoidal category* is a monoidal category with a family of isomorphisms for every pair of objects A, B,

$$\sigma_{A,B} : A \otimes B \to B \otimes A.$$

Hence $\sigma^{-1}_{A,B}$ exists, where

$$\sigma^{-1}_{A,B} : B \otimes A \to A \otimes B.$$

Two *hexagon axioms* are satisfied for every triple of objects A, B, C:

$$(id_B \otimes \sigma_{A,C}) \circ \alpha_{B,A,C} \circ (\sigma_{A,B} \otimes id_C) = \alpha_{B,C,A} \circ \sigma_{A,B\otimes C} \circ \alpha_{A,B,C}$$

and

$$(id_B \otimes \sigma^{-1}_{C,A}) \circ \alpha_{B,A,C} \circ (\sigma^{-1}_{B,A} \otimes id_C) = \alpha_{B,C,A} \circ \sigma^{-1}_{B\otimes C,A} \circ \alpha_{A,B,C}.$$

The first of these axioms is captured in the diagram below.

$$\begin{array}{ccc}
(B \otimes A) \otimes C & \xrightarrow{\alpha_{B,A,C}} & B \otimes (A \otimes C) \\
{\scriptstyle \sigma_{A,B}\otimes id_C}\uparrow & & \downarrow{\scriptstyle id_B\otimes\sigma_{A,C}} \\
(A \otimes B) \otimes C & & B \otimes (C \otimes A) \\
{\scriptstyle \alpha_{A,B,C}}\downarrow & & \uparrow{\scriptstyle \alpha_{B,C,A}} \\
A \otimes (B \otimes C) & \xrightarrow{\sigma_{A,B\otimes C}} & (B \otimes C) \otimes A
\end{array}$$

It follows that

$$\sigma_{A,B} \circ \sigma^{-1}_{A,B} = id_{A\otimes B}.$$

Graphical language is extended to picture braiding $\sigma_{A,B}$ and is represented by an under- (over-) crossing.

A *symmetric monoidal category* is a braided monoidal category where the braiding $\sigma_{A,B}$ is the inverse $\sigma^{-1}_{B,A}$. It is called symmetry and is graphically represented by a crossing.

For monoidal categories $\mathcal{C}$ and $\mathcal{D}$, a functor $\Phi : \mathcal{C} \to \mathcal{D}$ is called a *monoidal functor* if there are also morphisms $\phi_{A,B} : \Phi(A) \otimes \Phi(B) \to \Phi(A \otimes B)$ and

$\phi : I_{\mathcal{D}} \to \Phi(I_{\mathcal{C}})$, which preserve the tensor structure as follows. For every triple of objects A, B, C of $\mathcal{C}$,

$$\Phi(\alpha_{A,B,C}) \circ \phi_{A\otimes B,C} \circ (\phi_{A,B} \otimes id_{\Phi(C)}) = \phi_{A,B\otimes C} \circ (id_{\Phi(A)} \otimes \phi_{B,C}) \circ \alpha_{\Phi(A),\Phi(B),\Phi(C)},$$

$$\rho_{\Phi(A)} = \Phi(\rho_A) \circ \phi_{A,I} \circ (id_{\Phi(A)} \otimes \phi),$$

$$\lambda_{\Phi(A)} = \Phi(\lambda_A) \circ \phi_{I,A} \circ (\phi \otimes id_{\Phi(A)}).$$

For example, the last equation has the diagram:

$$\begin{array}{ccc}
I \otimes \Phi(A) & \xrightarrow{\lambda_{\Phi(A)}} & \Phi(A) \\
{\scriptstyle \phi\otimes id_{\Phi(A)}}\big\downarrow & & \big\uparrow{\scriptstyle \Phi(\lambda_A)} \\
\Phi(I) \otimes \Phi(A) & \xrightarrow{\phi_{I,A}} & \Phi(I \otimes A)
\end{array}$$

If the maps $\phi_{A,B}$ and ϕ are also invertible (isomorphisms), the functor is called a *strong monoidal functor*; if they are the identity maps, the functor is called a *strict monoidal functor*.

Given two monoidal categories, $\mathcal{C}$ and $\mathcal{D}$, and two strong monoidal functors from $\mathcal{C}$ to $\mathcal{D}$, Φ with ϕ^Φ and Ψ with ϕ^Ψ, a natural transformation $\mathcal{N} : \Phi \to \Psi$ with morphisms $\mu_A : \Phi(A) \to \Psi(A)$ is a *monoidal natural transformation* if for every pair of objects A, B of $\mathcal{C}$, we have

$$\mu_{A\otimes B} \circ \phi^{\Phi}_{A,B} = \phi^{\Psi}_{A,B} \circ (\mu_A \otimes \mu_B).$$

For braided monoidal categories $\mathcal{C}$ and $\mathcal{D}$, a monoidal functor $\Phi : \mathcal{C} \to \mathcal{D}$ is called a *braided monoidal functor* if it is compatible with braiding as follows. For every pair of objects A, B of $\mathcal{C}$,

$$\Phi(\sigma_{A,B}) \circ \phi_{A,B} = \phi_{B,A} \circ \sigma_{\Phi(A),\Phi(B)}.$$

An example of a symmetric monoidal category is the category of sets with functions as morphisms, with Cartesian product, and symmetry given by $\sigma_{A,B}(x, y) = (y, x)$. Another example of a symmetric monoidal category is the category of vector spaces with linear maps as morphisms, with tensor product, and symmetry given by $\sigma_{A,B}(x \otimes y) = y \otimes x$.

A monoidal category $\mathcal{C}$ is called *right autonomous* if every object A of $\mathcal{C}$ has a right dual, denoted by A^*, and there are two morphisms, the unit $\eta_A : I \to A^* \otimes A$ and the counit $\varepsilon_A : A \otimes A^* \to I$, which satisfy the following adjunction triangle equalities:

$$id_A = (\varepsilon_A \otimes id_A) \circ (id_A \otimes \eta_A),$$

$$(id_{A^*} \otimes \varepsilon_A) \circ (\eta_A \otimes id_{A^*}) = id_{A^*}.$$

A^*, ε_A, η_A and the first triangle equality are graphically represented as follows:

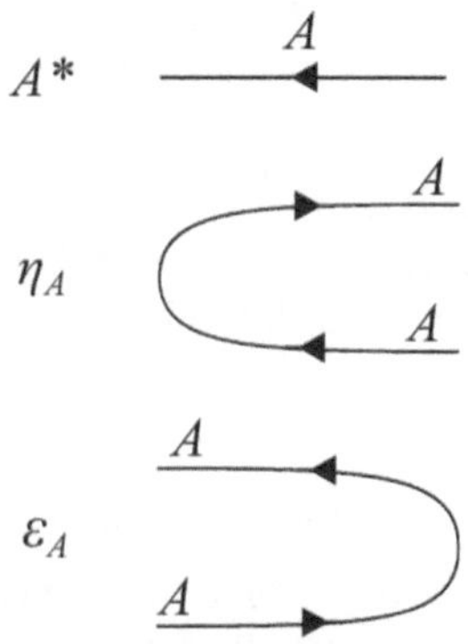

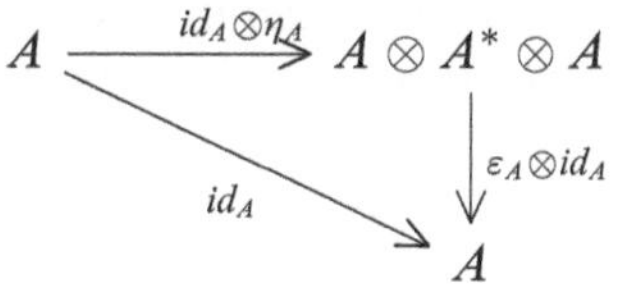

A *left autonomous* monoidal category is defined dually and a left dual of A is denoted by ${}^*A$. A monoidal category is *autonomous* if it is both right and left autonomous. In a braided right autonomous category, a right dual of A is also a left dual of A, so the category is autonomous. A *compact closed category* is a right autonomous symmetric monoidal category. A category of sets with binary relations as morphisms and direct product as tensor product and where $A^* = A$ is a compact closed category. The category of finite dimensional vector spaces (or finite dimensional Hilbert spaces) with tensor product and with A^* being the dual space of A is a compact closed category. On the other hand, if we allow infinite dimensional vector spaces, the categories of vector spaces and of Hilbert spaces are not autonomous.

§7. **Dagger categories.** A *dagger category* is a category $\mathcal{C}$ equipped with a contravariant functor $\dagger : \mathcal{C} \to \mathcal{C}$, which is identity on the objects and involutive on the morphisms. More specifically, to each morphism $f : A \to B$ a morphism $f^\dagger : B \to A$ is assigned such that

$$(f^\dagger)^\dagger = f \wedge id_A^\dagger = id_A,$$

and for every morphism $g : B \to C$,

$$(g \circ f)^\dagger = f^\dagger \circ g^\dagger.$$

Morphism $f^\dagger$ is called the *adjoint* of f. The adjoint is diagrammatically represented by reversing the location but not the direction of the wires and by marking the upper right corner (in contrast to the upper left corner) in the box. In general, the adjoint of a diagram is its mirror image.

The category of sets with binary relations as morphisms is a dagger category with relational inverse $R^\dagger$ as adjoint of R. The category of Hilbert spaces with bounded linear maps is a dagger category with the usual adjoints. A morphism f is called *Hermitian* if it is self-adjoint: $f^\dagger = f$. A morphism f is called *unitary* if it is an isomorphism and $f^{-1} = f^\dagger$. A *dagger functor* Φ between two dagger categories $\mathcal{C}$ and $\mathcal{D}$ is a functor that satisfies the following additional equality for every morphism f in $\mathcal{C}$:

$$\Phi(f^\dagger) = (\Phi(f))^\dagger.$$

A *dagger monoidal category* $\mathcal{C}$ is a category that is both monoidal and dagger and the two structures are compatible in the sense that the morphisms from the monoidal structure, $\alpha_{A,B,C}$, λ_A, ρ_A, are unitary and the following equality is satisfied for every pair of morphisms f, g:

$$(f \otimes g)^\dagger = f^\dagger \otimes g^\dagger.$$

A *dagger symmetric monoidal category* is a dagger braided monoidal category such that its symmetry (braiding) is unitary. A *dagger compact closed category* $\mathcal{C}$, also simply called *dagger compact category*, is a dagger symmetric monoidal category that is also compact closed, together with a relation to connect the dagger structure to the compact structure. Specifically, the dagger is used to connect the unit to the counit so that for all objects A in $\mathcal{C}$, we have:

$$\eta_A = \sigma_{A \otimes A^*} \circ \varepsilon_A^\dagger.$$

Dagger compact categories are of great importance for foundations of quantum information and computing. Selinger [4] proved a completeness and hence coherence result for dagger compact closed categories. That is, he established that an equation follows from the axioms of dagger compact closed categories if and only if it holds in finite dimensional Hilbert spaces. Thus, this coherence theorem allows us to use the diagrammatic calculus of dagger compact categories to precisely express and verify some fundamental quantum information notions and protocols.

REFERENCES

[1] S. Abramsky and B. Coecke, *A categorical semantics of quantum protocols*, ***Proceedings of the 19th Annual IEEE Symposium on Logic in Computer Science***, IEEE, 2004, pp. 415–425.

[2] A. Joyal and R. Street, *The geometry of tensor calculus I*, ***Advances in Mathematics***, vol. 88 (1991), pp. 55–112.

[3] P. Selinger, *A survey of graphical languages for monoidal categories*, ***New Structures for Physics*** (B. Coecke, editor), Lecture Notes in Physics, vol. 813, Springer, 2011, pp. 289–355.

[4] ———, *Finite dimensional Hilbert spaces are complete for dagger compact closed categories*, ***Logical Methods in Computer Science***, vol. 8 (2012), pp. 1–12.

DEPARTMENT OF MATHEMATICS
UNIVERSITY OF SAN FRANCISCO
SAN FRANCISCO, CA 94117
E-mail: jcchubb@usfca.edu

DEPARTMENT OF MATHEMATICS
GEORGE WASHINGTON UNIVERSITY
WASHINGTON, D.C. 20052
E-mail: harizanv@gwu.edu

COULD LOGIC BE EMPIRICAL? THE PUTNAM-KRIPKE DEBATE

ALLEN STAIRS

Abstract. Not long after Hilary Putnam published "Is Logic Empirical," Saul Kripke presented a critique of Putnam's argument in a lecture at the University of Pittsburgh. Kripke criticized both the substance of Putnam's version of quantum logic and the idea that one could "adopt" a logic for empirical reasons. This paper reviews the debate between Putnam and Kripke. It suggests the possibility of a "middle way" between Putnam and Kripke: a way in which logic could be broadly a priori but in which empirical considerations could still bear on our views about the logical structure of the world. In particular, considerations drawn from quantum mechanics might provide an example.

Some years ago, Hilary Putnam published a paper called "Is Logic Empirical?" [7] in which he argued that quantum mechanics provides an empirical case for revising our views about logic. (The paper was republished in his collected works as "The Logic of Quantum Mechanics". Page references will be to the reprinted version.) In 1974, Saul Kripke presented a talk at the University of Pittsburgh called "The Question of Logic," offering a detailed rebuttal of Putnam's case. As of this writing, almost 40 years later, Kripke's paper still hasn't appeared in print and apart from my 1978 dissertation and a paper I published 28 years later [9], very little has been written on the disagreement between Putnam and Kripke. This is unfortunate; the issues are well worth investigating. In my 2006 paper [9], I adopted the device of writing about Paul Kriske and Prof. Tupman out of deference to the fact that there is no published version of Kripke's talk. Here I'll simply write directly about Putnam and Kripke. If I get Kripke wrong, I hope he'll let us know.

As for the plan of the paper, we begin by reviewing Putnam's arguments; after that we move to Kripke's rebuttal. This will lead to a larger discussion of what logic and the empirical might have to do with one another.

§1. Putnam on quantum logic. We think of logical truths as a special case of necessary truths, but Putnam reminds us that we now reject certain claims about geometry that once seemed necessary. We would once have said that if two lines are straight and a constant distance apart over some portion of their span, they can't converge elsewhere. For anyone not familiar with

Logic and Algebraic Structures in Quantum Computing
Edited by J. Chubb, A. Eskandarian and V. Harizanov
Lecture Notes in Logic, 45
© 2016, Association for Symbolic Logic

non-Euclidean geometry, Putnam claims that this seems as intuitively clear as saying that there are no married bachelors, or that nothing can be scarlet all over and bright green all over at the same time. In the case of the lines, however, we've come to believe not just that the claim *might* be false but that in some instances it *is* false.

We might say that what Putnam describes applies to *geodesics*, but "geodesic" doesn't mean "straight line." However, Putnam insists that this won't do. On our intuitive conception, shortest paths are straightest and conversely. The notion of a geodesic preserves this, and lines that depart from geodesics will not seem straighter. One way to put it: if we say that geodesics behaving as Putnam describes aren't straight lines, we'll have to say that there can be points with no straight line between them. Putnam thinks we miss the significance of relativity if we represent its geometrical claims as mere changes of meaning. He writes:

> The important point is that ['straight line'] does not 'change meaning' in the trivial way one might at first suspect. Once one appreciates that something that was formerly literally unimaginable has indeed happened, then one also appreciates that the usual 'linguistic' moves only help to distort the nature of the discovery and not to clarify it. (p. 177)

Putnam argues that we've made a similar discovery about logic itself. We pair statements about quantum quantities with subspaces of Hilbert space and we can extend this map from simple statements to compound ones by associating "or" with subspace span ($p \vee q$), "and" with subspace intersection ($p \wedge q$), and "not" with orthocomplement ($p^{\perp}$). If we take the mapping seriously, however, we have a conflict with classical logic. Suppose the quantity A has two possible values a_1 and a_2, associated with rays α_1 and α_2. Suppose, likewise, that B has two values b_1 and b_2 associated with rays β_1 and β_2. Now consider the expressions

$$(A = a_1 \text{ or } A = a_2), (B = b_1 \text{ or } B = b_2)$$

and associate them with the subspaces

$$(\alpha_1 \vee \alpha_2), (\beta_1 \vee \beta_2).$$

Quantum mechanics, read as Putnam reads it, gives us cases where both disjunctions are true. That means the conjunction

$$(A = a_1 \text{ or } A = a_2) \text{ and } (B = b_1 \text{ or } B = b_2)$$

is also true. However each of the following pick out the null subspace of Hilbert space

$$(\alpha_1 \wedge \beta_1), (\alpha_1 \wedge \beta_2), (\alpha_2 \wedge \beta_1), (\alpha_2 \wedge \beta_2)$$

and so the corresponding conjunctions are false. Hence

$$(A = a_1 \wedge B = b_1) \vee (A = a_1 \wedge B = b_2) \vee (A = a_2 \wedge B = b_1) \vee (A = a_2 \wedge B = b_2)$$

is false but this discrepancy between the distributed and undistributed formulas is impossible classically. Putnam writes:

> *Conclusion*: the mapping is nonsense—*or, we must change our logic.* (p. 179)

On the other hand, if we do "adopt the heroic course of changing our logic" there's a straightforward way to proceed:

> ...*just read the logic off from the Hilbert space* $H(S)$. (p. 179)

The advantage, says Putnam, is that

> *all* so-called anomalies in quantum mechanics come down to the *non-standardness of the logic*. (p. 179)

and the anomalies go away if we change our logic.

Putnam offers several illustrations. "Complementarity," understood as the failure of quantum mechanics to specify joint values for noncommuting quantities comes down to logical incompatibility in quantum logic; the complementary quantities don't share eigenspaces. He also argues that quantum logic accounts for the two-slit experiment. To derive the incorrect classical probabilities, we have to distribute a proposition R about where the photon hits the screen over a disjunction of propositions A_1 and A_2 about which hole the photon passes through. If we treat '$R \wedge (A_1 \vee A_2)$' as equivalent to '$(R \wedge A_1) \vee (R \wedge A_2)$', then we end up with the wrong probabilities.

Putnam also claims that if we analyze barrier penetration quantum logically, we avoid explaining the effect by appeal to a supposedly mysterious 'disturbance by the measurement.' (p. 182) In fact, the account he gives (on p. 183) can't be right for any finite population of atoms (exercise for the reader; look especially at statement (8) and Putnam's comment on it) but let that pass. In classical physics, the state provides a complete description relative to the terms of the theory, of the system. In quantum theory, there are "states" or "state descriptions," but Putnam writes that

> A system has no complete description in quantum mechanics; such a thing is a logical impossibility (p. 185)

Quantum states are "logically strongest consistent statements" but they aren't states "in the sense of statements which imply every true proposition about S" (p. 185) This might suggest that quantum states are like statistical states in classical mechanics, and that their failure to provide a complete list of all the truths is a reflection of our epistemic situation. However, this isn't Putnam's view. Rather, he tells us that a quantum system has, e.g., a position by virtue of the truth of a disjunction of position statements and it also has a momentum by virtue of the truth of disjunction of momentum statements[1]. Here is Putnam articulating what we will call the *value-definiteness* thesis:

[1] Putnam knows that strictly, there are no position and momentum eigenstates; the oversimplification is merely for illustration.

1. For any such question as 'what is the value of $M(S)$ now?' where M is a physical magnitude, *there exists* a statement U_i which was true of S at t_0 such that *had I known* U_i was true at t_0, I could have predicted the value of $M(S)$ now, but
2. It is *logically impossible* to possess a statement U_i which was true of S at t_0 from which one could have predicted the value of *every* magnitude M now.

   We can predict any *one* magnitude, if we make an appropriate measurement, but we can't predict them *all*.

The advantage of giving up classical logic, according to Putnam, is this:

> These examples makes the principle clear. The only laws of classical logic that are given up in quantum logic are distributive laws . . . and every single anomaly vanishes once we give these up. (p. 184)

Putnam's argument for adopting quantum logic is that if we do, the interpretive puzzles of the theory dissolve. If we insist on classical logic, we have to say such supposedly objectionable things as that measurements create the values of the quantities measured or that there is a "cut between the observer and the observed" or that there are undetectable hidden variables. But Putnam says

> . . . I think it is *more likely that classical logic is wrong* than that there are either hidden variables or "cuts between the observer and the system", etc.

This completes the analogy with geometry. We could preserve Euclidean geometry, but only by paying the high intellectual price of admitting gratuitous universal forces. Likewise for classical logic: we can preserve it only by paying an unacceptable price in the coin of untoward claims about quantum systems.

§2. Kripke on Putnam. Kripke's critique of Putnam has two parts. One deals with the particulars of Putnam's argument. There Kripke's case is strong. However, granting that Kripke is right about Putnam's particular quantum logical proposal wouldn't show that logic isn't empirical, nor would it show that quantum mechanics doesn't give us a reason to change our views about logic. In the second part of his critique, Kripke argues that the very idea of changing logic for empirical reasons is confused.

In what follows, I quote at length from my partial transcript of Kripke's talk. The indirect debate between Kripke and Putnam was an important episode, and the reader will get a better sense of it if s/he reads Kripke's own words. To be sure, there is a matter of propriety here, but Kripke's own words do a better job than my paraphrase would of spelling out his view and therefore, it seems fairest to him to use those words.

2.1. Quantum logic and simple arithmetic. The first part of Kripke's argument is intended to show that if we follow Putnam, we have to agree to the untoward conclusion that $2 \times 2 \geq 5$. According to Putnam, if M has possible values $m_1, m_2, \ldots, m_n$ then there is a true statement ascribing one of these values to M. The statement

$$M = m_1 \vee M = m_2 \vee \cdots \vee M = m_n$$

is true, Putnam would say, and the summary of the value-definiteness thesis above makes clear what he means: one of the disjuncts really is true, and if we knew which, we could predict the outcome of an M-measurement. However, the logically strongest statement about the system may not tell us which disjunct is true.

Is it really clear that Putnam meant this? Here's a passage that would be hard to make sense of otherwise. S_z is a position state and T_1, T_2, etc. are momentum states. (Substitute eigenstates of different spin components if you prefer.) We suppose S_z to be known.

> The idea that momentum measurement 'brings into being' the value found arises very naturally if one does not appreciate the logic being employed in quantum mechanics. If I know that S_z is true, then I know that for *each* T_j the conjunction $S_z \cdot T_j$ is false. It is natural to conclude (smuggling in classical logic) that $S_z \cdot (T_1 \vee T_2 \vee \cdots \vee T_R)$ is false, and hence that we must reject $(T_1 \vee T_2 \vee \cdots \vee T_R)$—i.e., we must say 'the particle has no momentum'. Then one measures momentum, and one gets a momentum—so the measurement must have 'brought it into being'. However, the error was in passing from the falsity of $S_z \cdot T_1 \vee S_z \cdot T_2 \vee \cdots \vee S_z \cdot T_R$ to the falsity of $S_z \cdot (T_1 \vee T_2 \vee \cdots \vee T_R)$. This latter statement is *true* (assuming S_z) and so it is *true* that 'the particle has a momentum' ... and the momentum measurement merely *finds* this momentum (while disturbing the *position*); it does not create it, or disturb it in any way. It is as simple as that. (p. 186)

"Simple" or not, Kripke draws out an untoward consequence. Suppose we're given two quantities, A and B, each with two possible values 1 and 2. Thus the set $\{1, 2\}$ is the set of possible values of A and also of B. Putnam will say that

1. $A = 1 \vee A = 2$
2. $B = 1 \vee B = 2$

are both true. However, he will also say that each of the following are false:

1. $A = 1 \wedge B = 1$
2. $A = 1 \wedge B = 2$
3. $A = 2 \wedge B = 1$
4. $A = 2 \wedge B = 2$

But Kripke argues:

> The usual mathematical definition of multiplication is this: suppose we have two sets with two elements. Then the cardinality of their product is the cardinality of the Cartesian product of the two sets ... where x comes from the first set and y comes from the second set, so where x comes from $\{a, b\}$ and y comes from $\{c, d\}$ where $\{a, b\}$ and $\{c, d\}$ are our two two-element sets. We want to consider how many ordered pairs there are. So the classical arithmetician says "There are four, namely $\langle a, c\rangle, \langle a, d\rangle, \langle b, c\rangle, \langle b, d\rangle$ ... " But we can all see the fallacy in any conclusion that these are the only pairs.

The "fallacy" is that if x comes from $\{a, b\}$, then we have the disjunction

$$x = a \vee x = b$$

and similarly we have

$$y = c \vee y = d.$$

Now suppose that the set $\{a, b\}$ is the set of possible values of the quantity A above (i.e., $\{a, b\} = \{1, 2\}$) and $\{c, d\}$ is the set of values of the quantity B (i.e., $\{c, d\} = \{1, 2\}$ as well.) We'll let Kripke pick up the story:

> Now I claim that there is a fifth pair $\langle A, B\rangle$ where these are the two quantities mentioned by Putnam. Remember that Putnam does not think these are funny pseudo-numbers. The idea is that A was already one of the two numbers 1 and 2 [and] B was already one of the numbers 1 and 2. So A is certainly in the first set [i.e., $\{a, b\} = \{1, 2\}$—AS] because A is equal either to 1 or to 2. B is certainly in the second set [i.e., $\{c, d\} = \{1, 2\}$— AS] because B is either equal to 1 or to 2, though we may not have measured which. So the pair $\langle A, B\rangle$ is in our Cartesian product. But certainly we cannot say that $\langle A, B\rangle$ equals $\langle 1, 1\rangle$ if we adopt the usual criterion of identity of ordered pairs because that would mean that $A = 1$ and B equals 1, and that contradicts [the falsity of (3)]. Also, $\langle A, B\rangle$ does not equal $\langle 1, 2\rangle$ because it is false that A equals 1 and B equals 2. And $\langle A, B\rangle$ does not equal $\langle 2, 1\rangle$ [and] $\langle A, B\rangle$ does not equal $\langle 2, 2\rangle$ ... So there is a fifth and hitherto overlooked, I might say, ordered pair in the Cartesian product of these two finite sets.

Kripke's point, of course, is that this is absurd, but that it's where we end up if we follow Putnam.

There may be various ways Putnam could respond, but Kripke insists that one obvious rejoinder won't do: it won't do to accuse Kripke of begging the question. Kripke insists: he'd only be begging the question if he had assumed a premise that Putnam rejects. However, the distributive law isn't a premise in his argument. Kripke simply reasons from premises that Putnam accepts to the conclusion that if none of the pairs $\langle 1, 1\rangle$, $\langle 1, 2\rangle$, $\langle 2, 1\rangle$, $\langle 2, 2\rangle$ gives the joint values for A and B, then joint values require that there be another pair in the

Cartesian product. Since Putnam would claim that none of the four ordered pairs gives the joint values, and would also claim that both quantities really do have values, the untoward (and absurd) conclusion follows. As Kripke puts it in connection with a closely-related example

> ... if you say that I am begging the question then you yourself, I think, are begging the question, because only if my reasoning was invalid did I need any extra premise which I have begged against Putnam.

2.2. The impossibility of "adopting" a logic. Kripke is right, I believe: there's no convincing quantum-logical defense of the value-definiteness thesis. (See [8] for more discussion) and in what follows, we will assume that value-definiteness doesn't hold. Kripke's larger point is that there is a problem at the core of Putnam's view. Putnam, he thinks, believes that we could somehow decide to "adopt" a logic; Kripke insists that this is incoherent. We misunderstand logic if we think there are "logics" among which we could somehow choose. There is *reasoning*. Specific formal systems may or may not adequately capture aspects of correct reasoning. But there is no neutral place outside logic from which to decide what "logic" to adopt.

Whether Putnam really holds the view of logic that Kripke attributes to him isn't clear. That said, it's a useful foil for making Kripke's own view of logic clearer. Therefore, while we won't ignore the question of how well Kripke's criticisms fit Putnam, the exegetical question won't be our main concern.

Putnam remarked on our intuitive sense of contradiction when faced with his geometrical example. Kripke reads Putnam this way:

> Just as in the case of non-Euclidean geometry we throw intuition to the wind and adopt an axiomatic system as supposedly describing the real physical world ... so on every other domain we cannot rely on intuition. Once one has a rival system of axioms, the mere fact that an old system struck us as the only intuitively acceptable one should be given little weight. Once alternative geometries are under consideration, we abandon any mere intuitive preference for Euclidean geometry, and once alternative logics are under consideration, we abandon any mere intuitive preference for a particular system of logic.

Kripke thinks there is a deep confusion here. Formal systems are not *logic*. Formal systems may or may not faithfully reflect correct principles of *reasoning*, but we have no alternative to using "intuition," by which Kripke means reasoning, to assess the formal systems. If changing our formal system is supposed to entail changing the way we *reason*, then we have no place to stand outside of reasoning from which to do this. "Logics" *qua* formal systems aren't *logic*. As Kripke sees it, Putnam's fundamental error lies in missing this point.

Once we grasp this, the idea that we could change our logic in response to empirical considerations makes no sense.

Even if we grant that there's no place to stand outside reasoning, there's a more general phenomenon here. What William Alston called *doxastic practices* (see his [1, Ch. 4], for instance) typically have the sort of self-supposing quality that Kripke's point relies on. We can reconsider how to evaluate beliefs based on sensory input; when we do, we'll need to rely on at least some such beliefs and hence on the practice of forming beliefs based on sense evidence. We can consider what memory can and can't teach us; we can't avoid relying on at least some memories when we do. Equally important, these practices aren't insulated from one another. In considering what weight to give memory, for example, we'll make use of claims that we've accepted on the basis of the implicit and explicit rules/practices we use for assessing other kinds of empirical claims. We can also reason about how to reason, as Kripke would be the first to insist. Putnam may seem to be saying that we can evaluate logic without relying on logic broadly conceived (i.e., on logic *qua* reasoning) but it's not clear that he means this or needs to say it. In order to rebut a measured version of Putnam's view, Kripke would have to show that reasoning is the one doxastic practice to which the deliverances of other doxastic practices are irrelevant. Putnam's larger point would be made if sometimes what we discover empirically can *properly* enter into our deliberations about how to reason.

Be that as it may, Putnam's main argument seems to be that if we give up the distributive law, we'll be blocked from drawing untoward conclusions. Thus, we won't be able to argue that the probabilities in the two-slit experiment must fit a crude application of the law of total probability, and we won't need to say that measurement creates the values that it records. However, this is too quick. We might be able to avoid any number of unwelcome conclusions if we simply refused to reason in certain ways; that hardly makes a case for merely opportunistic "revisions" of logic. And while Putnam might judge that a failure of the distributive law is "more likely" than hidden variables or "cuts between observer and observed," Kripke can reply that without something more than a mere and tendentious cost-benefit analysis, we haven't been presented with an intelligible alternative. The distributive law *seems* to be a correct way to reason. Putnam hasn't shown us any deep problems with the idea that there are Bohmian-style hidden variables; he merely tells us that he finds them unlikely. He objects to the idea that measurement might bring the values it yields into being. However, his main objection seems to be that this is a strange notion of "measurement." This threatens to turn the argument into a mere quibble. The idea that the interactions we call measurements bring new states of affairs into being might be a reason to pick or invent a different word, but it doesn't count against the possibility that things really work this way.

We leave the vexed issue of measurement (or "measurement") aside and turn to a different part of Kripke's reply: his case that the very idea of "adopting a logic" makes no sense. Kripke takes his cue from Lewis Carroll's "What the Tortoise Said to Achilles" and from Quine's "Truth by Convention". He says:

> The basic problem is this: if logical truths are mere hypotheses ... and one can adopt them as one will, how, unless one has a logic in advance, can one possibly deduce anything from them?

Kripke develops the example of universal instantiation at greatest length. Imagine someone who doesn't see that from a universal claim, each instance follows. Imagine further that our poor reasoner is willing to accept Kripke's authority that all ravens are black and is also willing to accept Kripke's authority in more general logical matters. There's a raven, *J*, out of our subject's sight, but he doesn't see that believing this and accepting that all ravens are black commits him to accepting that *J* is black. Kripke tells the tale charmingly:

> So I say to him "Oh. You don't see that. Well let me tell you: from every universal statement, each instance follows." He will say "Oh. Yes. I believe you." So now I say to him, "Ah. So 'All ravens are black' is a universal statement and 'This raven is black' is an instance." "Yes. Yes" He agrees. So I say to him "All universal statements imply their instances. This particular statement that all ravens are black implies this particular instance." "Well, hmm, I'm not entirely sure," he will say. "I don't really see that I've got to accept that!"

The problem is clear. As Kripke puts it

> If he was not able to make the simple inference "All ravens are black, therefore *J* is black" where *J* is a particular raven, then giving him some super-premise like "Every universal statement implies each instance" won't help him either.

It won't help because he would already have to be in command of the principle to apply it; the idea that he could *adopt* it is incoherent. Kripke makes similar points about non-contradiction, adjunction and Lewis Carroll's modus ponens example. We can embody these principles in formal systems, but there's no sense to the idea that someone, so to speak, standing outside these principles could adopt them.

These are all cases where we couldn't adopt a particular principle unless we already grasped it intuitively. Perhaps that doesn't apply to all logical principles, and in any case Putnam's example had to do with giving up rather than adopting a principle. However, Kripke thinks this would miss the point. Here's what he says:

> ... I don't really mean that we adopt as basic just those things to which we can figure out that this argument applies, What I mean is this: you can't undermine intuitive reasoning in the case of logic and try to get everything on a much more rigorous basis. One has just

> to think not in terms of some formal set of postulates but intuitively. That is, one has to *reason*. One can't just adopt a formal system independently of any reasoning about it because if one tried to do so one wouldn't understand the directions for setting up the system itself. And so any comparison of logic to geometry which says that in the case of logic as in the supposed case of geometry, intuition can be thrown to the dogs—that is, any reasoning outside the system of postulates can be thrown to the dogs—must be wrong. One can only reason as we always did, independently of any special set of rules called "logic," in setting up a formal system or in doing anything else. And if proof by cases was part of our intuitive apparatus then there is no analogy to geometry which says that this should not be respected.

Kripke is surely right: logic isn't just a matter of formal systems. We can also agree that questions about how to reason have a special status among the various kinds of questions we can ask. We can agree further that for at least some logical principles there's no sense to be made of the idea that we might "adopt" them, and we can even concede that nothing could count as adopting a logic wholesale. Whether this scuttles the idea that empirical considerations could bear on logic is less clear, however.

2.3. Rejecting subalternation as a case of change in logic. To make progress, we need to look at what Kripke concedes about changes in logic and how he accounts for them. The most useful place to begin is with what he says about the principle of subalternation for universal categoricals—in particular, that "All *P* are *Q*" implies "Some *P* are *Q*."[2] Logicians once accepted this principle, and yet we no longer do. What Kripke says is surely right: if we accept subalternation, we overlook the case where '*P*' is empty. 'All deserters will be shot on sight' may be true, and that may be exactly the reason why there are no deserters. But if there are no deserters, it would be very odd to say that some deserters are shot. Intuitive reflection makes clear that something has to give. If we overlook empty terms, we'll be tempted to think subalternation is valid. We correct ourselves by mere reflection—by ordinary reasoning. However, we can ask if this is always so. When we discover that we've overlooked a case and accepted an incorrect logical principle as a result, is this always a matter of ordinary reasoning, or do empirical considerations sometimes come into play?

---

[2]Kripke talks briefly about cases where we see that an argument we once accepted is invalid. Here we change our beliefs about logic, but we do so simply and straightforwardly by *reasoning*. He also offers a cursory discussion of intuitionism. Here he claims that the intuitionists introduced new connectives, defined in terms of provability, and so the intuitionist's apparent rejection, e.g., of excluded middle isn't really in competition with the classical principle. Whether that's the best reading of intuitionists such as Brouwer I will leave to others to decide.

2.4. Future contingents, bivalence and the empirical. Consider a debate that Kripke doesn't mention but that has a long history: whether propositions about the future provide reasons to give up bivalence. Two sorts of views suggest that the answer might be yes. One is that some propositions about the future (e.g. "There will be a sea battle tomorrow" or "This atom will decay an hour from now") are contingent in a more-than-merely-logical sense. Another is the view that only the present exists, usually called "presentism" and the "growing block" view, according to which the present and the past but not the future are real. The difference between presentism proper and the growing block won't matter for our purposes, we'll use "presentism" for both.

Neither future contingent propositions nor presentism alone make the case against bivalence. Suppose some propositions about the future are contingent in the sense of not being determined by facts about the past. Suppose, that is, that determinism is false. If the so-called "block universe view" is correct, all events, past present and future, are ontologically on a par. In that case, the facts about the events making up the block entire settle the truth or falsity of future contingent propositions even if determinism is false. An event's being undetermined is a matter of its relationship to other events and to the laws of nature; whether we live in a block universe and whether the laws are deterministic are independent questions. On the other hand, suppose that presentism is correct. Then even though future states of affairs don't exist, deterministic laws plus the facts about the present could suffice to settle the truth or falsity of propositions about the future.

What, then, if presentism is true and determinism false? Perhaps bivalence about future contingent propositions can still be defended, though it's not clear how or why. What if it can't? One response is to abandon excluded middle—to claim that when 'P' is indeterminate, '$P \vee$ not-P' is likewise indeterminate (a view usually associated with Łukasiewicz.) However, there's a plausible objection: if 'P' is indeterminate, then it's not *true* that P, hence 'not-P' is true. If so, then even if 'P' is indeterminate, '$P \vee$ not-P' will be true by virtue of its second disjunct.[3]

Another familiar account of future contingents appeals to branching time and supervaluation (see [12]). On this approach, a statement about the future is true at the present moment just in case it holds on each branch or history passing through this moment, and false if it is false on each such branch. Contingent statements about the future will therefore be neither true nor false. However, this permits true disjunctions with no true disjuncts. Suppose $\{P_1, P_2, \dots, P_n\}$ is a set of future contingent propositions that are mutually exclusive, *not* logically exhaustive, but such that on each branch passing through

[3]Scope matters here; using 'F' as a future-tense operator, the claim is that when '$F(X)$' is indeterminate, 'not-$F(X)$' is true, even though '$F(\text{not-}X)$' is indeterminate. See Bourne [3, pp. 82 ff.] for useful discussion.

the present, one of them is true. An artificial example: suppose a coin will be tossed, that the outcome isn't determined, but that on each branch the outcome is either Heads or Tails. Then

The coin will come up heads or the coin will come up tails

is true at the present moment even though neither disjunct is.

Supervaluation preserves excluded middle and non-contradiction. Whether it preserves all classical logical truths might be more of an accounting issue than a substantive one. Even with excluded middle intact, the possibility of a true disjunction with no true disjuncts isn't part of logical business as usual. The novelty seems at first to sit comfortably with Kripke's view. Our belief that true disjunctions require true disjuncts came from overlooking a (complex) possibility: the combination of presentism and future contingents. However, further thought may seem to favor Putnam. This particular case for true disjunctions without true disjuncts depends on assumptions about the world: that the block universe view and determinism are both false. The overlooked possibility is a substantive one, and reasoning alone won't tell us if it holds. This suggests that matters of logic depend on the way things are, as Putnam's view would maintain.

The status of determinism is a contingent, empirical matter. However, as we noted above, even if determinism is false, this wouldn't be enough to undermine bivalence. The crucial additional assumption is presentism, and it might be argued that this is not an empirical matter; certainly the debate has often proceeded as though it's not. However, there are able defenders of the coherence of presentism *and* of the block universe. If both views are indeed coherent, empirical considerations plausibly bear on which is correct. Indeed, Putnam himself famously invoked special relativity to argue against presentism (albeit not under that name) in "Time and Physical Geometry." [6] His argument that past, present and future are equally "real" don't rest on general philosophical considerations; it depends on the structure of Minkowski space-time. It may be, then, that whether presentism is true depends on the facts about space-time. If so, it suggests that assessing the need for the logical revisions at issue in the debate over future contingents depends on contingent, empirical facts about the world.[4] The broad issue is whether claims about reality could have consequences for logic. Future contingent propositions give rise to a dilemma: if bivalence holds for such propositions, it's because of something about the world: the falsity of presentism or the truth of determinism. If bivalence fails, it's because presentism is true and determinism false. In either case, the claim is empirical. The question of determinism is *certainly* empirical and the question of presentism is at least arguably so. Thus, whether bivalence holds is an empirical matter, and that, it seems, is enough to make Putnam's larger point.

[4]Of course, not everyone agrees that Putnam's arguments are sound. See, for example, Stein [11] and Bourne [3]. To repeat, the point here is not to take sides in this debate.

The arguments above are skeletal and open to challenge, but suppose we grant them. There's a plausible Kripkean reply. Whether bivalence holds might be an empirical matter, but if so the correct conclusion is that bivalence is not a principle of *logic*. Furthermore, the conclusion that bivalence isn't a correct principle of logic is not an empirical one. We come to it by reflecting on the possibilities, and we discover that there is a genuine possibility we had overlooked: the possibility that there are no facts to ground the truth or falsity of certain propositions about the future. That this is *possible* remains so even if the possibility isn't realized.

2.5. ***Détente?*** What's just been said concedes something important to Kripke, but suggests a possibility for *détente*. Logic writ large (let's write bold-face '**logic**' for that) would remain a matter of reasoning, broadly understood. The logic of the actual world could still be a contingent matter. The analogy with geometry helps here. Suppose (unlikely, but science sometimes takes strange turns) we became convinced that the world is Euclidean after all. We would still know that the scenario Putnam describes is possible in a broad sense. It would just be that it's never actualized. The question of what the detailed geometry of a world *could* be would remain, broadly speaking, *a priori*; the question of what it is in fact would be empirical. That the world *could be* pseudo-Riemannian is not empirical knowledge. That it is or isn't pseudo-Riemannian is an empirical claim. Likewise, that bivalence *could* fail is arguably not empirical knowledge. That it *does* (or does not) fail in a particular way is arguably empirical. And though we won't try to give a general account of what counts as a question of logic, questions about the status of bivalence plausibly count.

This raises two questions. The first is whether there's a case of this sort to be made by appeal to quantum mechanics. We'll take that up in the following section. The second question will be raised but no more than raised: in light of what quantum mechanics teaches us, is it quite so clear that **logic** really is something we can know by a priori?

§3. Quantum logic reconsidered. Putnam's quantum logical proposal offered a formal structure and some interpretative principles and rationales. The structure is the lattice of subspaces of Hilbert space, but the beginnings of the disagreement with Kripke come from the interpretive overlay. Let '$S_z = +\frac{1}{2}$' say that the electron has spin $+1/2$ in direction z, and similarly for '$S_x = +\frac{1}{2}$' and '$S_x = -\frac{1}{2}$.' Putnam, as we know, would say that when

$$S_z = +\frac{1}{2} \wedge \left(S_x = +\frac{1}{2} \vee S_x = -\frac{1}{2}\right)$$

is true, one of the disjuncts in parentheses really is true, but that it's logically impossible for us to know which. However, there's another approach: treat

'($S_x = +\frac{1}{2} \vee Sx = -\frac{1}{2}$)' as a disjunctive fact—as a case of a disjunction that's true in spite of not having a true disjunct. We've already seen reasons of one sort for taking the idea of disjunctive facts seriously. Quantum mechanics gives reasons of a different sort.

What follows is intended merely as a sketch, and if the reader finds it hand-waving, that's because the author is waving his hands. The goal isn't to defend a view in detail (indeed, I am by no means certain that the view is correct) but simply to make its outlines clear enough to consider.

The paradigmatically curious quantum example is the case of two quantities—call them P and Q—that share no eigenstates. This is the heart of what Bohr called complementarity and it has two characteristic features. First, there's no arrangement that measures P and Q at the same time. Second, if we're certain what outcome a P-measurement would yield, we are *not* certain what outcome a Q-measurement would yield; all values of Q have at least some positive probability. The goal in this section is to see how we might move from here to something more clearly relevant to logic, and to do it in a way that doesn't stray far from what a typical physicist would find plausible. Note that we aren't following Putnam's approach. Putnam argued that if we adopt a strong set of logical claims, we solve the interpretive problems of quantum mechanics. There's no such goal here. We're trying to see what quantum mechanics might teach us about logic if we start from things that many physicists already believe.

The first point is simple: quantum mechanical quantities can have values. A system can have an energy or a spin in a particular direction. Few physicists would disagree.[5] The second point goes beyond ordinary common sense but not beyond the common sense of most physicists. Stick with our complementary quantities P and Q. When P has a value, Q does not. Thus: if there's a true statement

$$P = p_i$$

then there is no true statement

$$Q = Q_j.$$

No doubt most physicists believed this before no-hidden-variable proofs became widely known, but those proofs provide another reason. If we accept a handful of plausible constraints, then it's impossible for all quantum quantities to have values at the same time. Those constraints aren't beyond challenge, but our purpose isn't to make an iron-clad case. It's to make it *plausible* that quantum mechanics has consequences for logic.

The third point starts with a piece of physics common sense and then moves a bit beyond. It's that there are purely disjunctive truths about quantum

[5]Though *few* would disagree, this isn't the same as saying none would. Quantum Bayesians such as Carleton Caves, Christopher Fuchs and Rüdiger Schack are exceptions. See, for example, their [4]. For some relevant discussion see Stairs [10]

mechanical systems. To see why this is plausible, start with a special case of our first point: *degenerate* quantities can have values. For example: energy is often degenerate; the subspace that goes with

$$E = e$$

for some values e of the energy may not be one-dimensional. In spite of this, there's nothing strange about saying that the system really can have energy e—that '$E = e$' can be true. With that in mind, consider a simple but instructive example: a spin-one system whose z-spin is 0. The state $|z_0\rangle$ is a superposition of $|x_+\rangle$ and $|x_-\rangle$. On any orthodox account, the statement

$$S_x = 0$$

is definitely *false*; $|z_0\rangle$ and $|x_0\rangle$ are orthogonal. It's also of a piece with our second point to say that the system doesn't have a definite x-spin. Neither '$S_x = +1$' nor '$S_x = -1$' is true. But consider the degenerate quantity $(Sx)^2$—the square of the spin in the x direction. Again, on any orthodox account, *this* quantity has a value: $+1$. Few physicists would be shocked to be told that '$(Sx)^2 = +1$' is true when '$S_z = 0$' is true. But if the square of the spin is $+1$, it would be gratuitously peculiar to say that '$S_x = +1$' and '$S_x = -1$' are both *false*. Instead, we can say that for $(Sx)^2$ to take the value $+1$ and for

$$S_x = +1 \vee S_x = -1$$

to be true are one and the same fact: '$S_x = +1 \vee Sx = -1$' is true even though neither disjunct is. In short, bivalence fails, though for different reasons than in the case of future contingents, and we have a true disjunction with indeterminate disjuncts.[6] '$S_x = +1$' and '$S_x = -1$' stand in a different relationship to '$S_z = 0$' than '$S_x = 0$' does. '$S_z = 0$' excludes '$S_x = 0$' in an old-fashioned classical way: the two are contraries. The relationship between '$S_z = 0$' on the one hand and '$S_x = +1$' and '$S_x = -1$,' is not found in classical physics. For the states that go with these statements, the term is superposition, but there's no standard word for the relationship between the statements themselves. For present purposes, I propose *l-complementarity*.

In the language of Hilbert space, propositions are l-complementary when their associated projectors don't commute. But while that picks out the sorts of cases we're interested in, it doesn't make a connection with logic. It's also too restrictive: in principle l-complementarity is more general than Hilbert space non-commutativity. Kochen and Specker's [5] partial Boolean algebra approach is a better way to characterize l-complementarity formally. When X and Y are l-complementary they do not belong to a common Boolean

[6] Note that even if someone insisted that each disjunct is false, we'd still have a true disjunction with no false disjunct. Why anyone would insist on any such thing, however, is unclear to say the least.

subalgebra of the partial Boolean algebra.[7] However, this leaves the logical point unclear. The proposal on offer is that l-complementarity goes with a particular kind of failure of bivalence: if propositions X and Y are l-complementary, then there are possible states of affairs in which X is true but Y is neither true nor false.

With this in mind, consider distribution. In particular, consider

$$S_z = 0 \wedge (S_x = +1 \vee S_x = -1).$$

The proposal is that both conjuncts are true, but neither disjunct of the disjunction is true. That's why we can't distribute. The expression

$$(S_z = 0 \wedge S_x = +1) \vee (S_z = 0 \wedge S_x = -1)$$

either fails to pick out an element of the algebra of propositions (on the partial Boolean algebra approach) or picks out a statement that can't be true (on a lattice approach.) The distributive law fails, but not in a way that threatens looming arithmetical catastrophe; Kripke's "missing pair" is nowhere in the neighborhood.

This isn't what Putnam would say. He would say that the x-spin has a definite value, either $+1$ or -1 but that it's logically impossible to state this value along with the z-spin value. However, once we recognize the possibility of disjunctive facts, it's clear that Putnam's picture goes beyond saying that '$S_x = +1 \vee S_x = -1$' is true. We can assert the disjunction without accepting the value-definiteness thesis.

The proposal under consideration includes these points:

1. Quantum mechanical quantities sometimes have values, though not all quantities have values at once.
2. Bivalence fails; some statements about quantum systems are neither true nor false;
3. Disjunctions can be true even though none of their disjuncts are.
4. Unrestricted distribution of "and" over "or" fails.

Perhaps (1)–(4) fit quantum systems; perhaps not. What I hope to have made plausible is that they aren't shocking. A full discussion would call for much more detail (see Stairs [9] for some additional thoughts) but we turn to a different question: how well does the proposal meet Kripke's worries?

First, there's no question of "standing outside logic" and choosing a logic. This is a case of revision in light of finding an overlooked possibility: the possibility of l-complementary propositions. On the one hand, if l-complementarity is a genuine possibility, it's one that we came to by way of quantum mechanics, and quantum mechanics was an empirical discovery. However, grasping the

[7] A partial Boolean algebra is a family of Boolean algebras that share a common 0 and 1. $X \vee Y$ and $X \wedge Y$ are only defined when X and Y belong to a common member of the family.

implications for logic comes from reasoning about the theory and the conclusions about logic, and it would survive a change of physics. Recall the case of geometry. We can (dimly) imagine discovering that the best theory of space-time is Euclidean after all. However even if non-Euclidean geometry didn't fit *this* world, non-Euclidean space-time would be a genuine, albeit unrealized possibility. Reasoning won't tell us the actual geometrical structure of the world, but empirical discoveries won't tell us what the geometrical possibilities are. Similarly, for all we can say for sure, we'll find that the correct account of quantum phenomena is some version of Bohmian mechanics. If we do, physics would give us no reason to believe that the world exhibits l-complementarity, nor disjunctive facts, nor failures of distributivity. However, this wouldn't undermine the *possibility* of l-complementarity, nor the *possibility* of disjunctive facts, nor the *possibility* of a world where distributivity fails. The analogy with geometry is still apt: the *possible* structures, logical or geometrical, go beyond the actual. Empirical findings may prompt us to have thoughts we wouldn't have had otherwise, but the discovery that something is a non-actual possibility is not an empirical discovery. However, the structure that the world actually instantiates—logical or geometrical—is something we can only discover empirically.

§4. Coda: Some loose ends and some thoughts on logic and the limits of thought. A question that often comes up in discussions of quantum logic is whether it's meant to apply universally, so to speak—whether quantum logic is 'the true logic,' to use the phrase in Bacciagaluppi's "Is Logic Empirical?" [2] The point of view of this paper is that this is a misleading question. The proposal, rather, is that if quantum mechanics is true, the world embodies a logical relationship that hadn't been noticed before: the one we've called l-complementarity. If so, not all propositions are bivalent and distributivity fails in certain special circumstances. Even if l-complementarity is a genuine possibility, however, it doesn't apply to every set of propositions. Compare: suppose failures of bivalence are possible because it's possible that determinism and the block universe picture both fail. That admission wouldn't call for treating all propositions as neither true nor false, nor for saying that bivalence fails in every domain.[8] The point, rather, is that something we might have taken to hold in all cases—as a matter of **logic**—holds only in some.

What's been said also doesn't take issue with the idea that our knowledge of **logic** is a matter of reasoning. That's not because this is beyond dispute. It's because a central aim of the paper was to see where things stand if we concede to Kripke that what we've labeled **logic** is a matter of reasoning. We have argued that even if Kripke is right and **logic** is not empirical, there's still a

[8] In particular, to take one important example, it gives no reason at all to think that mathematical propositions are non-bivalent.

place for empirical considerations in thinking about logic. The empirical is not about what the logical possibilities are, but about which ones are realized.

That leaves a perplexing possibility that we'll raise but not unravel. The quantum logical story sketched here sees what we've called l-complementarity as a feature of the world. The world, so this story goes, has logical structure just as surely as it has geometrical structure; a bit too cutely, logic is empirical even if **logic** isn't. However, if this is correct it has an interesting implication: we might not be capable of grasping all of what **logic** encompasses. This, in turn *could* have the consequence that we are incapable of grasping the full logical structure of the actual world.

Go back to the case of geometry. Suppose space-time indeed has the structure of a pseudo-Riemannian manifold. In order to figure this out, we needed the capacity to grasp the relevant concepts. That wasn't inevitable; after all, there are individual people who lack that capacity. Even if we had all been unable to think the right thoughts, the world would still be pseudo-Riemannian. The same goes if the l-complementarity-based account of quantum mechanics gets the character of the world right. We are, collectively, lucky enough to be able to grasp the relevant structures and concepts; collective truth though this may be, it doesn't apply to everyone and need not have applied to anyone.

However, it *might* be that the actual geometrical structure *or* the actual logical structure of the world isn't what we think it is. And it *might* be that whatever that structure is lies beyond our cognitive reach. Logic would come unpinned from reasoning in a different way than the one Kripke argued against.

One might dismiss this as a silly kind of skepticism. That would be fair if the suggestion were that we might be deeply and radically ignorant about logic. However, that's not the thought. On the contrary (though we haven't discussed this) a full explication of l-complementarity assumes that propositions are sometimes related exactly as classical logic says they are. (A partial Boolean algebra, after all, is a family of Boolean algebras. Similar remarks apply to orthomodular lattices.)

The point, rather, is this. What quantum mechanics may well represent is a case in which we stumbled on a surprising exception to logical business as usual. However, a full account of l-complementarity calls for positing relationships among properties that we don't grasp easily. Studying, for example, partial Boolean algebras, as abstract mathematical structures is, of course, not the issue. The difficulty is in grasping what it means for states of affairs in the world to mirror that structure. One might fairly say that the persistent difficulty in understanding quantum mechanics has been understanding what it means for *the world* to have the structure that the mathematics seems to attribute to it. In light of this, the possibility that there might be yet more esoteric exceptions to business as usual doesn't seem quite so silly. A proper modesty suggests that there's no guarantee that we'll find them even if they exist. And a healthy suspicion about our limitations suggests there's no guarantee we would be able to recognize them even if they're there. **Logic** in its fullness just might be beyond our grasp.

REFERENCES

[1] W. ALSTON, ***Perceiving God***, Cornell University Press, Ithaca New York, 1991.

[2] G. BACCIAGALUPPI, *Is logic empirical?*, ***Handbook of Quantum Logic*** (D. Gabbay, D. Lehmann, and K. Engesser, editors), Elsevier, Amsterdam, 2009, pp. 49–78.

[3] C. BOURNE, ***A Future for Presentism***, Clarendon Press, Oxford, 2006.

[4] C. M. CAVES, C. A. FUCHS, and R. SCHACK, *Subjective probability and quantum certainty*, ***Studies in History and Philosophy of Modern Physics***, vol. 38 (2007), p. 255.

[5] S. KOCHEN and E. SPECKER, *The problem of hidden variables in quantum mechanics*, ***Journal of Mathematics and Mechanics***, vol. 17 (1967), pp. 59–87.

[6] H. PUTNAM, *Time and physical geometry*, ***Journal of Philosophy***, vol. 64 (1967), pp. 240–247. Reprinted in ***Mathematics, Matter and Method***, Cambridge University Press, 1975, pp. 198-205.

[7] ———, *Is logic empirical?*, ***Boston Studies in the Philosophy of Science*** (Robert S. Cohen and Marx W. Wartofsky, editors), vol. 5, D. Reidel, Dordrecht, 1968, pp. 216–241. Reprinted as *The logic of quantum mechanics* in ***Mathematics, Matter and Method***, Cambridge University Press, 1975, pp. 174-197.

[8] A. STAIRS, *Quantum logic, realism and value-definiteness*, ***Philosophy of Science***, vol. 50 (1983), pp. 578–602.

[9] ———, *Kriske, Tupman and Quantum Logic: the quantum logician's conundrum*, ***Physical Theory and its Interpretation*** (W. Demopoulos and I. Pitowsky, editors), Springer, 2006.

[10] ———, *A loose and separate certainty: Caves, Fuchs and Schack on quantum probability one*, ***Studies in History and Philosophy of Modern Physics***, vol. 42 (2011), pp. 158–166.

[11] H. STEIN, *On Einstein-Minkowski space-time*, ***The Journal of Philosophy***, vol. 65 (1968), pp. 5–23.

[12] R. H. THOMASON, *Indeterminist time and truth value gaps*, ***Theoria***, vol. 36 (1970), pp. 264–281.

DEPARTMENT OF PHILOSOPHY
UNIVERSITY OF MARYLAND
COLLEGE PARK, MD 20742
E-mail: stairs@umd.edu

THE ESSENCE OF QUANTUM THEORY FOR COMPUTERS

WILLIAM C. PARKE

Abstract. Quantum computers take advantage of interfering quantum alternatives in order to handle problems that might be too time consuming with algorithms based on classical logic. Developing quantum computers requires new ways of thinking beyond those in the familiar classical world. To help in this thinking, we give a description of the foundational ideas that hold in all of our successful physical models, including quantum theory. Our emphasis will be on the proper interpretation of our theories, and not just their statements. Our tact will be to build on the concept of information, which lies central to the operation of not just computers, but the Universe. For application to quantum computing, the essence of quantum theory is given, together with special precautions and limitations.

§1. Introduction. Having a grasp on the ideas behind a theory helps to apply it correctly, to understand its limitations, and to generate new ideas. Getting a firm hold on quantum theory is not an easy task, because our experiences and even our genetic predispositions have been developed in a world in which quantum effects are largely washed out.[1] Remarkably, our predilection for finding logic behind the behavior of what we observe,[2] including that of electrons and atoms, has led us to quantum theory, a description of nature that is hard for us to conceptualize, but is logical, accurate, and explains a wide variety of phenomena with only a few statements and input.

As background to quantum theory and quantum computing, an attempt is made here to give the primitive notions and essential observations that underlie current physical theories, so that foundational ideas are explicit, and a common language is established. In our description, information storage and transfer is made central.[3] A short description of quantum theory follows, and

[1]Although these days, macroscopic quantum effects can be seen in the actions of lasers and of quantum fluids.

[2]Our curiosity is enhanced by genetic selection, as there is advantage to being able to make sense of what goes on around us, so that we can anticipate what might happen next.

[3]Traditionally, energy transfer is used to characterize interactions in current theories. However, the concept of energy is several steps removed from more basic ideas. Moreover, information processing is not only the purpose of computers, but also lies underneath all natural processes.

Logic and Algebraic Structures in Quantum Computing
Edited by J. Chubb, A. Eskandarian and V. Harizanov
Lecture Notes in Logic, 45
© 2016, Association for Symbolic Logic

then applied to quantum computing, focusing on what the theory says, and particularly does not say, in areas where conceptual difficulties have arisen.

§2. Physical theory and reality. A physical theory is a logical model capable of making predictions of what we observe. It is judged by its accuracy in matching measurements, and by its economy, i.e. whether the proposed theory has only a few relationships and input data needed for its ability to explain observations over a wide realm.[4]

We should not, however, become too enamored with the auxiliary structures within a successful theory. Just as it is possible to transform, isomorphically, a logical structure into an equivalent one involving distinctly different relationships and symbols, it is also possible to so transform a physical theory. A good example is the transformation of Maxwellian electrodynamics into an action-at-a-distance form. The transformed theory, invented by Wheeler and Feynman,[5] no longer contains electric or magnetic fields. Even so, it makes the same predictions as Maxwell's theory.[6] A lesson from this example and others is that one should not endow physical meaning to all the symbols and relationships in a theory. Electric fields do not 'exist' in nature. They exist as symbols on paper and in our minds. But Maxwell's theory does make definite statements about observations using the electric field concept. Only those points in the theory that are stated as predictions can be connected to nature. In quantum theory, wave functions are clearly not physical; in general, they are complex numbers. They can also be transformed away in alternate but equivalent theories.[7] Rather, one should think of the symbols and relationships in a theory as *tools for making predictions*. Predictions are the touchstones in the theory. All else is ancillary.

Here is another caution: Predictions of pure counts are testable as either true or false, but predictions of continuous values will never be proved to match nature exactly, since our measuring instruments are finite. Theories which take space as continuous implicitly do so only down to the scale permitted by our instruments. There should be no implication that even continuity exists at finer scales.

---

[4]In information theory terms, the information contained in the independent data explained by a theory should be much larger than the information needed to express the theory.

[5]J. A. Wheeler and R. P. Feynman, *Classical electrodynamics in terms of direct interparticle action*, ***Reviews of Modern Physics***, vol. 21 (1949), pp. 425–433.

[6]We generally use Maxwell's theory to solve electrodynamics problems because the Wheeler-Feynman theory is a more complicated mathematical system.

[7]For example, Werner Heisenberg's formulation of quantum theory, shown by P. A. M. Dirac to be completely equivalent to Erwin Schrödinger's, uses no wave functions. Neither do various so-called hydrodynamic formulations, such as that of E. Madelung in *Quantentheorie in hydrodynamischer Form*, ***Zeitschrift für Physik***, vol. 40 (1927), pp. 322–326.

Our best physical theory so far is the so-called 'Standard Model',[8] which describes, with quantum field theory, all of the interactions yet detected, except for gravity. The Standard Model has made remarkable and now verified predictions and agrees with the most precise of measurements made to one part in a trillion. Even so, the theory is not tight, having many unexplained interaction strengths and masses. We expect new theories will give a deeper and simpler explanation of particles, of their interactions, and of the yet unexplored regions in nature.

In the next section, a set of tentative propositions and observations underlying all physical theories is proposed, building toward the foundations of quantum theory and application to quantum computing. Information storage and transfer will be seen to be fundamental to natural processes.

§3. Basic properties of physical systems. The natural world is divisible into a collection of observable subsystems. Each observable subsystem will be referred to as a *physical system*. If a physical system can be further divided, the parts may be called 'components' of the system. The number of divisions may reach a limit.

A physical system can store *information*, taken to be an additive quantity which grows with the number of distinct ways that the system may be configured under given physical constraints. The number of ways is called the system's '*multiplicity*', W.[9] To be additive across independent systems, the information I in a system must be proportional to $\ln W$.[10] With $I = \ln_2 W$, the information is given in 'bits'.[11] If the multiplicity W of a system decreases, we say the system has become more '*ordered*'.[12]

---

[8]For a personal perspective in the development of the Standard Model, see S. Weinberg's article, *The making of the standard model*, ***European Physical Journal C***, vol. 34 (2004), pp. 5–13.

[9]One of the many remarkable implications of quantum theory is that the count W can be performed over a denumerable number of quantum states of a system.

[10]If there were two independent systems of multiplicity W_1 and W_2, then the multiplicity of both together would be $W_1 W_2$. The condition $f(W_1 W_2) = f(W_1) + f(W_2)$ makes $f(W)$ proportional to $ln(W)$.

[11]If a given system subject to physical constraints cannot be re-configured, then that system has no information. If the system has two possible configurations, its reading transmits one bit of information, the equivalent of a yes or a no, but no more, and so forth.

[12]In the late nineteenth century, Ludwig Boltzmann introduced the number W ('Wahrschein-lichkeit'), connecting it to the disorder (Clausius' 'entropy', S) of a system with $S = k \ln W$, where k is Boltzmann's constant. Leó Szilárd showed that each bit of information we gather from a system and discard necessarily requires an increase in entropy of at least $k \ln 2$. (L. Szilárd, *On the decrease of entropy in a thermodynamic system by the intervention of intelligent beings*, ***Zeitschrift für Physik***, vol. 53 (1929), pp. 840–856.) Claude Shannon developed the formalism of information theory, including information transfer in the presence of noise. (C. E. Shannon, *A mathematical theory of communication*, ***Bell System Technical Journal***, vol. 27 (July & October, 1948), pp. 379–423 & 623–656.)

An *interaction* between two physical systems, by definition, exchanges *information* between them. An *open physical system* can interact with other systems. *Observation* is made by allowing two physical systems to interact, one of which is prepared as a measuring instrument. A *measuring instrument* is a physical system whose information gathered from an observed system is capable of being copied with a relatively high assurance. The copy will act as a *record* of the observation. Statements about a physical system are verified only by observations.[13] A statement about a physical system is *predictive* if it relates a number of observations of that system. A physical system is *isolatable* if the measurable effect of all interactions with other external systems can be made arbitrarily *small.*[14] An isolated physical system is said to be 'closed' when external interactions which might influence the results of intended measurements of that system are negligible.

If a set of observations of a system is found to repeat, that system can act as a *clock*, with *time defined* and measured by the number of repeats, each smallest repeating cycle called a *period* of the clock. If a large set of independent periodic systems, prepared in the same way, are found to consistently have the same number N of periods, these clocks are said to be 'good' to a precision of at least one part in $1/N$.

The *distance* between two interacting physical systems is defined, up to a selected constant factor, to be the minimum time needed for an observable change in one of those two systems to cause an observable effect on the other. *Space* is defined to be the set of available distances between all systems. Two systems with a finite distance between them are said to be spatially separated. If one isolatable system can be spatially separated from all others, it is *localizable*.[15] If N localizable systems can be spatially separated from each other by the same distance, then space has at least $N-1$ spatial dimensions. A system localizable in each spatial dimension can be referred to as a *body*. The *spatial coordinates* of a body are the minimal set of numbers that uniquely determine a definable *location* within the body. These coordinates are measured by one observer relative to an 'origin', a location used by that observer to coordinate a set of bodies. In an N dimensional space, a complete set of such coordinates for one location is denoted $\{x^1, x^2, \ldots, x^N\}$. An *event*, $\{x^0, x^1, x^2, \ldots, x^N\}$, specifies when and where an observation has occurred.

---

[13]This grounding is particularly poignant in quantum theory, wherein a quantum system is described by a set of interfering possible states for each observable, with only one such state realized by observation.

[14]We will use the term 'small' for a quantity which has the property that if made smaller, there would be no significant effect.

[15]Defining the localizability of zero-mass particles with spin greater than $1/2$ (in units of Planck's constant over 2π), such as the photon, is tricky. For a definition, and references back to Wolfgang Pauli, see M. HAWTON, *Photon position operator with commuting component*, ***Physical Review A***, vol. 59 (1999), no. 2, pp. 954–959.

A *frame of reference* characterizes how one observer records events. If the spatial separation between two bodies changes with the observer's time, we say they have relative motion. Bodies with no macroscopic motion relative to the observer are said to be stationary. The *velocity* of a body is its spatial change per unit observer's time along each of the independent spatial directions, and the *acceleration* is the change of velocity per unit time, each measurement made in a single frame of reference.

A *particle* is a localizable physical system with some identifiable intrinsic characteristics, i.e. quantities that are independent of how the observer measures them. A *fundamental particle* is a particle that suffers no measurable change in its intrinsic information even after engaging in all available interactions. A 'free particle' is a particle whose interactions with other systems can be neglected.

Recording a *complete set of observables* in a system determines (to the degree possible) the information present in that system at the time of measurement and before any further interaction with the system. The selection of observables is made such that the measurement of any one does not change the result that would be found for the measurement of any other in the selected set. Those observables that are time independent are called *conserved*.

The 'dynamics' of a physical system, i.e. a description of how interacting subsystems change over time, follows logical predictive schemes which reveal *cause and effect*. These schemes are most easily tested using isolatable 'simple systems', i.e. those with only a few discernible component subsystems and low information content. So far, all physical systems can be described by the interactions of fundamental particles in space-time.

Systems with many interacting components, called 'complex systems' or 'macrosystems', have been successfully described when those components can be tracked, or when statistical likelihood arguments become meaningful. Systematics in the behavior of complex systems make global properties referred to as *emergent relationships*. Those of thermodynamics and statistical mechanics are examples. Rules for optimal dynamics in biosystems[16] form others.

Some systems, through the mutual interactions of their particles, will form bound bodies, i.e. systems that retain their localized character provided external interactions are sufficiently weak. A *confined* system is one which, when initially localized in a certain volume with zero average velocity, and then left alone, will have a non-zero lower bound on the probability of being found in the initial volume later in time. The ability to create bound systems gives preference to the evolution of differentiated systems and to condensation into locally ordered subsystems. With a sufficient variety of particles and interactions,

[16]A 'biosystem' is a physical system whose activities support life. A life system is one which is capable of self-replication by interactions with external systems, using information stored within the life system.

the evolution of complexity in open subsystems is natural,[17] including the evolution of life.

The *Universe* is defined as the collection of everything that can be observed.

§4. Space-time as background to quantum theory. The Universe appears to have existed in a finite number of current clock periods, and the volume of our Universe apparently is also finite. There is a limit to the greatest separation between bodies. The dimension of our space is at least three.[18] The distance between widely separated bodies has been growing relative to the size of the smallest bodies since time started.

Observation shows that, to a good approximation, there exist *inertial frames* in which an isolated body nearby and initially stationary relative to the observer will continue to be nearly stationary. We will use the term 'inertial observer' for an observer in an inertial frame. In inertial frames, an interaction experienced by one body can always be associated with the effect of other local bodies. At small scales, the relationships between local events can be expressed in a form that is independent of the observer's position, orientation, or motion relative to the events. This is the grand '*Principle of Relativity*'.[19] The Principle of Relativity allows for the existence of a finite *universal limiting speed* for all bodies.[20] Examination shows that our Universe has a finite limiting speed. To the precision of current measurement, the interactions due to electromagnetism and gravity carry information between bodies at the universal speed c.

[17] L. Onsager, *Reciprocal relations in irreversible processes*, I & II, ***Physical Review***, vol. 37 & 38 (1931), pp. 405–426, 2265–2279; I. Prigogine, ***Introduction to Thermodynamics of Irreversible Processes***, Interscience, New York, 1955.

[18] Three dimensions is also the minimum dimension needed to build a computer or brain having more than four devices with mutual connections. At present, there is no evidence for higher extended dimensions than three. The strong experimental support of our conservation laws in three dimensions suggests that if higher dimensions of space existed, matter and energy would have had extreme difficulty passing into or out of it.

[19] Radiation from distant galaxies and radiation left over from the hot big bang do establish a unique frame of reference, but these are taken as part of the initial conditions in dynamics and so do not vitiate the relativity principle. In our Universe, the residual effects of these initial conditions on present observations of local events are often small.

[20] As demonstrated by H. Poincaré in *L'état actuel et l'avenir de la physique mathématique*, St. Louis Conference, ***Bulletin des sciences mathématiques***, vol. 28 (1904), pp. 302–324. Einstein's second postulate, the constancy of the speed of light, is not needed. Relativity alone, under reasonable assumptions about how events are measured in close by inertial frames with relative motion, initially aligned, allows only one relationship between their space-time coordinates. That relationship is the Lorentz transformation, containing a fixed universal speed called c. Explicitly, if the second inertial frame moves at a speed v away along the positive x-axis of the first, then $x_2 = (x_1 - vt_1)/\sqrt{1-(v/c)^2}$, $y_2 = y_1$, $z_2 = z_1$, $t_2 = (t_1 - (v/c^2)x_1)/\sqrt{1-(v/c)^2}$. The Galilean transformation is approached when the universal speed in the Lorentz transformation is taken much larger than the relative speeds of the observed bodies. This makes $t_2 \cong t_1$, so that time becomes universal in this limit.

In Relativity, one observer's measure of spatial separation between two bodies is related to a combination of space and time coordinates of another observer moving relative to the first. This makes the concept of space and time inseparable, and gives utility to the idea of a *four-vector* using the coordinates in space and time for a pair of close by events, in the form $\{dx^\mu\} = \{dx^0, dx^1, dx^2, dx^3\}$, where $x^0 \equiv ct$. Any other ordered set of four quantities forms a four-vector if they transform by coordinate transformations just like $\{dx^\mu\}$ does.

Relativity makes the small interval between two events, $ds = \sqrt{g_{\mu\nu} dx^\mu dx^\nu}$ invariant,[21] i.e. independent of the observer's frame of reference. The set of quantities $\{g_{\mu\nu}\}$ form what is called the *metric tensor*. Each infinitesimal space-time region within any inertial frame can be covered by an orthogonal coordinate grid, so that the metric tensor is well approximated by $\{g_{\mu\nu}\} \approx diag\{1, -1, -1, -1\}$. A vector 'dual' to $\{dx^\mu\}$ can be defined by $dx_\mu \equiv g_{\mu\nu} dx^\nu$, so that $dx_\mu dx^\mu$ is a 'scalar', i.e. a number who value is independent of the frame of reference of the observer. The sum $A_\mu B^\mu$ defines the scalar product of the two vectors, and the length of A is $\sqrt{A_\mu A^\mu}$. An important example of a four-vector is a particle's four-momentum, $\{p^\mu\}$, with cp^0 being the energy of the particle and $\vec{p}$ its spatial momentum. The length of $\{p^\mu\}/c$ is the mass of the particle.[22]

A general coordinate transformation between frames of reference, $x' = f(x)$, becomes a 'Poincaré transformation' when $x'^\mu = a^\mu_{\ \nu} x^\nu + b^\mu$ and the coefficients $\{a^\mu_{\ \nu}\}$ satisfy $g_{\mu\nu} a^\mu_{\ \kappa} a^\nu_{\ \lambda} = g_{\kappa\lambda}$.[23] Rotations, Lorentz transformations, and displacements are included. The set $\{a^\mu_{\ \nu}, b^\kappa\}$ forms the so-called 'Poincaré group', with the product rule $\{a''^\mu_{\ \nu}, b''^\kappa\} = \{a'^\mu_{\ \lambda} a^\lambda_{\ \nu}, (a'^\mu_{\ \lambda} b^\lambda + b'^\kappa)\}$.[24]

A body initially stationary in an inertial frame, but acted on by one other body some distance away, will accelerate. If a duplicate of the first body is weakly bound to the first, and the experiment repeated, then the acceleration of the pair will be half the rate of the single one. We say the pair has twice the '*inertial mass*' of the single body. The inertial mass of a particle is an intrinsic property.

The observation of the effects on the motion of bodies due to the acceleration of the observer's frame with respect to an inertial frame is locally indistinguishable from the effects of gravity. This is Einstein's *Equivalence Principle*. Einstein's Equivalence Principle makes inertial mass the same as '*gravitational*

---

[21] By convention, repeated indices, one upper, one lower, should be summed from 0 to 3.

[22] The energy and momentum of a system are best defined, in our successful theories, through the generators of time and space translations, with a scale determined by gravity. These ideas will be presented shortly in the context of Noether's Theorem and Einstein's General Relativity Theory.

[23] Note that this relation makes the metric components an 'invariant tensor', in that the components take the same values after a coordinate transformation.

[24] Reflections are excluded by imposing $det\,|a| = 1$. Then the transformations are called 'proper'.

mass', which is the intrinsic property of a body that determines the strength of its gravitational influence on nearby systems.[25] The mass of any localized system (including the equivalent mass of any associated localized field energy) can be measured by using the gravitational pull that system creates on a distant mass.

The Equivalence Principle, together with the Principle of Relativity, requires that the distance measure of space-time in the presence of a gravitating body be non-Euclidean, i.e. there will be intrinsic curvature to the space-time around a body with mass, and the metric tensor $\{g_{\mu\nu}\}$ can no longer be transformed by a coordinate choice to the form $\{g_{\mu\nu}\} = diag\{1, -1, -1, -1\}$ in any finite region of the space near the body. However, even in the presence of mass, inertial observers will still find an approximate flat metric in their infinitesimal neighborhood.

Einstein showed that the effects of gravity due to masses could be found from conditions on the Riemannian curvature of space-time. Curvature can be characterized by the behavior of vectors as they are moved from one point to another across space. Infinitesimal changes in any vector that are observed while transporting that vector along a path define the 'covariant derivative': $D_\kappa A^\mu = \partial_\kappa A^\mu - \Gamma^\mu_{\nu\kappa} A^\nu$. The changes due to the underlying geometry come from the 'connections' $\Gamma^\mu_{\nu\kappa}$ in the space. In Riemannian geometry, the connections are determined by gradients of the metric tensor.[26] The vector $A^\mu(x_0) - \int_{x_0}^{x} \Gamma^\mu_{\nu\kappa} A^\nu \, dx^\kappa$ is said to be the components of the 'parallel transport' of the original vector at x_0 along a particular path to x. The change $\delta_\Box A^\mu$ in the components of any vector field, $A^\mu(x)$, by carrying the vector in parallel transport around an infinitesimal closed loop, must be proportional to the area of the loop and the size of the original vector field. The proportionality constants in each small patch of space-time defines the curvature tensor $\{\mathcal{R}^\mu{}_{\nu\kappa\lambda}\}$ in that patch, to wit: $\delta_\Box A^\mu = \mathcal{R}^\mu{}_{\nu\kappa\lambda} A^\nu \, dx^\kappa dy^\lambda$, where the loop is given orthogonal sides dx^μ and dy^μ.

Einstein's General Theory of Relativity[27] is the simplest of a class of theories that incorporate the Equivalence Principle and the Principle of Relativity.[28] Einstein discovered that in empty space, the condition on the metric curvature tensor[29] given by $R^\mu{}_{\kappa\mu\lambda} = 0$ numerically predicts: Newtonian gravitational fields when the effects of gravity differ little from flat space; The size of the extra perihelion precession of Mercury's orbit; The amount of the gravitational

[25]The Equivalence Principle also means that mass m can be measured in distance units by giving Gm/c^2, where G is Newton's gravitational constant that determines the strength of gravity.

[26]In the form $g_{\mu\lambda}\Gamma^\lambda_{\nu\kappa} = (1/2)(\partial_\kappa g_{\mu\nu} + \partial_\nu g_{\mu\lambda} - \partial_\mu g_{\nu\kappa})$.

[27]A. Einstein, *Die Grundlage der allgemeinen Relativitätstheorie*, ***Annalen der Physik***, vol. 49 (1916), pp. 50–205.

[28]More general theories can be constructed using higher derivatives of the metric tensor in the field equations than the second.

[29]The metric curvature tensor $\{R^\mu{}_{\nu\mu\lambda}\}$ is that part of the local curvature tensor $\{\mathcal{R}^\mu{}_{\nu\kappa\lambda}\}$ due solely to changes in the metric across space-time.

deflection of light, and; The interval for the slowing of clocks in a gravitational field. All these and more have been confirmed to the precision of current instruments.[30]

In both the Special and the General Theory of Relativity, time is not universal. If two good clocks are synchronized in one frame of reference, and one is set in motion relative to the other, they may differ in the number of periods each had when they are brought back together.[31]

In General Relativity, bodies acted on by gravity follow a 'geodesic', i.e. a path that makes the invariant four-dimensional distance $\int ds$ along the path between fixed initial and final points of the motion extreme. Free particles that travel at the ultimate speed c also follow geodesics, are necessarily massless, carry no charge, and cannot spontaneously decay.[32]

Einstein's General Relativity Theory describes how the classical field $\{g_{\mu\nu}\}$ should vary over space-time. All 'dynamical fields', to be consistent with quantum theory, must have corresponding quanta.[33] We expect that the quantum aspects of gravity will be important near the 'Planck scale' [34] $\sqrt{\hbar G/c^3} \approx 1.6 \times 10^{-35}\ m$. Although this is far smaller than the regions we can explore with current accelerators, the very rarely detected ultra-high energy cosmic rays may be scattered by this quantum granularity of space.

§5. Quantum theory.

5.1. The essence of quantum theory. Boiled down to its essence, quantum theory follows from a prescription due to Feynman:[35]

---

[30] Calculations of position on Earth using Global Positioning Satellites at height h and speed v over an Earth of mass M_E and radius R_E, have Special Relativity corrections included to order v^2/c^2 for the relativistic Doppler shift and General Relativity corrections included to order $GM_E h/(c^2 R_E^2)$ for clock slowing in a gravitational field. Without these, errors in positions would be unacceptable!

[31] This leads to the 'Twin Paradox', that one twin can end up younger than the other, yet each sees the other move away and then come back. The resolution came from Einstein using his General Theory of Relativity. The difference in the time elapsed by the clocks will be the difference between the values of $\int \sqrt{|g_{\mu\nu}\, dx^\mu dx^\nu|}/c$, integrated along the path of each clock from the common starting point to the common endpoint.

[32] In relativistic quantum theory, no localizable charge can be carried by a massless particle with spin greater than 1/2, nor can there be a localizable flow of energy and momentum for massless particles with spin greater than 1. See S. Weinberg and E. Witten, *Limits on massless particles*, ***Physics Letters***, vol. B 96 (1980), no. 1–2, pp. 59–62.

[33] See, for example, M. P. Bronstein, *Quantentheorie schwacher Gravitationsfelder*, ***Physikalische Zeitschrift der Sowjetunion***, vol. 9 (1936), pp. 140–157. Generally, a dynamical field varies both over space and in time. Formally, fields which have a kinetic energy term in the Lagrangian for the system are dynamic.

[34] M. Planck, *Über irreversible Strahlungsvorgänge. Fünfte Mitteilung*, ***Sitzungsberichte der Königlich Preussischen Akademie der Wissenschaften zu Berlin*** (1899), pp. 440–480.

[35] Feynman began thinking of these ideas in 1942. They are described in: R. P. Feynman and A. R. Hibbs, ***Quantum Mechanics and Path Integrals***, McGraw Hill, 1965.

For each particle that was initially observed at A and later observed at B, construct a complex number, called the *transition amplitude*, as a sum of unimodular complex numbers according to:

$$\langle B|A\rangle = N \sum_{paths} \exp(2\pi i S/h)\,. \tag{1}$$

The factor N will be fixed by a 'normalization' condition, introduced shortly. Each exponential term in the sum has a phase given by $2\pi S/h$. The number h is called Planck's constant. The quantity S is called the action, defined by a time-integration from A to B of a function L:

$$S = \int_A^B L\,dt\,. \tag{2}$$

The Feynman sum Eq. (1) is carried over all distinct paths between A and B.[36]

The function L, called the Lagrangian, depends on the particle coordinates and time changes of coordinates for the possible paths between A and B. The Lagrangian is presumed known, and often can be expressed as the particle's kinetic energy minus its potential energy. Helping to strongly limit the possible Lagrangians is the imposition of the symmetries we observe, such as the Poincaré symmetry of Relativity.

By reversing the order of the time limits in the action integral Eq. (2), the phases of the Feynman amplitudes change sign, so that time reversal of a transition amplitude is equivalent to taking its complex conjugate: $\langle B|A\rangle = \langle A|B\rangle^*$. Let B range over all possible states into which A may evolve. Then $\sum_B \langle A|B\rangle\,\langle B|A\rangle$ gives the amplitude for the state A to explore all possible alternatives but then return to itself. We can take this amplitude to be unity and thereby fix the magnitude of the normalization constant N. We will then have

$$\sum_B \langle A|B\rangle\,\langle B|A\rangle = \sum_B |\,\langle B|A\rangle\,|^2 = 1. \tag{3}$$

This relation makes it possible to interpret the magnitude square of the Feynman amplitude as a probability for a given transition. Doing so creates quantum theory. That's it. All of quantum mechanics follows.

In contrast to the determinism of Newtonian theory,[37] quantum theory gives probabilities for the result of each measurement of a system. These probabilities are not simply the result of statistics applied to events. In quantum theory, a system can be in an interfering combination of possible realizable events before

[36] For an excellent description on how Feynman paths are constructed, see H. KLEINERT, ***Path Integrals in Quantum Mechanics, Statistics, Polymer Physics, and Financial Markets***, 5th edition, World Scientific, Singapore, 2009.

[37] The assumption that systems have a definite state of existence between interactions would follow from having only a 'single' path dominate the Feynman sum over paths.

one of these events is determined by interactions with another system such as by measurement.[38]

If one takes field quantities as a set of equivalent particle oscillators in each infinitesimal volume of space, with the field amplitudes as the particle displacements, then quantum field theory follows.

5.2. The classical limit. Note that the summation of unit complex numbers with wildly different phases will tend to cancel (think of adding unit vectors in a plane with arbitrary angles between them), while a collection of such complex numbers with almost the same phase tend to add coherently. This observation applied to the Feynman path sum shows how to take the classical limit, in that those paths causing the least change in the action S relative to the size of h contribute the most to the probability. Classical physics includes only those paths between two events that minimize S. This is the famous 'Principle of Least Action', from which Newton's laws and Maxwell's electrodynamics can be derived, after the appropriate choice of L.[39] When compared with quantum theory, Newtonian theory for particles, Maxwell's electrodynamics, and statistics applied to Newtonian systems with a large number of particles are together in a realm called 'Classical Physics'.

A *'classical computer'* is a dedicated physical system which transforms a prepared initial state into a desired output state by applying the equivalent of Boolean logic in one or more steps between input and output.

5.3. Superposition. From the observation that the action satisfies $S_{BA} = S_{AC} + S_{CB}$, it follows from Eq. (1) that

$$\langle B|A\rangle = \sum_C \langle B|C\rangle \langle C|A\rangle \ . \tag{4}$$

Quantum amplitudes contain a linear superposition of possible intermediate states. If the allowed Feynman paths from A to B are restricted to only those that pass through two small intermediate regions, say C_1 and C_2, there will be interference of the amplitudes constructed to pass through C_1 with those constructed to pass through C_2. This interference can be completely

[38]The fact that certain predictions of quantum theory have intrinsic probabilistic character and that the possible realizable states of a system retain strange correlations over arbitrarily long distances between particles, greatly disturbed Einstein. But John von Neumann showed that quantum theory cannot be trivially subsumed into a bigger deterministic theory. See J. von Neumann, ***Mathematical Foundations of Quantum Mechanics***, Princeton University Press, 1955, Chapter 4. For more recent work, see R. Colbeck and R. Renner, *No extension of quantum theory can have improved predictive power*, ***Nature Communications***, vol. 2 (2011), pp. 411–416. So far, all careful observations are consistent with quantum theory, even ones that Einstein called 'spooky action at a distance'.

[39]That non-relativistic quantum mechanics has Newtonian theory as a limit is an example of the 'correspondence limit' which we impose on any new theory in order to sustain the verified predictions of earlier observations. After all, Newton's theory predicts natural processes quite well for massive slowly moving bodies, like baseballs, moons, and spacecraft.

destructive, so that repeated searches for a particle at B that was launched from A come up practically empty. This effect is observed, and has *no* explanation in classical particle theory. Yes, you might say, but isn't the particle a wave? No, we never observe particles as waves. We never find a particle 'spread out'. Rather, the *probability* of finding a particular particle *somewhere* can be spread out over space. Individual particles are *always* found localized. Quantum theory lets us calculate these new kinds of probabilities. New, because these probabilities are found by first adding complex amplitudes, a formulation for probabilities unheard of before the second decade of the 1900s. Addition of amplitudes allows for interference effects, even for a single particle. This makes the resultant probabilities an intrinsic property of the theory, and not just due to ignorance of states in a more deterministic theory.

5.4. Wave functions and quantum states. The Feynman transition amplitude for a particle to leave any earlier location A with coordinates x_0 at time t_0 and arrive at B having the location x at time t is called the *wave function* for that particle over the spatial coordinates x at the time t:

$$\psi(x,t) = \langle B(x,t)|A(x_0,t_0)\rangle \ . \tag{5}$$

From Eq. (3), $\int \psi^*(x,t)\psi(x,t)\,dx = 1$. The symbol dx in the integral is to be interpreted as the volume element in space. We see that $\psi^*(x,t)\psi(x,t)dx$ is the probability of finding the particle within the volume dx. Dirac recognized that wave functions may be considered a projection of the 'state of the system' described by a vector denoted $|\psi\rangle$ onto a specific state ('eigenstate') of position: $\psi(x,t) = \langle x|\,\psi(t)\,\rangle$. Each '*quantum state*' $|\psi\rangle$ can be considered a vector in a Hilbert space.[40] Superposition allows us to expand the quantum state into a complete set of basis states:

$$|\psi\rangle = \sum_a |a\rangle\,\langle a|\,\psi\rangle \, .$$

(The sum over 'a' may be given continuous regions as an approximation to discrete sums which are dense in those regions.)

From the Feynman path sums, the state of a system evolves in time according to a linear transformation

$$|\psi(t)\rangle = U(t,t_0)\,|\psi(t_0)\rangle \ , \tag{6}$$

where, to keep the total probability of finding the particle anywhere unity, the operation U must be unitary: $U^\dagger U = 1$. (The 'dagger' here performs a transpose-complex-conjugate operation, rather than just complex-conjugation, to include cases in which ψ is taken to have components.) The Feynman path summation divided into small time steps means we can write $U = \exp(-i\int H\,dt/\hbar)$. (The sign in the exponent is conventional. The constant $\hbar = h/(2\pi)$.) The operator H, called the Hamiltonian, satisfies the 'Hermiticity

[40] Essentially a vector space with lengths and angles defined, but possibly infinite dimensional.

condition' $H = H^\dagger$. In the language of Lie groups, H is a generator of time translations. For small shifts in time, ψ satisfies a linear equation:

$$i\hbar\partial_t\psi(x,t) = H\psi(x,t)\,. \tag{7}$$

This is a wave equation, which formed the basis of the dynamics of quantum theory originated by Schrödinger.

5.5. Particles in relativistic quantum theory. Our present quantum theory incorporates Einstein's Special Theory of Relativity.[41] P. A. M. Dirac, recognizing that Relativity requires that physical laws be expressible with space and time on an equal footing, wrote the Hamiltonian as a linear operator in the generators of space translation, so that the wave equation took the form[42]

$$\sum_{\mu=0}^{3}\gamma^\mu(i\hbar\partial_\mu - (e/c)A_\mu)\psi = mc\psi\,. \tag{8}$$

When the fields A_μ vanish, there are plane wave solutions $\psi \propto \exp(-ip_\mu x^\mu)/\hbar)$, so that $g_{\mu\nu}i\hbar\partial^\mu i\hbar\partial^\nu\psi = p_\mu p^\mu\psi = m^2c^2\psi$, and the γ's must satisfy

$$\gamma^\mu\gamma^\nu + \gamma^\nu\gamma^\mu = 2g^{\mu\nu}I\,.$$

If we add the assumption of reflection symmetry, the γ's are square matrices with even dimension at least four. Taking the γ's to be dimension four, and the fields A_μ as the electromagnetic vector potentials due to other charges, the Dirac equation very accurately describes electrons in the field of other charges, and therefore atomic structure and, in principle, all of chemistry and molecular biology. The components of electron wave function can be decomposed into two pairs, each pair corresponding to the two possible intrinsic spin directions measurable, and the combined pair corresponding to the electron carrying positive or negative energy. As an indication of the profound reach gained by merging quantum theory and Relativity, Dirac was able to show that the electron spin and its magnetic moment followed from relativistic quantum theory, and that antimatter must exist, a prediction before anyone dreamed of the concept.

The possibility that fundamental particles can be created and destroyed is included into quantum theory by taking the particle wave functions and interacting fields as quantum fields, entering into the action S with their own dynamics. We find that if disturbed, particle pairs can even 'bubble' out of empty space. The time-and-space-reversed wave function for a particle describes the forward progression of a corresponding antiparticle. This becomes the

[41] A. Einstein, *Zur Elektrodynamik bewegter Körper*, ***Annalen der Physik***, vol. 17 (1905), pp. 891–921.

[42] P. A. M. Dirac, *The Quantum Theory of the Electron*, Part I & II, ***Proceedings of the Royal Society of London***, vol. A 117 & A 118 (1928), pp. 610–624 & pp. 351–361.

'CPT Theorem' in quantum field theory, referring to the operations of charge conjugation, parity transformation, and time reversal.

Quantum field theory distinguishes particles with half-odd integer spin, called 'fermions', from those with integer spin, called 'bosons'.[43] The quantum field in a three-dimensional space and associated with a pair of identical particles will undergo a phase change when those two particles are exchanged: $|\psi(1,2)\rangle = (-1)^{2s}\,|\psi(2,1)\rangle$. If the particles are fermions ($s = 1/2, 3/2, \dots$) the phase change is -1, while no phase change occurs if the particles are bosons ($s = 0, 1, 2, \dots$). This means no two fermions of the same type (such as electrons in atoms) can occupy the same quantum state. This is the 'Pauli Exclusion Principle'. Any number of bosons of the same type can be in the same quantum state (e.g. photons in lasers).

The fundamental particles making up the structure of materials currently appear to be three generations of the doublet electron-neutrino,[44] and three generations of a doublet of quarks, all fermions. The family of electrons and neutrinos are called 'leptons'. Each generation of quark comes in one of three distinct varieties according to their 'color charge'. The bound state of a 'red', 'green', and 'blue' quark and any other 'color-neutral' combination of an odd number of quarks generates a 'baryon', such as the familiar proton and neutron. A zoo of more fleeting particles exist, including 'mesons' coming from bound color-neutral quark-anitiquark systems. The large family of baryons and mesons, all strongly interacting particles, are called 'hadrons'. In the Standard Model, leptons have no direct strong interactions.

5.6. Interactions in quantum theory. All the observed interactions of one particle with another can be categorized by the so-called strong, electromagnetic, weak, and gravitational forces.[45]

The numerical strength of a particle's interactions with other particles is always associated with an intrinsic property called its 'charge'. For each category of interaction, there is one or more corresponding charges. If the total charge of a closed physical system is preserved during a sequence of

---

[43] Particles must have quantized spin with length $\sqrt{s(s+1)}\hbar$ and projection along some measurement axis of $\mu\hbar$, where s is either a half or whole integer, and $-s \le \mu \le s$. It is conventional to use the label s to characterize the particle spin, as in "The electron has spin $1/2$". Particles that move at the speed of c have only two projections of their spin, called their 'helicities', either along their momentum, or in the opposite direction. The characteristic properties of particles following from relativistic quantum theory were first described by Eugene Wigner in *On unitary representations of the inhomogeneous Lorentz group*, ***Annals of Mathematics***, vol. 40 (1939), no. 1, pp. 149–204.

[44] Our observations of the sky together with General Relativistic cosmology seem not to allow more than three generations.

[45] The electromagnetic and weak interactions were linked, principally by the work of Salam, Glashow, Weinberg, Higgs, 't Hooft, and Veltman, from 1964 to 1975.

interactions within that system, we say the charge has been 'conserved'. In nature, all charges are quantized, i.e., they come from a countable set.[46]

The existence of conserved and localizable charges means one can always define an *interaction field* that has those charges as its source, using the following argument: If $\{j^\mu\} = \{\rho c, \rho\vec{v}\}$ represents the charge density and current density for a set of charges, then the local conservation of the total charge, $Q \equiv \int \rho\, d^3x$, can be read from $\partial_\mu j^\mu = 0$. But this implies the existence of an *'interaction field'* $\{F^{\mu\nu}\}$, antisymmetric in its indices, satisfying $\partial_\kappa F^{\kappa\nu} \propto j^\nu$. An associated field, $F^*{}_{\mu\nu} \equiv (1/2)\varepsilon_{\mu\nu\kappa\lambda}F^{\kappa\lambda}$ defines a 'dual' conserved charge with current $j^{*\nu} \propto \partial_\mu F^{*\mu\nu}$.[47] If no such dual charge exists in a region of space, then the field $\{F^{\mu\nu}\}$ can be expressed in terms of a vector field $\{A^\mu\}$ by $F^{\mu\nu} = \partial^\mu A^\nu - \partial^\nu A^\mu$. The field $\{A^\mu\}$ is called the *'gauge field'* going with the corresponding charge. Gauge fields are not uniquely determined, but may be transformed into new fields $\{A'^\mu\}$ which have the same interaction field $\{F^{\mu\nu}\}$ by adding a gradient: $A'^\mu = A^\mu + \partial^\mu\Lambda$. The choice of the 'gauge function' $\Lambda(x)$ is open, provided $\oint \partial_\mu\Lambda dx^\mu$ vanishes for all closed loops in regions where the gauge field acts. Theories whose predictions are independent of the choice of gauge have *'gauge symmetry'*.[48]

Conventional theory describes particle interactions by introducing interaction fields which 'mediate' the effect of one charge on another. We say each particle with a charge of some kind 'creates' an interaction field in the space around it, and that field acts on other particles having the same kind of charge. In the case of electromagnetic interactions, the interaction field is $\{F^{\mu\nu}\}$ with components that are the electric and magnetic field, while the gauge field $\{A^\mu\}$ is called the electromagnetic vector potential. Maxwell's equations, $\partial_\kappa F^{\kappa\nu} = (4\pi/c)j^\nu$ and $\partial_\kappa F^{*\nu\kappa} = 0$, then express two conditions: Electric charge is conserved locally, and there is no observable local magnetic charge. How particles react to other charges requires knowledge of the dynamics for those particles. Dynamics is incorporated into quantum theory.

In quantum theory, a second kind of gauge transformation occurs when the phase of particle wave functions are shifted. A constant shift has no observable effect. But making a shift in phase which depends on location will introduce a relative phase between wave components. If those component waves converge, their interference is observable in the associated particle probability. Now,

[46] Dirac showed that if magnetic monopoles exist, then electric charge must be quantized. See P. A. M. Dirac, *Quantised singularities in the electromagnetic field*, ***Proceedings of the Royal Society of London***, vol. A 133 (1931), pp. 60–71.

[47] The $\{\varepsilon_{\mu\nu\kappa\lambda}\}$ is the completely antisymmetric tensor in four dimensions, with $\varepsilon_{0123} = 1$, called the Levi-Civita symbol. Like $g_{\mu\nu}$, its components are invariant under a proper Poincaré transformation.

[48] The use of 'gauge symmetry' was introduced by Herman Weyl in his consideration of theories with invariance in the scale of length. (H. Weyl, *Gravitation und Elektrizität*, ***Sitzungsberichte der Königlich Preussischen Akademie der Wissenschaften zu Berlin*** (1918), pp. 465–480.)

if, along with the phase shift, a shift in the derivatives of the wave function occurs, one can make the combined shifts cancel. This is the property built into *'gauge symmetric quantum theories'*. In fact, all the interactions among fundamental particles have been found to follow from theories which satisfy gauge symmetry!

Another property of our current dynamical theories can be called the *principle of quasi-local interactions*: The known interactions of one particle with another can be described by 'quasi-local' effects, castable into a form that requires only knowledge of the fields of other particles in a small local space-time neighborhood of the affected particle. These fields are the gauge fields described above. Consider the free-electron Dirac equation $\hbar\gamma^{\mu} i\partial_{\mu}\psi = mc\psi$. A gauge transformation of the second kind on the wave function can be expressed as $\psi'(x) = \exp\left[-i(e/(\hbar c))\Lambda(x)\right]\psi(x)$. The free-particle wave equation becomes $\gamma^{\mu}(i\hbar\partial_{\mu} - (e/c)\partial_{\mu}\Lambda)\psi' = mc\psi'$. Gauge symmetry can be enforced by adding to the derivative term a gauge field A_{μ} which undergoes a gauge transformation of the first kind: $A'^{\mu} = A^{\mu} + \partial^{\mu}\Lambda$. We arrive at the full Dirac equation (8). This technique for introducing interactions is referred to as the 'minimal coupling principle'.[49]

A marvelous theorem was derived by Emmy Noether,[50] who showed that symmetries of our theories based on continuous groups of transformations, such as the Poincaré group and the gauge transformations, lead to conservation laws. In the case of the Poincaré symmetry, the conserved quantities are total energy-momentum, total angular momentum, and the velocity of the center-of-energy. An important example is the symmetry under time translation: If experiments done now with a given system have the same set of results as those done at any time later, then the system's energy is fixed. Symmetry under constant phase shifts of a lepton or baryon wave function makes lepton and baryon number conservation. Gauge symmetry makes the corresponding charges conserved.

As the gauge fields have their own dynamics, quantum theory requires that gauge fields be quantized. That means that the interactions between material particles occur only by the exchange of quanta. These quanta are necessarily bosons. For electromagnetic interactions, the gauge-field quantum

---

[49] Gauge symmetry in quantum theory can be re-expressed in terms of the action of 'covariant' derivatives $D_{\mu} = \partial_{\mu} + i(e/(\hbar c))A_{\mu}$, acting on a quantum state for a particle. In this interpretation, the interactions arise from the behavior of quantum states by parallel transport across space. When the gauge fields themselves are taken to be operators on the internal components of a quantum state, the gauge group elements may not commute. These kind of 'non-abelian' gauge fields were introduced by C. N. Yang and R. Mills (*Conservation of isotopic spin and isotopic gauge invariance*, ***Physical Review***, vol. 96 (1954), no. 1, pp. 191–195) and are used in the Standard Model to describe interactions between fundamental particles grouped into families. For example, the quark color charge follows from an SU_3 gauge symmetry.

[50] E. Noether, *Invariante Variationsprobleme*, ***Nachr. d. König. Gesellsch. d. Wiss. zu Göttingen, Math.-phys. Klasse*** (1918), pp. 235–257.

is the photon. The photon at present appears to travel at the maximum speed in Relativity, has unit spin, and carries no electric charge.[51] For the strong interactions between quarks, the quanta of the field are called gluons. Gluons also have unit spin but they carry various 'color' charges. By having charge, gluons can directly interact with themselves, making their dynamics more complicated than for photons. For example, the gluon fields, through their self-interaction, can form flux tubes between quarks.

5.7. Prepared states and measurement. Each possible quantum state of a system is referred to as a 'pure quantum state', as contrasted to a 'mixed quantum state', for which we may only know probabilities for the system to be in given quantum states. A pure quantum state made from a superposition of component states is called a 'coherent quantum state' when all the phases between its various component states are known to be fixed.

A quantum system is prepared by first selecting a physical system, isolating the system from unwanted interactions, determining its initial configuration, and then stimulating or allowing the system to approach a desired initial state.

Isolating a system and determining its initial configuration are often daunting tasks. The state of most macrosystems will be practically impossible to completely specify. Some interactions, such as those from stray fields or background radiation, may be difficult or impossible to eliminate. Helping the effort are quantum states with unusual stability. These stable states are changed only by the input of energies larger than typically available, so they are effectively isolated until such energies enter the system.[52] After isolation, a system will evolve by quantum dynamics following a unitary transformation, and may eventually become a 'steady state', i.e. one with no change in probability densities for its particles, if these were observed.

Consider the expansion of a pure quantum state into component states which together span the system's Hilbert space:

$$|\psi\rangle = \sum_i \alpha_i \, |\phi_i\rangle \,. \tag{9}$$

If the phases between two or more components of the quantum state are related, then these components are said to be in '*coherence*'. Quantum interference between various possible outcomes of a measurement requires some coherence in a quantum state. The states $|\phi_1\rangle$ and $|\phi_2\rangle$ might be two possible interfering states of a single electron, or even a trapped atom. The two states of the atom

[51] A particle with charge will carry energy associated with the field of that charge, and therefore, if it can be separated from other particles, must have mass, and must move slower than the universal speed c.

[52] Such stability is a pure quantum effect, since, if the energy states in a bound subsystem were not quantized, any small energy could excite the subsystem. Subsystems such as bound electrons, bound atoms, nuclei, and topologically constrained subsystems, are all stable at sufficiently low energy arriving from the outside.

might have opposite motions, so that the wave function for each state can oscillate back and forth across the trap. Then the probability distribution of finding the atom at a specific location within the trap shows an interference pattern.[53] This quantum effect, however, is no different in principle than that seen as an interference pattern made by the bright spots of light on the surface of a phosphor plate, those spots produced by electrons passing through two slits in a screen, one electron at a time, and then hitting the phosphor.

Starting with a set of identically prepared systems in a coherent state represented by Eq. (9), measurements of the observable A will have an average $\sum_{ij} \alpha_i^* \alpha_j \langle\phi_i| A |\phi_j\rangle$. Interference will arise from terms for which $\langle\phi_i| A |\phi_j\rangle$ are not zero for $i \neq j$. However, if a quantum system interacts with another system or with a measuring device, some or all of the components of residual quantum states for that system may be left with no well-defined phase relationships. This is a process of *'decoherence'*. During the measurement, information is transferred between the system and the measuring device, and some may be lost to the environment.

One of the important measurements of a system locates the position of particles. After a number of such measurements in each small region dx of space, we find a distribution of positions. For one particle, the wave function determines the probability density for position across space, so the distribution of measured positions is predicted to be an approximation of $\psi(x)^*\psi(x)dx$. The average position over all space is predicted to be $\langle\psi| x |\psi\rangle$. More generally, each distinct measurement of a property of the system can be associated with a Hermitian operator A that acts on wave functions for the system ψ as follows: The average value of A will be

$$\langle A\rangle = \int \psi^\dagger(x,t) A(x,\hat{p}_x)\psi(x,t)\, dx\,, \tag{10}$$

wherein $\hat{p}_x$ is taken proportional to the space translation operator in accord with Noether's Theorem.[54] The operators A may also act on the spin and other components of the wave function.

Those states $|a\rangle$ satisfying

$$A\,|a\rangle = a\,|a\rangle$$

---

[53]This game was played using a Beryllium atom by Dr. Christopher Monroe and colleagues at the National Institute of Standards and Technology, Boulder, Colorado. See C. MONROE et al., *A 'Schrödinger cat' superposition of an atom*, ***Science***, vol. 272 (May 24, 1996), pp. 1131–1136. Some members of the press mis-represented the observation as indicating that one atom can be found in two places at once. For example, see M. W. BROWNE's article *Physicists put atom in 2 places at once*, published in the New York Times, May 28, 1996.

[54]The proportionality constant is fixed by noting that if a free particle is left unobserved, then within some bounded region its wave function becomes a 'plane' wave $\phi \propto \exp((ip_x x - iEt)/\hbar)$, for which $\hat{p}_x\phi = -i\hbar\partial_x\phi$. The order of non-commuting operators in A must be determined by physical arguments.

are called the 'eigenstates' of A and a an 'eigenvalue'. For Hermitian operators A, i.e. $A^\dagger = A$, the eigenvalues a will be real numbers, and therefore each is a value which may result from a measurement. The elements of the set $\{|a\rangle\}$ for distinct values a will be 'orthogonal', i.e. $\langle a'|a\rangle = \delta_{a'a}$, and 'complete', i.e. they span the space of possible states, expressible as $\sum_a |a\rangle\langle a| = I$ by reading from $|\psi\rangle = \sum_a |a\rangle\langle a|\psi\rangle$. The measured values of A will have an uncertainty defined by $\Delta A \equiv \sqrt{\langle(A - \langle A\rangle)^2\rangle}$. This means that after measurement of A for a number of identically prepared systems, the observed values will be distributed around the average with a 'width' of ΔA.

After a single measurement of the observable A for a system in a pure quantum state $|\psi\rangle$ that has A as one of its observables, one of the eigenvalues of A, say a, will be found, and the system will be left in the state $|a\rangle$. The effect of measurement can be represented by a *'projection operation'*: $P_a \equiv |a\rangle\langle a|$. The measurement of A has 'collapsed' the quantum state to $|a\rangle \propto P_a|\psi\rangle$. The collapse evidently does not preserve unitarity for the system, expressed by $|\psi(t)\rangle = U(t, t_0)|\psi(t_0)\rangle$, unless the system was already in an eigenstate of A. Quantum unitarity applies to isolated systems. The measurement process has involved another interacting system which reduced the system's available states in a subsequent measurement.

The interaction (called the 'coupling') between two systems during a measurement may cause one or more of the phase differences between the components in the resulting quantum states to become indeterminate, especially likely if the measuring device is macroscopic. After the measurement, the system may be left in a 'mixed' state, for which only the probabilities p_k for any particular pure quantum state $|\psi_k\rangle$ are known. Then a subsequent measurement of the observable A will have an average value of $\langle A\rangle = \sum_k p_k \langle\psi_k|A|\psi_k\rangle$. This expression can be usefully re-written in terms of a 'density operator', defined by $\hat{\rho} \equiv \sum_k p_k |\psi_k\rangle\langle\psi_k|$, so that $\langle A\rangle = Tr(\hat{\rho}A)$. In this way, the choice of the mixed state is left implicit. A pure state can then be simply characterized by $\hat{\rho}^2 = \hat{\rho}$.

5.8. Entangled states. An *'entangled quantum system'*, by definition, has two or more particles in a quantum state which cannot be factorized into states for each particle.[55] For example, if we let the quantum state $|\mu_\kappa\rangle$ represent the electron labeled by κ and having a spin projection along the z-axis of $(\mu - 1/2)\hbar$, then one of the possible entangled two-electron states can be written

$$|\psi_0\rangle = (1/\sqrt{2})(|0_1\rangle|1_2\rangle - |1_1\rangle|0_2\rangle),$$

which happens to have a total spin of zero.

[55]There is a special caution for quantum states describing photons, in that the number of photons is not fixed, but rather has an uncertainty which increases as the phase of the electromagnetic wave becomes more definite.

The outcome of a measurement of one of the electrons in an entangled pair will be correlated with the outcome of measurement of the other, even when they are far apart. This kind of correlation also occurs under classical conditions. Suppose you put a jack in one envelope and a queen in another. Now send one of the envelopes to one friend, the second envelope to a second friend. If one friend opens your envelope and finds a jack, then your other friend must find a queen even before hearing from your first friend. However, there is a twist in the quantum world. Take the case when a pair of electrons is prepared in a zero total spin state along a z-axis expressed above, and then the electrons are allowed to move far apart. Next, while the electrons are in flight, have one of the distant observers rotate her electron-spin measuring apparatus away from the z-axis direction to an angle of her choosing, i.e. the first distant observer makes a *'delayed-choice experiment'*.[56] If this first observer finds an electron aligned along her new axis, then the second observer, far away, will find the other electron aligned along the negative direction of the *new* axis constructed by the first observer. Now we, on first hearing and with our classical thinking, *should* be surprised! Even so, this is the way nature acts. The result does *not* mean that the pair interacted after traveling apart, nor was there 'superluminal' transmission of information.[57] This suggestion of faster-than-light signaling is a misinterpretation of quantum theory, and such information transfer has not been seen. Rather, those who say so are likely to have been tripped up by picturing each unobserved electron as being localized between observations!

5.9. Non-classical interactions. There are interactions predicted by quantum theory without classical explanation. Yakir Aharonov and David Bohm[58] showed that a single electron wave which never enters a region of electric or magnetic field could never-the-less have a measurable shift in the probability of finding that electron after an electric or magnetic field changes in the excluded region. The effect occurs, for example, when the electron passes on either side but does not enter a tube where a magnetic field is confined. The difference in phase of that wave when followed around a closed loop is given by $e \oint A^\mu dx_\mu/(\hbar c)$, where $\{A^\mu\}$ is the electromagnetic potential. This is the

[56]If the decision on how a component of a system is measured comes after that system has had sufficient time to cause interference between quantum alternatives for that component, then this becomes a delayed-choice experiment as introduced by John Wheeler in ***Mathematical Foundations of Quantum Theory***, edited by A. R. Marlow, Academic Press (1978).

[57]Information transmitted by a wave disturbance that started at a certain time cannot be transferred faster than the outgoing wavefront from that disturbance. In Special Relativity, the speed of the wavefront, also called the signal speed, is always less than or equal to the universal limiting speed, c. There is no such restriction on the group velocity or the phase velocity of the wave.

[58]Y. Aharonov and D. Bohm, *Significance of electromagnetic potentials in the quantum theory*, ***Physical Review***, vol. 115 (1959), pp. 485–491.

flux of magnetic field somewhere inside the loop.[59] The shift in the observed interference pattern produced by the electrons when the magnetic flux is turned on has no explanation in classical physics.

Measurement of a system may disturb the system. If the measurement process transfers complete information about a system, that system will no longer contain entangled states. This effect leads to the 'no-cloning theorem',[60] the statement that a general quantum system for which we have no prior knowledge cannot be identically copied. If a copy of a quantum state could be made, then we could defeat the interfering effect of measurement by first making a copy, and then measuring the copy, leaving the original system undisturbed.

The wave function for a particle confined to a fixed region of space and initially localized to a much smaller part of that region and then left with no external interaction will diffuse outward in space as time progresses. The wave for an unobserved particle will spread over the entire allowed region, and eventually the probability for finding the particle in any small location will have no measurable change in time, and its quantum wave function will be steady.[61]

A localized and isolated physical system will have denumerable ('quantized') possible values for its measurable energies and momenta. Periodicities of the wave function also enforce quantization if there is a simply-connected closed path over which the corresponding particle can move. For example, periodicity in the azimuthal angle in the wave function makes the measured values of the projection of the orbital angular momentum along a measurement axis become denumerable.

Suppose two observables A and B for a given system in the state $|\psi\rangle$ are measured in a certain time order. If these two measurements are repeated for identically prepared systems, a change in the order of measurement may change the probability for finding a given value for the second observable. In general, one can show that the uncertainties satisfy $\Delta A\,\Delta B \geq (1/2)\,|\langle AB - BA\rangle|$. This is called the uncertainty principle of Heisenberg. If the 'commutator' $[A, B] \equiv (AB - BA)$ vanishes, then the observables A and B may be measured 'simultaneously', i.e. without the measurement of one affecting the results of measuring the second. The state of a physical system can be labeled by a set of measured values for a maximal set of mutually commuting observables that are also conserved over time.

---

[59] Mandelstam re-expressed the local interaction with $\{A^\mu\}$ as a non-local effect of the electric and magnetic fields, i.e. a topological effect of fields over space-time. See S. MANDELSTAM, *Quantum electrodynamics without potentials*, ***Annals of Physics***, vol. 19 (1962), pp. 1–24.

[60] W. ZUREK, *A single quantum cannot be cloned*, ***Nature***, vol. 299 (1982), pp. 802–803; D. DIEKS, *Communication by EPR devices*, ***Physics Letters***, vol. A 92 (1982), no. 6, pp. 271–272.

[61] Steady wave functions necessarily have a sinusoidal time dependence through a factor of the form $\exp(-i\omega t)$, making the probability $\psi^\dagger\psi dx$ time independent.

5.10. Quantum theory for complex systems. After a relaxation time for a system containing a large number of interacting particles, the most likely distribution of particles in the available quantized energy states will be those that tend to maximize, under the physical constraints, the multiplicity W, simply because as the system evolves through various configurations, it will spend most of its time in those configurations which have many ways to be constructed. One can then show[62] that if the number of particles n_i in each quantum state labeled by the index i is also large, then $n_i \propto \exp(-\varepsilon_i/(kT))$, where ε_i is the energy of the corresponding quantum state, and T is the temperature of the system. Interactions from the outside can change the total energy, $E = \sum_i n_i \varepsilon_i$, of a system either by changing the 'occupation numbers' $\{n_i\}$, and/or by changing the energies $\{\varepsilon_i\}$ of the quantum states. The first kind of change is heat transfer and the second is work transfer. By increasing the multiplicity of the system, putting heat into a system is a 'disordering process'. Work involves changing the particle energies by changing the volume of the system, without moving particles between quantum states.[63] These ideas incorporate the first and second law of thermodynamics.

In terms of information, the second law of thermodynamics implies that *if two systems interact, each with fixed volumes, then that system of the two which has the smaller variation in its information content as its total energy changes will tend to spontaneously transfer information into the second system.*[64]

These are important concepts for quantum computers, as there is an intimate connection between entropy, information, decoherence, wave function collapse, and heat from memory loss.

§6. Quantum computation. Feynman[65] considered the possibility that we might take advantage of quantum systems to perform computations quicker than so-called classical computers. Modern classical computers use bistable systems to store information, and logical gates to perform Boolean operations on sets of ones and zeros.[66] For some problems involving numbers with

[62]L. Boltzmann, *Über die Beziehung zwischen dem zweiten Hauptsatze der mechanischen Wärmetheorie und der Wahrscheinlichkeitsrechnung respektive den Sätzen über das Wärmegleichgewicht*, ***Sitzungsberichte der Akademie der Wissenschaften zu Wien***, vol. 76 (1877), pp. 373–435.

[63]If particles remain in their quantum states, no heat is transferred, and the process is called 'adiabatic'.

[64]If a system near thermal equilibrium is held at fixed volume and a small amount of energy dE is put in, causing an increase in its information content by dI, then the ratio dE/dI turns out to be proportional to the temperature of that system. The spontaneous flow of information, i.e. non-forced flow, results from statistical likelihood.

[65]R. Feynman, *Simulating physics with computers*, ***International Journal of Theoretical Physics***, vol. 21 (1982), pp. 467–488.

[66]These discrete-level computers are often referred to as 'digital', in contrast to 'analog' computers that use internal signals that are assumed to vary smoothly with time. Mechanical computers, which work by the movement and interaction of shaped objects, and molecular

n digits and that may require solution times that rise exponentially with n when performed on computers using only Boolean logic, the computation on a quantum computer may take times that rise no faster than a power of n. Below are some of the special consequences of quantum theory for quantum computers and communications:

The simplest system for the storage of information gives only two possible values by a measurement. These values can be taken as 0 or 1, in which case the states are called $|0\rangle$ and $|1\rangle$. Classically, such a system stores one bit of information. A quantum system can be constructed that has only these two values for the outcome of a measurement, but whose quantum state is a linear combination of the two possible outcome states $|0\rangle$ and $|1\rangle$:

$$|q\rangle = \alpha\,|0\rangle + \beta\,|1\rangle\,.$$

This state is called a 'qubit', where α and β are complex numbers satisfying $|\alpha|^2 + |\beta|^2 = 1$. An alternative parameterization takes $\alpha = \cos(\theta/2)$ and $\beta = \exp(i\phi)\sin(\theta/2)$. Evidently, the possible qubit states can be pictured as points on a unit sphere (called the 'Bloch sphere') with $|0\rangle$ at the north pole and $|1\rangle$ at the south. Two-valued qubit states are easily realized in nature: The electron spin has only two possible projection values $\pm\, 1/2\hbar$, and the photon has only two possible helicity values $\pm\, 1\hbar$.

As it is always possible to expand an arbitrary quantum state into a basis set for that state's Hilbert space, N-particle states in a quantum computer can be made by constructing these quantum states from a linear combination of the states for each of the N particles. Taking these particles to have only two internal quantum states, the state of the computer is expressible by

$$\begin{aligned} |\psi_N\rangle &= \sum_{\{i_k=0,1\}} \beta_{i_1 i_2 i_3,\ldots i_N}\, |i_1\rangle_1\, |i_2\rangle_2\, |i_3\rangle_3 \ldots |i_N\rangle_N \\ &= \sum_{i=1}^{2^N} \beta_i\, |i_1 i_2 i_3 \ldots i_N\rangle \text{ with } \sum_i |\beta_i|^2 = 1\,. \end{aligned} \tag{11}$$

In the second line of the equation, the product base state is represented in a shortened form, in which the order of the 0's and 1's corresponds to

computers, that work through molecular interactions and transformations, are a mixed breed. The phrase digital computer, referring to counts base ten, can now mean any device which manipulates information by discrete changes. These days, the changes are made in systems which can flip between *off* and *on* in a specified clock time, i.e. a binary coding. By using such switching to encode information, digital computers can be more tolerant of a small amount of noise than analog devices. Shannon and Hartley showed that the maximum number of bits per second that can be transmitted from one storage location to another is given by $B\log_2(1 + S/N)$, where B is the bandwidth (in cycles per second), S is the average signal power, and N is the average noise power. See R. V. L. HARTLEY, *Transmission of information*, **Bell System Technical Journal** (July 1928); C. E. SHANNON, *Communication in the presence of noise*, **Proc. Institute of Radio Engineers**, vol. 37 (January 1949), no. 1, pp. 10–21.

the labeling of each of the separate qubits, and $i = i_1 i_2 i_3 \ldots i_N$ is a binary number constructed from the i's. If the quantum state $|\psi_N\rangle$ cannot be factorized, it harbors entanglement. Quantum computation takes advantage of entanglement within those states.[67] It follows that a useful initial state of a quantum computer has at least a subset of particles prepared in one of the *maximally entangled states*, i.e. states with equal probability for all possible configurations of its component particles, making it also that state which has maximum information content.[68] The maximally entangled states made from qubits as in Eq. (11) will have all $|\beta_i|^2 = 1/N$, leaving $(2^N - 1)$ free relative phases between the basis states.[69]

To sustain coherence, quantum computers must operate on the input information stored in quantum states by unitary transformations. In the following, the substage of a quantum computation holding the intermediate state of a calculation will be called $|\psi(k)\rangle$, where k labels a particular intermediate state, with $k = 0$ labeling the initial state. For a given quantum computer, a solution to a solvable problem is a unitary transformation U_S that carries the input quantum state $|\psi(0)\rangle$ encoding the required initial data into an output quantum state that carries the information about the solution, at least in probabilistic terms. To be a non-classical computer, at least some the intermediate states must be entangled. It is possible that U_S can be decomposed into a finite product of simpler or more universal unitary operations: $U_S = \prod_i U_{g_i}$, where the set $\{U_{g_i}\}$ are called 'quantum gates', a generalization of classical logic gates. Each term in the product acts on the state $|\psi(k)\rangle$ left by the previous operation labeled by k and produces $|\psi(k+1)\rangle$.

Since a general unitary transformation will contain continuous parameters, U_S might only be approximated by a finite sequence of quantum gates. In the classical case, all Boolean operations on a set of bits can be performed by a combination of NAND gates. This makes NAND gates universal for classical computing. The same is true of NOR gates. In the quantum case, there are 'universal' sets of simple gates that can be used to build arbitrarily close representations of a general unitary transformation, such as U_S. (Arbitrarily close here means that if V_S is the approximation, then $|\langle\psi|\,(U_S - V_S)\,|\psi\rangle|^2$ is a number that can be made arbitrarily small for all $|\psi\rangle$ by increasing the number of universal gates used in V_S.)

Quantum gates acting on a single qubit can all be represented by a general unitary transformation $U_{\vec{\theta}}$ which is an arbitrary rotation in Hilbert space:

$$U_{\vec{\theta}} = \exp\left(i\theta\, \hat{n}\cdot\vec{\sigma}/2\right) = \cos\left(\theta/2\right) I + i\hat{n}\cdot\vec{\sigma}\sin\left(\theta/2\right),$$

[67] See, for example, R. Jozsa and N. Linden, *On the role of entanglement in quantum computational speed-up*, ***Proceedings of the Royal Society A: Mathematical, Physical and Engineering Sciences***, vol. 459 (2002), no. 2036, pp. 2011–2032.

[68] One learns most when the outcomes are least predictable!

[69] These maximally entanglement states are also called 'generalized Bell states'.

where θ is an angle of rotation around an axis fixed by the direction $\hat{n}$, and the $\{\sigma_i\ ;\ i = 1, 2, 3\}$ are the Pauli matrices,

$$\sigma_x = \begin{bmatrix} 0 & 1 \\ 1 & 0 \end{bmatrix}, \quad \sigma_y = \begin{bmatrix} 0 & -i \\ i & 0 \end{bmatrix}, \quad \sigma_z = \begin{bmatrix} 1 & 0 \\ 0 & -1 \end{bmatrix}$$

which act on the base states $|0\rangle = \begin{bmatrix} 0 \\ 1 \end{bmatrix}$ and $|1\rangle = \begin{bmatrix} 1 \\ 0 \end{bmatrix}$. For an initial state consisting of the many qubits (perhaps realized by many particles capable of being in two distinct quantum states), such as $|i_1 i_2 i_3 \ldots i_N\rangle$, a $2N$ dimensional unitary transformation would be implemented to carry out one step of a computation.

If noise or other spurious interactions occur in the system, quantum coherence may be degraded or lost, and there will be both a 'coherence time' and a 'coherence length' over which the system retains a semblance of its coherence. A fault-tolerant quantum computer uses states that have long coherence times, quantum entangled states with long life times, and/or error correcting schemes. Systems for transferring qubits over long distances require long coherence lengths.[70]

A new measurement acting on a quantum state generally causes some decoherence, so that a number of components of the wave function may have their phase become stochastically indeterminate. The observation of the state of a particle in a multi-particle entangled state removes the entanglement of that particle. As we have seen, measurement of an observable is the equivalent of projecting out a subspace of the initial state. Such a projection into a proper subspace is irreversible and non-unitary. The resulting state of the system no longer holds information about the projected state.

In quantum theory, all processes preserve the condition that the probability of finding any of the possible states of the system add to unity. Formally, quantum states evolve by a unitary transformation. In the 'Copenhagen' view, the act of measurement causes the wave function for the system to 'collapse'. A collapse of a quantum state from a superposition of substates to one such substate violates unitarity, and therefore is outside the formalism of quantum theory. This produces a paradox: The measuring instrument is also a physical system, so that the larger system that contains the observed system and the measuring devices, left unobserved, should evolve by a unitary transformation, and no wave function 'collapse' should occur. There is no easy way out of this paradox.

[70]Transferring qubits across space was first described by C. H. Bennett et al. in *Teleporting an unknown quantum state via dual classical and Einstein-Podolsky-Rosen channels*, ***Physical Review Letters***, vol. 70 (1993), pp. 1895–1899. Note that transferring a qubit from one system to another does not violate the no-cloning theorem, because the initial qubit is destroyed in the process, and that the transfer is cannot be superluminal, as two classical bits must be sent from the first system to the second before reconstruction of the qubit can take place.

The measuring devices in the larger combined system must introduce interactions that do not project out quantum substates in the combined system, but rather redistribute the amplitudes for various quantum states, making the observed state highly probable, and the other possible states in the observed system left with very small amplitudes. Being unitary for the combined system of the observed and the measuring device, such a measurement process is, in principle, 'reversible'. The entanglement of an observed system with a measurement instrument and subsequent restoration of the original quantum state has been demonstrated for simple systems and measuring devices with highly restricted interactions.[71] But for a multitude of interactions, restoration after interactions is typically unfeasible with our current resources.[72] It is also possible that nature does not just scatter information so much that we cannot easily put systems such as broken eggs back together again, but rather actually does lose information over time. This possibility is outside the realm of quantum theory.

The same ideas apply to quantum computers. In quantum theory, even the measurement of a final state after a computer calculation is a reversible process for the computer, the measurement device and the surrounding interacting systems. In principle, no information is lost. But if the information transferred by erasing a quantum memory state produces heat in the environment, some information is practically lost.

If one of the particles in an entangled state is sent to a second observer as a form of communication, then attempts to intercept that particle will degrade or destroy the entanglement, and therefore will be detectable. This opens the possibility of 'absolute' security in transmission lines, particularly since macroscopically long coherence lengths have been realized with laser beams.

If a set of identical particles are restricted to a two-dimensional surface, or the space is not simply connected, the quantum state representing two particles may gain a phase factor of $\exp(2i\pi p)$ when the two particles are exchanged, where p need not be integer or half-integer.[73] If the phase factor p is not $n/2$ (where n is integer), the particles are called 'anyons'. For three particles, if the order in which the particles are exchanged produces a different wave function phase, the group of such exchanges is non-abelian. This consideration may be important in the construction of quantum computers through the storing of information in the topological braiding of non-abelian anyons as they progress in space-time.

Topological structures have been shown to be important in quantum theory. For example, the continuity condition for the wave function describing particles

---

[71] See, for example, N. Katz et al., *Reversal of the weak measurement of a quantum state in a superconducting phase qubit*, ***Physical Review Letters***, vol. 101 (2008), p. 200401.

[72] This difficulty is related to the ergotic hypothesis in classical mechanics, and the development of entropy concept in statistical thermodynamics.

[73] By contrast, in a connected region of three dimensional space, there is a space transformation that will 'untangle' the pair, and make p an integer multiple of $1/2$.

adds significance to global space-time topology. In some models, particle charges come from topological structures. A variety of promising systems for quantum computers take advantage of the difficulty of breaking topological structures in order to preserve quantum coherence, making the system more 'fault immune' against the effect of noise and other external interactions.

§7. Limits to computing.

7.1. Practical limits to computing and information storage. Classical computers have practical limitations in density. Gates and memory elements smaller than nanoscale will suffer quantum fluctuations, with growing uncertainties in bit structures and Boolean transformations as the size of the elements are reduced. Even our DNA code can be mutated by quantum tunneling. If the system has a certain level of noise, classical correction schemes can eliminate errors, at a cost of size. The techniques to control heat buildup also require volume in the ancillary heat sinks or channels for radiative cooling. Taking systems at the nanoscale and finding technology that minimizes heat production toward the Szilárd value of $kTln2$ per bit lost gives an upper limit to computer density made from materials. Memory and gates based on information in light beams have corresponding limits due to pulse duration and wave length uncertainties.

Quantum computers require coherence within the involved quantum states of the computer during computation. Working against us are physical limitations. For example, the quantum states being used to store information typically have finite lifetimes through spontaneous decay, resulting in the collapse of the employed coherent states. Uncontrollable interactions both within and from the outside a quantum computer will tend to collapse coherent states. After sufficient time, coupling to the environment will cause decoherence and disentanglement within a quantum system.

Coherence can be maintained for some period of time by using quantum states which have some intrinsic stability and suffer little debilitating interactions with adjacent systems or with the environment. Explorations to find strategies which minimize the limitations are ongoing. Evidently, each quantum gate must act within the shortest coherence time. Some mixing and degradation in quantum states can be tolerated by using repeated calculations and/or Implement error corrections which can reconstruct, with some assurance, a degraded quantum state. Overall, even though we can anticipate severe practical difficulties to building a quantum computer which can outperform its classical cousin, we see no fundamental limitation, unless our ambitions reach across the cosmos.

7.2. Cosmological limits. Strong gravitational fields exist near black holes, which are predicted by Einstein's General Theory of Relativity to occur when the density of an object of mass m exceeds about $3c^6/(2^5\pi G^3m^2)$. Such black holes got their name because no form of radiation can escape from the hole if it starts out within a region around the hole bounded by a surface called the 'horizon'. For a non-spinning hole without charge, this surface has the 'Schwarzschild radius'[74] $R_S \equiv 2Gm/c^2$. Astronomers have found stellar-mass black holes in binary systems by analyzing the orbits of companion stars. Nearby large galaxies are known to contain one or more super-massive black holes at their center, and we suspect all large galaxies do.

Using quantum theory, Hawking showed[75] that the fluctuations in particle fields near but outside the horizon of a black hole can produce particle pairs with some of the positive energy particles having sufficient kinetic energy to reach large distances away, while the negative energy particles fall into the black hole. Thus, quantum theory requires that black holes evaporate, with a mass loss rate inversely proportional to the square of the hole mass m $\left[dm/dt = -\hbar c^4/(3 \cdot 5 \cdot 2^{10}\pi G^2 m^2)\right]$. The flux of photons emitted is close to that of a hot body at a temperature inversely proportional to m $\left[T = \hbar c^3/(8\pi k G m)\right]$.

However, to be consistent with quantum theory, a system initially containing an object and a black hole, with the object destined to disappear into the black hole, with no other interaction but gravity, cannot lose information: The quantum state of the hole and the object evolves unitarily. One resolution of this paradox is to have the object's information transferred to a region close to the horizon of the black hole.[76] In this way, Hawking radiation can carry the stored information back out (so the radiation is not perfectly thermal). Even before Hawking proposed that black holes evaporate, Jacob Berkenstein[77] conjectured that the entropy of a black hole, which is also the information storage capacity, is proportional to the area of the hole's horizon, $4\pi R_S^2$, and inversely proportional to the square of Planck's length. Hawking then calculated the proportionality constant to be $k/4$, where k is Boltzmann's constant.

General Relativity limits the density of a computer, and concurrently the density of information storage. As a computer becomes larger in a given volume, its density eventually forces the computer to collapse into a black hole. This leads to the idea that the limiting density of information storage may be effectively two dimensional, with each bit stored in a Planck-size area. Some (as yet untested) theories even have the information of the whole Universe reflected by a kind of holographic image in one less dimension.

---

[74]K. Schwarzschild, *Über das Gravitationsfeld eines Massenpunktes nach der Einsteinschen Theorie*, ***Sitzungsberichte der Königlich Preussischen Akademie der Wissenschaften***, vol. 1 (1916), pp. 189–196.

[75]S. W. Hawking, *Black hole explosions?*, ***Nature***, vol. 248 (1976), no. 5443, pp. 30–31.

[76]It is even possible that the volume surrounded by a black hole horizon is completely empty, even of any space-time structure, with any infalling matter ending up just outside the horizon.

[77]J. D. Bekenstein, *Black holes and entropy*, ***Physical Review D***, vol. 7 (1973), pp. 2333–2346.

A cosmological limitation on computation also comes from the fact that we appear to live in a finite Universe. A computer can be no larger than the Universe itself. Any smaller computer cannot hold the data of the Universe at one time, which is needed to unambiguously project the Universe's future. In addition, being that the computer is within the Universe, it cannot predict both itself and the Universe. Our current theories do not incorporate these kinds of limitations, although there are propositions that connect the very small to the very large.

§8. Conclusions. Quantum computers take advantage of quantum operations in physical systems in order to solve well-posed problems. Quantum theory describes these operations based on how nature processes information. Space and time are important primitives in quantum theory, and active participants in both information transfer and information storage. While we formulate how nature handles information, we should recognize that our physical theories are always tentative. Each covers a limited realm and has a limited accuracy. Also, since each theory has a variety of equivalent formulations, with their own language, our main focus should be on the predictions of a theory. Even though very successful, quantum theory makes some rather non-intuitive and thought-provoking predictions. Correspondingly, there are a variety of precautions to which we should be attentive when applying and interpreting the theory. Reflecting on the underlying ideas central to quantum theory should help us in the exploration of possibilities for future quantum computers.

DEPARTMENT OF PHYSICS
COLUMBIAN SCHOOL OF ARTS & SCIENCES
THE GEORGE WASHINGTON UNIVERSITY
WASHINGTON, D. C. 20052
E-mail: wparke@gwu.edu

FIBER PRODUCTS OF MEASURES AND QUANTUM FOUNDATIONS

ADAM BRANDENBURGER AND H. JEROME KEISLER

Abstract. With a view to quantum foundations, we define the concepts of an *empirical model* (a probabilistic model describing measurements and outcomes), a *hidden-variable model* (an empirical model augmented by unobserved variables), and various properties of hidden-variable models, for the case of infinite measurement spaces and finite outcome spaces. Thus, our framework is general enough to include, for example, quantum experiments that involve spin measurements at arbitrary relative angles. Within this framework, we use the concept of the *fiber product of measures* to prove general versions of two determinization results about hidden-variable models. Specifically, we prove that: (i) every empirical model can be realized by a deterministic hidden-variable model; (ii) for every hidden-variable model satisfying locality and λ-independence, there is a realization-equivalent hidden-variable model satisfying determinism and λ-independence.

§1. **Introduction.** Hidden variables are extra variables added to the model of an experiment to explain correlations in the outcomes. Here is a simple example. Alice's and Bob's computers have been prepared with the same password. We know that the password is either p2s4w6r8 or 1a3s5o7d, but we do not know which it is. If Alice now types in p2s4w6r8 and this unlocks her computer, we immediately know what will happen when Bob types in one or other of the two passwords. The two outcomes—when Alice types a password and Bob types a password—are perfectly correlated. Clearly, it would be wrong to conclude that, when Alice types a password on her machine, this somehow causes Bob's machine to acquire the same password. The correlation is purely informational: It is our state of knowledge that changes, not Bob's computer. Formally, we can consider an r.v. (random variable) X for Alice's password, an r.v. Y for Bob's password, and an extra r.v. Z. The r.v. Z takes the value z_1 or z_2 according as the two machines were prepared with the first or the second

We are grateful to Samson Abramsky, Bob Coecke, Amanda Friedenberg, Barbara Rifkind, Gus Stuart, and Noson Yanofsky for valuable conversations, to John Asker, Axelle Ferrière, Tobias Fritz, Elliot Lipnowski, Andrei Savochkin, participants at the workshop on Semantics of Information, Dagstuhl, June 2010, and participants at the conference on Advances in Quantum Theory, Linnaeus University, Växjö, June 2010, for helpful input, to a referee and the volume editors for very important feedback, and to the Stern School of Business for financial support.

Logic and Algebraic Structures in Quantum Computing
Edited by J. Chubb, A. Eskandarian and V. Harizanov
Lecture Notes in Logic, 45
© 2016, Association for Symbolic Logic

password. Then, even though X and Y will be perfectly correlated, they will also be independent (trivially so), conditional on the value of Z. In this sense, the extra r.v. Z explains the correlation.

Of course, even in the classical realm, there are much more complicated examples of hidden-variable analysis. But, the most famous context for hidden-variable analysis is quantum mechanics (QM). Having started with von Neumann [23, 1932] and Einstein, Podolosky, and Rosen [13, 1935], the question of whether a hidden-variable formulation of QM is possible was re-ignited by Bell [3, 1964], whose watershed no-go theorem gave conditions under which the answer is negative. The correlations that arise in QM—for example, in spin measurements—cannot be explained as reflecting the presence of hidden variables.

Let us specify a little more what we mean by an experiment. We imagine that Alice can make one of several measurements on her part of a certain system, and Bob can make one of several measurements on his part of the system. Each pair of measurements (one by Alice and one by Bob) leads to a pair of outcomes (one for Alice and one for Bob). We can build an *empirical model* of the experiment by choosing appropriate spaces for the sets of possible measurements and outcomes, and by specifying, for each pair of measurements, a probability measure over pairs of outcomes. An associated *hidden-variable* (henceforth *h.v.*) *model* is obtained by starting with the empirical model and then appending to it an extra r.v.

We can define various types of h.v. model, according to what properties we ask of the model. One property is *locality* (Bell [3, 1964]), which can be decomposed into *parameter independence* and *outcome independence* (Jarrett [18, 1984], Shimony [20, 1986]). Another property is *λ-independence* (the term is due to Dickson [12, 2005]), which says that the choices of measurement by Alice and Bob are independent of the process determining the values of any h.v.'s. Bell [5, 1985, p. 95] describes this as the condition that "the settings of instruments are in some sense free variables." We will use the term "free variables" below.

Here are two basic types of h.v. question one can ask:

(i) *The existence question.* Suppose we are given a certain physical system and an empirical probability measure e on the observable variables of the system. Can we find an extended space that includes h.v.'s, and a probability measure p on this space, where p satisfies certain properties (as above) and realizes (via marginalization) the empirical probability measure e?

(ii) *The equivalence question.* Suppose we are given an empirical probability measure e on the observable variables of a system, and an h.v. model, with probability measure p that satisfies certain properties and realizes e. Can we find another h.v. model, with probability measure q, where q satisfies other stipulated properties and also realizes e?

Bell's Theorem is the most famous negative answer to (i), obtained when the physical system is quantum and the properties demanded are locality and λ-independence.

In this chapter we will focus on positive results for questions of both types (i) and (ii). These positive results involve yet another property of h.v. models: The (strong) *determinism* property says that for each player, the h.v.'s determine 'non-probabilistically' (formally: almost surely) the outcome of any measurement. As we will see in Section 4, determinism implies locality. We consider the following positive results on questions (i) and (ii):

(i) *First determinization result.* Every empirical model (whether generated by a classical or quantum or even superquantum system) can be realized by an h.v. model satisfying determinism.

(ii) *Second determinization result.* Given an h.v. model satisfying locality and λ-independence, there is a realization-equivalent h.v. model that satisfies determinism and λ-independence.

Put together, these two results tell us a lot about Bell's Theorem. The first determinization result says that for every empirical model, an h.v. model with determinism is possible. It is also true that for every empirical model, an h.v. model with λ-independence is possible. (This is a trivial construction, which we note in Remark 5.1.) As usually stated, Bell's Theorem asks for an h.v. model satisfying locality and λ-independence. In light of the second determinization result, Bell's Theorem can be equivalently stated as asking for determinism and λ-independence. Thus, Bell's Theorem teaches us that: *It is possible to believe that Nature (in the form of QM) is deterministic, or it is possible to believe that measurement choices by experimenters are free variables, but it is not possible to believe both.*

The goal of this chapter is to prove the two determinization results at a general measure-theoretic level (Theorems 5.2 and 5.3). Bell [4, 1971] mentioned the idea of the first determinization result. Fine [14, 1982] produced the first version of the second determinization result. Both results have been (re-)proved for various formulations in the literature. A notable aspect of our formulation is that we allow for infinite measurement spaces. Thus, our set-up is general enough to include, for example, experiments that involve spin measurements at arbitrary relative angles. We assume that outcome sets are finite (such as spin up or spin down).

Our treatment uses the concept of the *fiber product of measures*. The construction of these objects comes from Shortt [21, 1984]. The fiber product generalizes independence in probability theory, and has in turn been generalized in several directions in the literature (e.g., see Adler [2, 2009] and Ben Yaacov and Keisler [7, 2009] in model theory, Dawid and Studený [11, 1999] in graph theory, and Flori and Fritz [15, 2013] in category theory. Fiber products of measures turn out to be well suited to the questions in quantum foundations which we study in this chapter.

§2. Empirical and hidden-variable models. Alice has a space of possible measurements, which is a measurable space $(Y_a, \mathcal{Y}_a)$, and a space of possible outcomes, which is a measurable space $(X_a, \mathcal{X}_a)$. Likewise, Bob has a space of possible measurements, which is a measurable space $(Y_b, \mathcal{Y}_b)$, and a space of possible outcomes, which is a measurable space $(X_b, \mathcal{X}_b)$. Throughout, we will restrict attention to bipartite systems. (We will comment later on the extension to more than two parts.) There is also an h.v. space, which is an unspecified measurable space $(\Lambda, \mathcal{L})$. Write

$$\begin{aligned}
(X, \mathcal{X}) &= (X_a, \mathcal{X}_a) \otimes (X_b, \mathcal{X}_b), \\
(Y, \mathcal{Y}) &= (Y_a, \mathcal{Y}_a) \otimes (Y_b, \mathcal{Y}_b), \\
\Psi &= (X, \mathcal{X}) \otimes (Y, \mathcal{Y}), \\
\Omega &= (X, \mathcal{X}) \otimes (Y, \mathcal{Y}) \otimes (\Lambda, \mathcal{L}).
\end{aligned}$$

DEFINITION 2.1. An *empirical model* is a probability measure e on Ψ.

We see that an empirical model describes an experiment in which the pair of measurements $y = (y_a, y_b) \in Y$ is randomly chosen according to the probability measure $\text{marg}_Y e$, and y and the joint outcome $x = (x_a, x_b) \in X$ are distributed according to e.

DEFINITION 2.2. A *hidden-variable (h.v.) model* is a probability measure p on Ω.

DEFINITION 2.3. We say that an h.v. model p *realizes* an empirical model e if $e = \text{marg}_\Psi p$. We say that two h.v. models, possibly with different h.v. spaces, are *(realization-)equivalent* if they realize the same empirical model.

An h.v. model is an empirical model which has an extra component, viz., the h.v. space, and which reproduces a given empirical model when we average over the values of the h.v. The interest in h.v. models is that we can ask them to satisfy properties that it would be unreasonable to demand of an empirical model. Thus, in the example we began with, the property we ask for is conditional independence—which we would only expect once the extra r.v. Z is introduced. We will come to other properties in Section 4.

§3. Products and fiber products of measures. We first introduce notation and recall some well-known facts about product measures. For background on the relevant measure theory, see e.g. Billingsley [8, 1995].

Recall that by a product $(X, \mathcal{X}) \otimes (Y, \mathcal{Y})$ of two measurable spaces $(X, \mathcal{X})$ and $(Y, \mathcal{Y})$ is meant the (Cartesian) product space $X \times Y$ equipped with the σ-algebra generated by the measurable rectangles $J \times K$, where $J \in \mathcal{X}$ and

$K \in \mathcal{Y}$. We use the following two conventions. First, when p is a probability measure on $(X, \mathcal{X}) \otimes (Y, \mathcal{Y})$ and $q = \text{marg}_X p$, then for each $J \in \mathcal{X}$ we write

$$p(J) = p(J \times Y) = q(J),$$

and for each q-integrable $f : X \to \mathbb{R}$ we write

$$\int_J f(x)\, dp = \int_{J \times Y} f(x)\, dp = \int_J f(x)\, dq.$$

Thus, in particular, a statement holds for p-almost all $x \in X$ if and only if it holds for q-almost all $x \in X$.

Second, when p is a probability measure on a product space $(X, \mathcal{X}) \otimes (Y, \mathcal{Y}) \otimes (Z, \mathcal{Z})$, $J \in \mathcal{X}$, and $z \in Z$, we write $p[J||\mathcal{Z}]$ for the conditional probability of J given z. Here, we refer to the concept of conditional probability given a sub σ-algebra; see Billingsley [8, 1995, Section 33] for a presentation. Formally, $p[J||\mathcal{Z}]$ denotes a function from Z into [0, 1] such that

$$p[J||\mathcal{Z}]_z = p[J \times Y \times Z | \{X \times Y, \emptyset\} \otimes \mathcal{Z}]_{(x,y,z)}.$$

(Note that $\{X \times Y, \emptyset\}$ is the trivial σ-algebra over $X \times Y$, so that the right-hand side does not depend on (x, y).)

We use similar notation for (finite) products with factors to the left of $(X, \mathcal{X})$ or to the right of $(Z, \mathcal{Z})$. Note that if $q = \text{marg}_{X \times Z} p$, then $q[J||\mathcal{Z}] = p[J||\mathcal{Z}]$. We will also need the concept of conditional expectation given a sub σ-algebra (Billingsley [8, 1995, Section 34]), and we will use an analogous notation. Thus, given an integrable function $f : X \to \mathbb{R}$, and $z \in Z$, we define $\mathrm{E}[f||\mathcal{Z}]$ by:

$$\mathrm{E}[f||\mathcal{Z}]_z = \mathrm{E}[f \circ \pi | \{X \times Y, \emptyset\} \otimes \mathcal{Z}]_{(x,y,z)},$$

where we write π for the projection from $X \times Y \times Z$ to X.

LEMMA 3.1. *The mapping $z \mapsto p[J||\mathcal{Z}]_z$ is the p-almost surely unique $\mathcal{Z}$-measurable function $f : Z \to [0, 1]$ such that for each set $L \in \mathcal{Z}$,*

$$\int_L f(z)\, dp = p(J \times L).$$

PROOF. Existence: Let $f(z) = p[J||\mathcal{Z}]_z$. Using the definition of $p[J||\mathcal{Z}]$, we see that

$$\int_L f(z)\, dp = \int_{X \times Y \times L} \mathrm{E}[1_{J \times Y \times Z} | \{X \times Y, \emptyset\} \otimes \mathcal{Z}]\, dp =$$
$$\int_{X \times Y \times L} 1_{J \times Y \times Z}\, dp = p((X \times Y \times L) \cap (J \times Y \times Z)) = p(J \times L),$$

as required.

Uniqueness: If $p(J) = 0$, then $f(z) = g(z) = 0$ p-almost surely. Suppose $p(J) > 0$. Let f and g be two such functions and let $L = \{z : f(z) < g(z)\}$.

Then $L \in \mathcal{Z}$. If $p(J \times L) > 0$, then $p(L) > 0$, and

$$0 < \int_L g(z)\,dp - \int_L f(z)\,dp = \int_L (g(z) - f(z))\,dp = 0,$$

a contradiction. Therefore $p(J \times L) = 0$, so $p(L) = 0$ and hence $f(z) \geq g(z)$ p-almost surely. Similarly, $g(z) \geq f(z)$ p-almost surely, so $f(z) = g(z)$ p-almost surely. ⊣

Corollary 3.2. *Let q be the marginal of p on $X \times Z$. Then, for each $J \in \mathcal{X}$, we have $p[J||\mathcal{Z}] = q[J||\mathcal{Z}]$ q-almost surely.*

Lemma 3.3. *If $p[J||\mathcal{Z}] \in \{0, 1\}$ p-almost surely, then $p[J||\mathcal{Y} \otimes \mathcal{Z}] = p[J||\mathcal{Z}]$ p-almost surely.*

Proof. Let $L_0 = \{z \in Z : p[J||\mathcal{Z}]_z = 0\}$ and $L_1 = \{z \in Z : p[J||\mathcal{Z}]_z = 1\}$. Then $L_0, L_1 \in \mathcal{Z}$ and $p(L_0 \cup L_1) = 1$. By Lemma 3.1,

$$\int_{L_0} p[J||\mathcal{Z}]_z\,dp = 0 = p(J \times L_0),$$

$$\int_{L_1} p[J||\mathcal{Z}]_z\,dp = p(L_1) = p(J \times L_1).$$

By Lemma 3.1 again,

$$\int_{Y \times L_0} p[J||\mathcal{Y} \otimes \mathcal{Z}]_{(y,z)}\,dp = p(J \times Y \times L_0) = p(J \times L_0) = 0,$$

so

$$p[J||\mathcal{Y} \otimes \mathcal{Z}]_{(y,z)} = 0 = p[J||\mathcal{Z}]_z \ \forall\, (y, z) \in Y \times L_0.$$

Similarly,

$$\int_{Y \times L_1} p[J||\mathcal{Y} \otimes \mathcal{Z}]_{(y,z)}\,dp = p(J \times Y \times L_1) = p(J \times L_1) = p(L_1),$$

so

$$p[J||\mathcal{Y} \otimes \mathcal{Z}]_{(y,z)} = 1 = p[J||\mathcal{Z}]_z \ \forall\, (y, z) \in Y \times L_1,$$

as required. ⊣

When $x \in X$, we write $p[x||\mathcal{Z}]_z = p[\{x\}||\mathcal{Z}]_z$. For the particular case of finite X, we get, by the properties of probability measures, that $\sum_{x \in X} p[x||\mathcal{Z}]_z = 1$ p-almost surely.

Given probability measures p on $(X, \mathcal{X}) \otimes (Y, \mathcal{Y})$ and r on $(Y, \mathcal{Y})$, we say that p is an *extension* of r if $r = \text{marg}_Y p$. We say that two probability measures p and q on $(X, \mathcal{X}) \otimes (Y, \mathcal{Y})$ *agree on* Y if $\text{marg}_Y p = \text{marg}_Y q$.

Given probability spaces $(X, \mathcal{X}, q)$ and $(Y, \mathcal{Y}, r)$, the product measure $p = q \otimes r$ is the unique probability measure p on $(X, \mathcal{X}) \otimes (Y, \mathcal{Y})$ such that q and r are independent with respect to p, that is,

$$p(J \times K) = q(J) \times r(K)$$

for all $J \in \mathcal{X}$ and $K \in \mathcal{Y}$. Note that p is a common extension of q and r.

Remark 3.4. Let $(X, \mathcal{X}, q)$ and $(Y, \mathcal{Y}, r)$ be as above and let p be a common extension of q and r on $(X, \mathcal{X}) \otimes (Y, \mathcal{Y})$. The following are equivalent:

(i) $p = q \otimes r$.

(ii) The σ-algebras $\mathcal{X} \otimes \{Y, \emptyset\}$ and $\{X, \emptyset\} \otimes \mathcal{Y}$ are independent with respect to p, that is,

$$p(J \times K) = p(J) \times p(K)$$

for all $J \in \mathcal{X}$ and $K \in \mathcal{Y}$.

(iii) $p[J||\mathcal{Y}]_y = p(J)$ p-almost surely for all $J \in \mathcal{X}$.

We next introduce the notion of a fiber product of measures. For the remainder of this section we let $\boldsymbol{X} = (X, \mathcal{X}), \boldsymbol{Y} = (Y, \mathcal{Y}), \boldsymbol{Z} = (Z, \mathcal{Z})$ be measurable spaces.

DEFINITION 3.5. Let q and r be probability measures on $\boldsymbol{X} \otimes \boldsymbol{Z}$ and $\boldsymbol{Y} \otimes \boldsymbol{Z}$, respectively. Assume that q and r have the same marginal s on $\boldsymbol{Z}$. We say that a probability measure p on $\boldsymbol{X} \otimes \boldsymbol{Y} \otimes \boldsymbol{Z}$ is a *fiber product* of q and r over Z, in symbols $p = q \otimes_Z r$, if

$$p(J \times K \times L) = \int_L q[J||\mathcal{Z}]_z \times r[K||\mathcal{Z}]_z \, ds$$

for all $J \in \mathcal{X}$, $K \in \mathcal{Y}$, and $L \in \mathcal{Z}$.

Intuitively, the fiber product $q \otimes_Z r$ is the common extension of q and r with respect to which q and r are as independent as possible given that they have the same marginal on Z. There are examples where a fiber product does not exist (see Swart [22, 1996] and Dawid and Studený [11, 1999]). But it is easily seen that if a fiber product $q \otimes_Z r$ does exist, then it is unique. Next is a characterization of the fiber product in terms of conditional probabilities and extensions.

LEMMA 3.6. *Let q and r be as in Definition 3.5, and let p be a common extension of q, r on $\boldsymbol{X} \otimes \boldsymbol{Y} \otimes \boldsymbol{Z}$. Then the following are equivalent:*

(i) $p = q \otimes_Z r$.

(ii) $p[J \times K||\mathcal{Z}]_z = q[J||\mathcal{Z}]_z \times r[K||\mathcal{Z}]_z$ *p-almost surely, for all $J \in \mathcal{X}$ and $K \in \mathcal{Y}$.*

(iii) $p[J \times K||\mathcal{Z}]_z = p[J||\mathcal{Z}]_z \times p[K||\mathcal{Z}]_z$ *p-almost surely, for all $J \in \mathcal{X}$ and $K \in \mathcal{Y}$.*

(iv) $p[J||\mathcal{Y} \otimes \mathcal{Z}]_{(y,z)} = p[J||\mathcal{Z}]_z$ *p-almost surely, for all $J \in \mathcal{X}$.*

PROOF. It is clear that (i), (ii), and (iii) are equivalent. Consider any $J \in \mathcal{X}, K \in \mathcal{Y}$, and $L \in \mathcal{Z}$. Assume (i). To prove (iv), it is enough to show that

$$\int_{K \times L} p[J||\mathcal{Z}] \, dp = p(J \times K \times L).$$

We have

$$\int_{K\times L} p[J||\mathcal{Z}]\,dp = \int_{Y\times L} p[J||\mathcal{Z}] \times 1_K\,dp.$$

By the rules of conditional expectations,

$$\mathrm{E}[p[J||\mathcal{Z}] \times 1_K||\mathcal{Z}] = p[J||\mathcal{Z}] \times \mathrm{E}[1_K||\mathcal{Z}] = p[J||\mathcal{Z}] \times p[K||\mathcal{Z}].$$

Therefore

$$\int_{Y\times L} p[J||\mathcal{Z}] \times 1_K\,dp = \int_L p[J||\mathcal{Z}] \times p[K||\mathcal{Z}]\,dp$$
$$= \int_L q[J||\mathcal{Z}] \times r[K||\mathcal{Z}]\,dp.$$

By (i), this is equal to $p(J \times K \times L)$, which shows that (i) implies (iv).

Now assume (iv). Then

$$p(J \times K \times L) = \int_{K\times L} p[J||\mathcal{Y}\otimes\mathcal{Z}]\,dp = \int_{K\times L} p[J||\mathcal{Z}]\,dp$$
$$= \int_{Y\times L} p[J||\mathcal{Z}] \times 1_K\,dp.$$

As in the preceding paragraph,

$$\int_{Y\times L} p[J||\mathcal{Z}] \times 1_K\,dp = \int_L q[J||\mathcal{Z}] \times r[K||\mathcal{Z}]\,dp,$$

and condition (i) is proved. ⊣

A version $g(J,z)$ of the conditional probability $q[J||\mathcal{Z}]_z$ is *regular* if $g(\cdot, z_0)$ is a probability measure on $\boldsymbol{X}$ for each fixed $z_0 \in Z$. It is well known that when $\boldsymbol{X}$ and $\boldsymbol{Z}$ are both Polish spaces, then $q[J||\mathcal{Z}]_z$ has a regular version. It is also easily seen that when X is finite and $\boldsymbol{Z}$ is any measurable space, then $q[J||\mathcal{Z}]_z$ has a regular version. This is the case we will need in this chapter. The next lemma is from Swart [22, 1996]:

LEMMA 3.7. *Let q and r be as in Definition* 3.5. *If $q[J||\mathcal{Z}]_z$ has a regular version, then the fiber product $q \otimes_Z r$ exists.*

COROLLARY 3.8. *Let q and r be as in Definition* 3.5. *If the space X is finite, then the fiber product $q \otimes_Z r$ exists.*

§4. Properties of hidden-variable models.

We can now formulate the various properties of h.v. models which we listed in the Introduction (we will not repeat their sources) and establish some relationships among them. At this point, we adopt:

Assumption: The outcome spaces X_a and X_b are finite, and $\mathcal{X}_a$ and $\mathcal{X}_b$ are the respective power sets.

Also, whenever we write an equation involving conditional probabilities, it will be understood to mean that the equation holds p-almost surely. By the term "measure" we will always mean "probability measure." Fix an h.v. model p. We will often make use of the following notation:

$$p_a = \mathrm{marg}_{X_a \times Y \times \Lambda} p, \quad p_b = \mathrm{marg}_{X_b \times Y \times \Lambda} p,$$
$$q_a = \mathrm{marg}_{X_a \times Y_a \times \Lambda} p, \quad q_b = \mathrm{marg}_{X_b \times Y_b \times \Lambda} p,$$
$$r = \mathrm{marg}_{Y \times \Lambda} p,$$
$$p_Y = \mathrm{marg}_Y p, \quad p_\Lambda = \mathrm{marg}_\Lambda p.$$

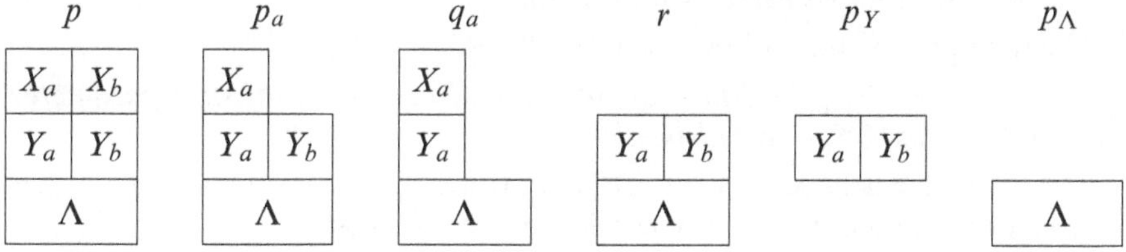

All expressions below which are given for Alice have counterparts for Bob, with a and b interchanged.

Definition 4.1. The h.v. model p satisfies *locality* if for every $x \in X$ we have

$$p[x||\mathcal{Y} \otimes \mathcal{L}] = p[x_a||\mathcal{Y}_a \otimes \mathcal{L}] \times p[x_b||\mathcal{Y}_b \otimes \mathcal{L}].$$

Definition 4.2. The h.v. model p satisfies *parameter independence* if for every $x_a \in X_a$ we have

$$p[x_a||\mathcal{Y} \otimes \mathcal{L}] = p[x_a||\mathcal{Y}_a \otimes \mathcal{L}].$$

Here is a characterization of parameter independence in terms of fiber products.

Corollary 4.3. *p satisfies parameter independence if and only if $p_a = q_a \otimes_{Y_a \times \Lambda} r$ and $p_b = q_b \otimes_{Y_b \times \Lambda} r$.*

Proof. By Lemma 3.6, $p_a = q_a \otimes_{Y_a \times \Lambda} r$ if and only if

$$p_a[x_a||\mathcal{Y} \otimes \mathcal{L}] = p_a[x_a||\mathcal{Y}_a \otimes \mathcal{L}]$$

for all $x_a \in X_a$. Since p is an extension of p_a, this holds if and only if

$$p[x_a||\mathcal{Y} \otimes \mathcal{L}] = p[x_a||\mathcal{Y}_a \otimes \mathcal{L}]$$

for all $x_a \in X_a$. Similarly, $p_b = q_b \otimes_{Y_b \times \Lambda} r$ if and only if

$$p[x_b||\mathcal{Y} \otimes \mathcal{L}] = p[x_b||\mathcal{Y}_b \otimes \mathcal{L}]$$

for all $x_b \in X_b$. The result follows. ⊣

DEFINITION 4.4. The h.v. model p satisfies *outcome independence* if for every $x = (x_a, x_b) \in X$ we have

$$p[x||\mathcal{Y} \otimes \mathcal{L}] = p[x_a||\mathcal{Y} \otimes \mathcal{L}] \times p[x_b||\mathcal{Y} \otimes \mathcal{L}].$$

The following corollary characterizes outcome independence in terms of fiber products.

COROLLARY 4.5. *p satisfies outcome independence if and only if*

$$p = p_a \otimes_{Y \times \Lambda} p_b.$$

PROOF. This follows easily from Lemma 3.6. ⊣

The next proposition follows Jarrett [18, 1984, p. 582].

PROPOSITION 4.6. *p satisfies locality if and only if it satisfies parameter independence and outcome independence.*

PROOF. It is easily seen from the definitions that if p satisfies parameter independence and outcome independence, then p satisfies locality.

Suppose that p satisfies locality. We have

$$\{x_a\} \times X_b = \bigcup_{x_b \in X_b} \{(x_a, x_b)\},$$

so

$$\begin{aligned} p[x_a||\mathcal{Y} \otimes \mathcal{L}] &= p[\{x_a\} \times X_b||\mathcal{Y} \otimes \mathcal{L}] \\ &= \sum_{x_b \in X_b} p[x_a, x_b||\mathcal{Y} \otimes \mathcal{L}] \\ &= \sum_{x_b \in X_b} (p[x_a||\mathcal{Y}_a \otimes \mathcal{L}] \times p[x_b||\mathcal{Y}_b \otimes \mathcal{L}]) \\ &= p[x_a||\mathcal{Y}_a \otimes \mathcal{L}] \times \sum_{x_b \in X_b} p[x_b||\mathcal{Y}_b \otimes \mathcal{L}] \\ &= p[x_a||\mathcal{Y}_a \otimes \mathcal{L}] \times 1 = p[x_a||\mathcal{Y}_a \otimes \mathcal{L}]. \end{aligned}$$

Similarly,

$$p[x_b||\mathcal{Y} \otimes \mathcal{L}] = p[x_b||\mathcal{Y}_b \otimes \mathcal{L}].$$

It follows that p satisfies parameter independence.

Again, supposing that p satisfies locality, we have

$$p[x_a, x_b||\mathcal{Y} \otimes \mathcal{L}] = p[x_a||\mathcal{Y}_a \otimes \mathcal{L}] \times p[x_b||\mathcal{Y}_b \otimes \mathcal{L}],$$

and hence

$$p[x_a, x_b||\mathcal{Y} \otimes \mathcal{L}] = p[x_a||\mathcal{Y} \otimes \mathcal{L}] \times p[x_b||\mathcal{Y} \otimes \mathcal{L}],$$

so p satisfies outcome independence. ⊣

We immediately get a characterization of locality in terms of fiber products.

COROLLARY 4.7. *p satisfies locality if and only if*

$$p = p_a \otimes_{Y \times \Lambda} p_b, \quad p_a = q_a \otimes_{Y_a \times \Lambda} r, \quad p_b = q_b \otimes_{Y_b \times \Lambda} r.$$

PROOF. By Proposition 4.6 and Corollaries 4.3 and 4.5. ⊣

DEFINITION 4.8. The h.v. model p satisfies *λ-independence* if for every event $L \in \mathcal{L}$,

$$p[L \| \mathcal{Y}]_y = p(L).$$

Remark 4.9. We observe:

(i) The λ-independence property for p depends only on r.
(ii) Any h.v. model p such that Λ is a singleton satisfies λ-independence.

By Remark 3.4, we have:

LEMMA 4.10. *The following are equivalent*:

(*i*) *p satisfies λ-independence.*
(*ii*) *The measure r is the product $p_Y \otimes p_\Lambda$.*
(*iii*) *The σ-algebras $\mathcal{Y}$ and $\mathcal{L}$ are independent with respect to p, i.e.,*

$$p(K \times L) = p(K) \times p(L)$$

for every $K \in \mathcal{Y}, L \in \mathcal{L}$.

The distinction between strong and weak determinism in the next two definitions is from Brandenburger and Yanofsky [10, 2008]. Strong determinism is the notion discussed in the Introduction.

DEFINITION 4.11. The h.v. model p satisfies *strong determinism* if for each $x_a \in X_a$ we have

$$p[x_a \| \mathcal{Y}_a \otimes \mathcal{L}]_{(y_a, \lambda)} \in \{0, 1\}.$$

This says that the set $Y_a \times \Lambda$ can be partitioned into sets $\{A_{x_a} : x_a \in X_a\}$ such that $p[x_a \| A_{x_a}] = 1$ for each $x_a \in X_a$.

DEFINITION 4.12. The h.v. model p satisfies *weak determinism* if for each $x \in X$ we have

$$p[x \| \mathcal{Y} \otimes \mathcal{L}]_{(y, \lambda)} \in \{0, 1\}.$$

This says that the set $Y \times \Lambda$ can be partitioned into sets $\{A_x : x \in X\}$ such that $p[x \| A_x] = 1$ for each $x \in X$.

LEMMA 4.13. *The following are equivalent*:

(*i*) *p satisfies weak determinism.*
(*ii*) *For each $x_a \in X_a$ we have*

$$p[x_a \| \mathcal{Y} \otimes \mathcal{L}]_{(y, \lambda)} \in \{0, 1\}.$$

PROOF. It is clear that (ii) implies (i).

Assume (i). Then for p-almost all (y, λ) there is an $x \in X$ such that $p[x||\mathcal{Y} \otimes \mathcal{L}]_{(y,\lambda)} = 1$, and hence

$$p[x_a||\mathcal{Y} \otimes \mathcal{L}]_{(y,\lambda)} = 1$$

for each $x_a \in X_a$. Therefore (ii) holds. ⊣

PROPOSITION 4.14. *If p satisfies strong determinism then it satisfies weak determinism.*

PROOF. Suppose p satisfies strong determinism. By Lemma 3.3, we have

$$p[x_a||\mathcal{Y}_a \otimes \mathcal{L}] = p[x_a||\mathcal{Y} \otimes \mathcal{L}]$$

p-almost surely, and therefore

$$p[x_a||\mathcal{Y} \otimes \mathcal{L}] \in \{0, 1\},$$

so p satisfies weak determinism by Lemma 4.13(ii). ⊣

PROPOSITION 4.15. *If p satisfies weak determinism then it satisfies outcome independence.*

PROOF. Suppose p satisfies weak determinism. By Lemma 4.13, we have

$$p[x_a||\mathcal{Y} \otimes \mathcal{L}] \in \{0, 1\}.$$

Therefore

$$p[x||\mathcal{Y} \otimes \mathcal{L}] = p[x_a||\mathcal{Y} \otimes \mathcal{L}] \times p[x_b||\mathcal{Y} \otimes \mathcal{L}],$$

as required. ⊣

PROPOSITION 4.16. *p satisfies strong determinism if and only if it satisfies weak determinism and parameter independence.*

PROOF. Suppose p satisfies strong determinism. By Lemma 3.3,

$$p[x_a||\mathcal{Y}_a \otimes \mathcal{L}] = p[x_a||\mathcal{Y} \otimes \mathcal{L}],$$

so p satisfies parameter independence. By Proposition 4.14, p satisfies weak determinism.

For the converse, suppose p satisfies weak determinism and parameter independence. Fix $x_a \in X_a$. By weak determinism and Lemma 4.13,

$$p[x_a||\mathcal{Y} \otimes \mathcal{L}]_{(y,\lambda)} \in \{0, 1\}.$$

By parameter independence,

$$p[x_a||\mathcal{Y} \otimes \mathcal{L}] = p[x_a||\mathcal{Y}_a \otimes \mathcal{L}].$$

Therefore

$$p[x_a||\mathcal{Y}_a \otimes \mathcal{L}]_{(y,\lambda)} \in \{0, 1\},$$

so p satisfies strong determinism. ⊣

COROLLARY 4.17. *p satisfies strong determinism if and only if it satisfies weak determinism and locality.*

PROOF. By Propositions 4.6, 4.15, and 4.16. ⊣

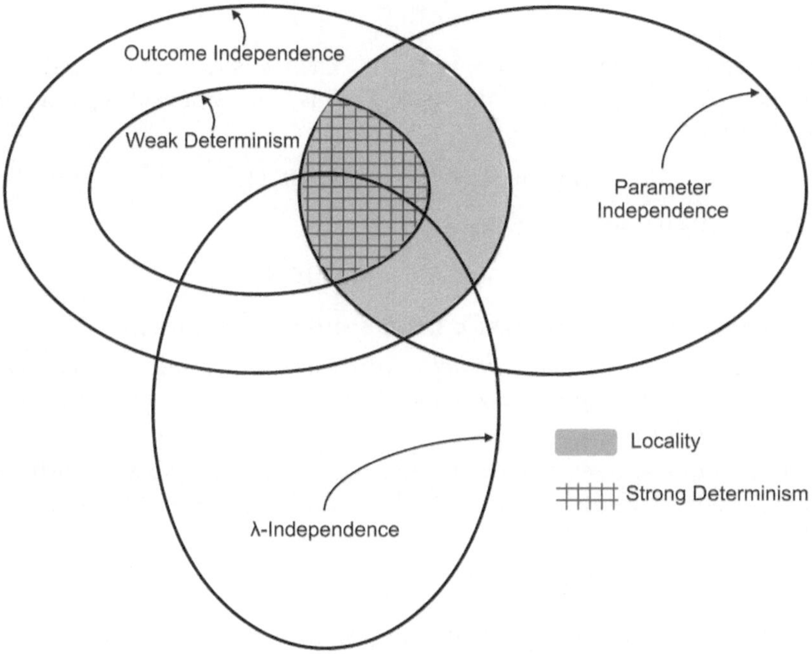

We can summarize the properties we have considered and the relationships among them in the above Venn diagram.

§5. Determinization theorems. Given an h.v. model p, we call the probability space $(\Lambda, \mathcal{L}, p_\Lambda)$ the *h. v. space of* p.

Remark 5.1. Every empirical model e can be realized by an h.v. model p where p satisfies λ-independence and the h.v. space of p has only one element.

PROOF. For every probability space $(\Lambda, \mathcal{L}, p_\Lambda)$, the product measure $p = e \otimes p_\Lambda$ is an h.v. model that realizes e and satisfies λ-independence. In particular, we can take Λ to be a one-element set and take $(\Lambda, \mathcal{L}, p_\Lambda)$ to be the trivial probability measure. ⊣

We now state and prove our determinization results.

THEOREM 5.2. *Every empirical model e can be realized by an h.v. model p where p satisfies strong determinism and the h.v. space of p is finite.*

PROOF. Let $s = \mathrm{marg}_X e$. Build an h.v. space $(\Lambda, \mathcal{L}, s)$ where Λ is a copy of X and $\mathcal{L}$ is the power set of X. Build a probability measure d on $X \times \Lambda$ so that, for each $x \in X$ and $x' \in \Lambda$,

$$d(x, x') = \begin{cases} s(x) & \text{if } x = x', \\ 0 & \text{otherwise.} \end{cases}$$

Note that d is an extension of s.

Let p be the fiber product $p = d \otimes_X e$. Then p is realization-equivalent to e. Since Λ is a copy of the finite space X, Λ is finite. For each $x_a \in X_a$ and $x' \in \Lambda$, we have

$$p[x_a \| x'] = d[x_a \| x'] \in \{0, 1\}.$$

By Lemma 3.3, for each x_a we have

$$p[x_a \| \mathcal{Y}_a \otimes \mathcal{L}]_{(y_a, x')} \in \{0, 1\}$$

p-almost surely. This shows that p satisfies strong determinism. ⊣

THEOREM 5.3. *Given an h.v. model p satisfying locality and λ-independence, there is a realization-equivalent h.v. model $\bar{p}$ that satisfies strong determinism and λ-independence.*

PROOF. Suppose p satisfies locality and λ-independence. We will construct a new h.v. model $\bar{p}$ whose h.v. space $(\bar{\Lambda}, \bar{\mathcal{L}}, \bar{p}_{\bar{\Lambda}})$ will be the product of $(\Lambda, \mathcal{L}, p_\Lambda)$ and the Lebesgue unit square

$$([0, 1]_a, \mathcal{U}_a, u_a) \otimes ([0, 1]_b, \mathcal{U}_b, u_b).$$

Here, $[0, 1]_a$ is a copy of the real unit interval, $\mathcal{U}_a$ is the set of Borel subsets of $[0, 1]_a$, and u_a is Lebesgue measure on $\mathcal{U}_a$; similarly for b.

Let $X_a = \{x_a^1, \dots, x_a^A\}$. For each $y_a \in Y_a$ and $\lambda \in \Lambda$, partition $[0, 1]_a$ into A consecutive intervals

$$I_a(x_a^1, y_a, \lambda), I_a(x_a^2, y_a, \lambda), \dots, I_a(x_a^A, y_a, \lambda),$$

where, for each $x_a \in X_a$, $I_a(x_a, y_a, \lambda)$ has length

$$u_a(I_a(x_a, y_a, \lambda)) = p[x_a \| \mathcal{Y}_a \otimes \mathcal{L}]_{(y_a, \lambda)}.$$

Note that the boundary point between the ith and $(i + 1)$th intervals is the $(\mathcal{Y}_a \otimes \mathcal{L})$-measurable function

$$\sum_{i=1}^{n} p[x_a^i \| \mathcal{Y}_a \otimes \mathcal{L}]_{(y_a, \lambda)}.$$

We carry out the same construction with b in place of a.

Let $\bar{r} = r \otimes u_a \otimes u_b$. Since p satisfies λ-independence, $r = p_Y \otimes p_\Lambda$, and thus $\bar{r} = p_Y \otimes \bar{p}_{\bar{\Lambda}}$. Let s_a be the unique probability measure on

$$(X_a, \mathcal{X}_a) \otimes (Y_a, \mathcal{Y}_a) \otimes (\Lambda, \mathcal{L}) \otimes ([0, 1]_a, \mathcal{U}_a)$$

such that for each $x_a \in X_a$, $K_a \in \mathcal{Y}_a$, $L \in \mathcal{L}$, and $U_a \in \mathcal{U}_a$, we have

$$s_a(\{x_a\} \times K_a \times L \times U_a) = \int_{K_a \times L \times U_a} 1_{I_a(x_a, y_a, \lambda)}(\alpha)\, d\bar{r},$$

where we write α for a typical element of $[0,1]_a$. Define s_b in a similar way. Now define $\bar{p}_a$, $\bar{p}_b$, and $\bar{p}$ as the fiber products

$$\bar{p}_a = s_a \otimes_{Y_a \times \Lambda \times [0,1]_a} \bar{r}, \ \bar{p}_b = s_b \otimes_{Y_b \times \Lambda \times [0,1]_b} \bar{r}, \ \bar{p} = \bar{p}_a \otimes_{Y \times \bar{\Lambda}} \bar{p}_b.$$

We see that the h.v. model $\bar{p}$ is a common extension of s_a, s_b, and $\bar{r}$. It also satisfies λ-independence because $\bar{r} = p_Y \otimes \bar{p}_{\bar{\Lambda}}$. By Lemma 3.1,

$$s_a[x_a || \mathcal{Y}_a \otimes \mathcal{L} \otimes \mathcal{U}_a] = 1_{I_a(x_a, y_a, \lambda)} \in \{0, 1\}.$$

By Lemma 3.3,

$$s_a[x_a || \mathcal{Y}_a \otimes \bar{\mathcal{L}}] \in \{0, 1\}.$$

Similarly for s_b. Therefore $\bar{p}$ satisfies strong determinism.

It remains to prove that $\bar{p}$ is an extension of p. By Fubini's Theorem,

$$\begin{aligned}
s_a(\{x_a\} \times K_a \times L) &= \int_{K_a \times L \times [0,1]_a} 1_{I_a(x_a, y_a, \lambda)}(\alpha)\, d\bar{r} \\
&= \int_{K_a \times L} \int_0^1 1_{I_a(x_a, y_a, \lambda)}(\alpha)\, du_a\, dr \\
&= \int_{K_a \times L} u_a(I_a(x_a, y_a, \lambda))\, dr \\
&= \int_{K_a \times L} q_a[x_a || \mathcal{Y}_a \otimes \mathcal{L}]_{(y_a, \lambda)}\, dr \\
&= q_a(\{x_a\} \times K_a \times L).
\end{aligned}$$

Thus s_a is an extension of q_a. Similarly, s_b is an extension of q_b.

Since both p and $\bar{p}$ satisfy locality, and $\bar{p}$ extends $\bar{r} = r \otimes u_a \otimes u_b$, by Fubini's Theorem we have

$$\begin{aligned}
\bar{p}(\{x\} \times K \times L) &= \int_{K \times L \times [0,1]_a \times [0,1]_b} \bar{p}[x || \mathcal{Y} \otimes \bar{\mathcal{L}}]\, d\bar{r} \\
&= \int_{K \times L \times [0,1]_a \times [0,1]_b} s_a[x_a || \mathcal{Y}_a \otimes \bar{\mathcal{L}}] \times s_b[x_b || \mathcal{Y}_b \otimes \bar{\mathcal{L}}]\, d\bar{r} \\
&= \int_{K \times L} \int_0^1 \int_0^1 s_a[x_a || \mathcal{Y}_a \otimes \bar{\mathcal{L}}] \times s_b[x_b || \mathcal{Y}_b \otimes \bar{\mathcal{L}}]\, du_a\, du_b\, dr \\
&= \int_{K \times L} q_a[x_a || \mathcal{Y}_a \otimes \mathcal{L}] \times q_b[x_b || \mathcal{Y}_b \otimes \mathcal{L}]\, dr \\
&= p(\{x\} \times K \times L).
\end{aligned}$$

Thus $\bar{p}$ is an extension of p, and hence $\bar{p}$ is realization-equivalent to p. This completes the proof. ⊣

All the results in Section 4 ("Properties of hidden-variable models"), and Theorems 5.2 and 5.3 in this section, extend immediately to multipartite systems. The only adjustment needed is that parameter independence must now be stated in terms of sets of parts instead of individual parts. Interestingly, outcome independence and locality do not need to be restated.

§6. Endnote. To keep things simple, we assumed in this chapter that the outcome spaces X_a and X_b are finite. However, the only result in this chapter that requires this assumption is Theorem 5.2. We show in [9, 2012] that all of the results in Section 4 hold for arbitrary outcome spaces X_a and X_b. Also, the arguments in [9, 2012] can be adapted to show that Theorem 5.3 holds assuming only that the outcome spaces have countably generated σ-algebras of events $\mathcal{X}_a$ and $\mathcal{X}_b$.

It would be of interest to extend the methods in this chapter to formulate other properties that have usually been studied only for the case of finite sets of measurements. For finite probability spaces, Abramsky and Brandenburger [1, 2011] establish a strict hierarchy of three properties: non-locality (à la Bell) is strictly weaker than possibilistic non-locality (exhibited by the Hardy [17, 1993] model), which is strictly weaker than strong contextuality (exhibited by the Greenberger, Horne, and Zeilinger [16, 1989] model). (In this language, the Kochen-Specker Theorem [19, 1967] is a model-independent proof of strong contextuality.) Extending these latter properties to the general measure-theoretic setting appears to be an open direction.

REFERENCES

[1] S. Abramsky and A. Brandenburger, *The sheaf-theoretic structure of non-locality and contextuality*, ***New Journal of Physics***, vol. 13 (2011), p. 113036.

[2] H. Adler, *A Geometric Introduction to Forking and Thorn-Forking*, ***Journal of Mathematical Logic***, vol. 9 (2009), pp. 1–21.

[3] J. Bell, *On the Einstein-Podolsky-Rosen Paradox*, ***Physics***, vol. 1 (1964), pp. 195–200.

[4] ———, *Introduction to the hidden-variable question*, ***Foundations of Quantum Mechanics, Proceedings of the International School of Physics 'Enrico Fermi,' Course IL*** (New York), Academic Press, 1971, (Reprinted in [6, 1987, pp. 29–39],), pp. 171–181.

[5] ———, *An exchange on local beables*, ***Dialectica***, vol. 39 (1985), pp. 85–96.

[6] ———, ***Speakable and Unspeakable in Quantum Mechanics***, Cambridge University Press, 1987.

[7] I. Ben Yaacov and H. J. Keisler, *Randomizations of models as metric structures*, ***Confluentes Mathematici***, vol. 1 (2009), pp. 197–223, also available at http://www.math.wisc.edu/keisler.

[8] P. Billingsley, ***Probability and measure***, 3rd ed., Wiley, 1995.

[9] A. Brandenburger and H. J. Keisler, *A canonical hidden-variable space*, (2012), available at http://www.adambrandenburger.com and http://www.math.wisc.edu/keisler.

[10] A. Brandenburger and N. Yanofsky, *A classification of hidden-variable properties*, ***Journal of Physics A: Mathematical and Theoretical***, vol. 41 (2008), p. 425302.

[11] A. P. Dawid and M. Studený, *An Alternative Approach to Conditional Independence*, ***Artificial Intelligence and Statistics 99, Proceedings of the 7th Workshop*** (D. Heckerman and J. Whittaker, editors), Morgan Kaufmann, 1999, pp. 32–40.

[12] W. M. Dickson, ***Quantum Chance and Non-Locality: Probability and Non-Locality in the Interpretations of Quantum Mechanics***, Cambridge University Press, 2005.

[13] A. Einstein, B. Podolsky, and N. Rosen, *Can quantum-mechanical description of physical reality be considered complete?*, ***Physical Review***, vol. 47 (1935), pp. 777–780.

[14] A. Fine, *Hidden variables, joint probability and the Bell inequalities*, ***Physical Review Letters***, vol. 48 (1982), pp. 291–295.

[15] C. Flori and T. Fritz, *Compositories and Gleaves*, (2013), available at http://arxiv.org/abs/1308.6548.

[16] D. M. Greenberger, M. A. Horne, and A. Zeilinger, *Going beyond Bell's theorem*, ***Bell's Theorem, Quantum Theory and Conceptions of the Universe*** (M. Kafatos, editor), Kluwer, 1989, pp. 69–72.

[17] L. Hardy, *Nonlocality for two particles without inequalities for almost all entangled states*, ***Physical Review Letters***, vol. 71 (1993), pp. 1665–1668.

[18] J. Jarrett, *On the physical significance of the locality conditions in the Bell arguments*, ***Noûs***, vol. 18 (1984), pp. 569–589.

[19] S. Kochen and E. Specker, *The problem of hidden variables in quantum mechanics*, ***Journal of Mathematics and Mechanics***, vol. 17 (1967), pp. 59–87.

[20] A. Shimony, *Events and processes in the quantum world*, ***Quantum Concepts in Space and Time*** (R. Penrose and C. Isham, editors), Oxford University Press, 1986, pp. 182–203.

[21] R. Shortt, *Universally measurable spaces: An invariance theorem and diverse characterizations*, ***Fundamenta Mathematicae***, vol. 121 (1984), pp. 169–176.

[22] J. Swart, *A conditional product measure theorem*, ***Statistics & Probability Letters***, vol. 28 (1996), pp. 131–135.

[23] J. von Neumann, ***Mathematische Grundlagen der Quantenmechanik***, Springer-Verlag, 1932, (Translated as ***Mathematical Foundations of Quantum Mechanics***, Princeton University Press, 1955.).

STERN SCHOOL OF BUSINESS
NEW YORK UNIVERSITY
NEW YORK, NY 10012
E-mail: adam.brandenburger@stern.nyu.edu
URL: www.adambrandenburger.com

DEPARTMENT OF MATHEMATICS
UNIVERSITY OF WISCONSIN-MADISON
MADISON, WI 53706
E-mail: keisler@math.wisc.edu
URL: www.math.wisc.edu/~keisler

OPERATIONAL THEORIES AND CATEGORICAL QUANTUM MECHANICS

SAMSON ABRAMSKY AND CHRIS HEUNEN

Abstract. A central theme in current work in quantum information and quantum foundations is to see quantum mechanics as occupying one point in a space of possible theories, and to use this perspective to understand the special features and properties which single it out, and the possibilities for alternative theories. Two formalisms which have been used in this context are *operational theories*, and *categorical quantum mechanics*. The aim of the present paper is to establish strong connections between these two formalisms. We show how models of categorical quantum mechanics have representations as operational theories. We then show how *non-locality* can be formulated at this level of generality, and study a number of examples from this point of view, including Hilbert spaces, sets and relations, and stochastic maps. The local, quantum, and no-signalling models are characterized in these terms.

§1. **Introduction.** A central theme in current work in quantum information and quantum foundations is to see quantum mechanics as occupying one point in a space of possible theories, and to use this perspective to understand the special features and properties which single it out, and the possibilities for alternative theories.

Two formalisms which have been used in this context are *operational theories* [48, 41, 52, 47], and *categorical quantum mechanics* [6, 7].

- Operational theories allow general formulations of results in quantum foundations and quantum information [11, 12, 10]. They also play a prominent rôle in current work on axiomatizations of quantum mechanics [36, 19, 49, 25].
- Categorical quantum mechanics enables a high-level approach to quantum information and quantum foundations, which can be presented in terms of string-diagram representations of structures in monoidal categories [7]. This has proved very effective in providing a conceptually illuminating and technically powerful perspective on a range of topics, including quantum protocols [6], entanglement [24], measurement-based quantum computing [29], no-cloning [1], and non-locality [22].

Logic and Algebraic Structures in Quantum Computing
Edited by J. Chubb, A. Eskandarian and V. Harizanov
Lecture Notes in Logic, 45
© 2016, Association for Symbolic Logic

The aim of the present paper is to establish strong connections between these two formalisms. We shall begin by reviewing operational theories. We then show how a proper formulation of *compound systems* within the operational framework leads to a view of operational theories as representations of monoidal categories of a particular form. We call these *operational representations*.

We then review some elements of categorical quantum mechanics, and show how *monoidal dagger categories*, equipped with a trace ideal, give rise to operational representations. Thus there is a general passage from categorical quantum mechanics to operational theories.

We go on to show how *non-locality* can be formulated at this level of generality, and study a number of examples from this point of view, including Hilbert spaces, sets and relations, and stochastic maps. The local, quantum, and no-signalling models are characterized in these terms.

We shall assume some familiarity with the linear-algebraic formalism of quantum mechanics, and with the first notions of category theory. To make the paper reasonably self-contained, we include an appendix which reviews the basic definitions of monoidal categories, functors and natural transformations.

We also include another appendix which proves a number of technical results on trace ideals. These are mathematically interesting, but would break up the flow of ideas in the main body of the paper.

§2. Why operational theories? Before proceeding to a formal description of operational theories, it may be useful to discuss the motivation for studying them.

As we see it, operational theories have the following attractions:

- Firstly, they focus on the empirical content of theories, and the means by which we can gain knowledge of the microphysical world. Any viable theory must account for this content.
- By focussing on this empirical and observational content, operational theories allow meaningful results to be formulated and proved about the 'space of theories' as a whole. At a stage in the development of physics where the next step is far from clear, this is a useful perspective, which may prove useful in finding 'deeper' theories.
- Indeed, the operational framework has proved fruitful as a basis for general results, e.g. on the information processing capabilities of theories under various assumptions [11, 12, 10]; and provides the setting for recent work on axiomatic reconstructions of quantum mechanics [36, 19, 49, 25].

On the debit side, operational theories attract criticism on philosophical grounds. They are seen as linked to an 'instrumentalist' or 'epistemic' view of physics, as opposed to a 'realistic' approach. From our perspective, the fact that we study operational theories does not indicate any such philosophical commitment. Rather, they are pragmatically useful for the reasons already mentioned, and can be seen as expressing some irreducible minimum of empirical content, which will have to be accounted for by any presumptive 'deeper' theory.

§3. **Operational theories formalized.** An operational theory is formulated in terms of directly accessible 'operations', which can be performed e.g. in a laboratory. We assume there are several different types of system, A, B, C, etc. For each system type A, the theory specifies the following:

- A set of *preparations* P_A which produce systems of that type.
- A set T_A of *transformations* which may be performed on systems of type A. More generally, we can consider transformations $T_{A,B}$ which can be performed on systems of type A to produce systems of type B.
- A set of *measurements* M_A which can be performed on systems of that type.

Each measurement has a set of possible *outcomes*. In this paper, we shall only consider 'finite-dimensional' theories, or parts of theories. This means that each measurement has only finitely many possible outcomes. For convenience, we shall assume a fixed *infinite* set of outcomes O, which will apply to all measurements. Any measurement with a finite set of outcomes $O' \subseteq O$ can be represented using O, where those outcomes outside O' have zero probability of occurring.

The empirical predictions of the theory are given by its *evaluation rule*, which is a function

$$v_A \colon P_A \times M_A \times O \to [0, 1]$$

which assigns a probability $v_A(p, m, o)$ to the event that a system of type A, prepared by p, yields outcome o when measurement m is performed on it.

For each choice of p and m, the function $v_A(p, m, -)$ defines a probability distribution on outcomes. We shall use the function

$$d_A \colon P_A \times M_A \to \mathcal{D} \qquad d_A(p, m) \colon o \mapsto v_A(p, m, o)$$

where $\mathcal{D}$ is the set of probability distributions of finite support on O.

3.1. Compound systems. An important additional ingredient is to give an account of *compound systems*, i.e. putting systems, possibly space-like separated, together.

This leads to the following additional requirements.

- For each pair of system types A, B, a compound system type AB.
- Ways of combining preparations, measurements, etc. on A and B to yield corresponding operations on the compound system AB.

Moreover, these operations should be subject to axioms yielding a coherent mathematical structure on these notions.

Rather than trying to develop such 'meta-operations' and axioms from first principles, we see the essential elements as provided by *monoidal categories*, which have been developed extensively as a setting for quantum mechanics and quantum information in the categorical quantum mechanics programme [6, 7].

We shall therefore proceed by giving a precise formulation of operational theories with compound system structure as a certain class of representations of monoidal categories, which we call *operational representations*.

3.2. Operational representations: concrete description. Before giving the 'official' definition of operational representation, which is mathematically elegant but a little abstract, we shall give a more concrete account, which shows the naturalness of the ideas, and also indicates why guidance from category theory is helpful in finding the right structural axioms.

For each system type A, we can gather the relevant data provided by an operational theory into a single structure

$$(P_A, M_A, d_A : P_A \times M_A \to \mathcal{D}).$$

This immediately suggests the notion of *Chu space* [14, 20], which has received quite extensive development [54], and was applied to the modelling of physical systems in [2]. Indeed, it can be seen as a generalization of the notion of model of a physical system proposed by Mackey in his influential work on the foundations of quantum mechanics [48].

There is a natural equivalence relation on preparations: p is equivalent to p', where $p, p' \in P_A$, if for all $m \in M_A$:

$$d_A(p, m) = d_A(p', m).$$

This is exactly the notion of extensional equivalence in Chu spaces [2]. We can regard *states* operationally as equivalence classes of preparations [51].

In an entirely symmetric fashion, there is an equivalence relation on measurements. We define m to be equivalent to m', where $m, m' \in M_A$, if for all $p \in P_A$:

$$d_A(p, m) = d_A(p, m').$$

We can regard *observables* operationally as equivalence classes of measurements.

Quotienting an operational system (P_A, M_A, d_A) by these equivalences corresponds to the *biextensional collapse* of a Chu space [2].

Having identified operational systems with Chu spaces, we now turn to morphisms. A transformation in $T_{A,B}$ induces a map $f_* : P_A \to P_B$. That is, preparing a system of type A according to preparation procedure p, and then subjecting it to a transformation procedure t resulting in a system of type B, is itself a procedure for preparing a system of type B.

Such a transformation can also be seen as a procedure for converting measurements of type B into measurements of type A: given a measurement $m \in M_B$, to apply it to a state prepared by $p \in P_A$, we apply the transformation

t to obtain a preparation of type B, to which m can be applied. Thus we can also associate a map $f^*\colon M_B \to M_A$ to the transformation t. The formal relationship that links the two maps f_* and f^* is that, whether we measure $f_*(p)$ with m, or p with $f^*(m)$, we should observe the same probability distribution on outcomes:

$$d_B(f_*(p), m) = d_A(p, f^*(m)). \tag{1}$$

This can be seen as an abstract form of the relationship between the Schrödinger and Heisenberg 'pictures' of quantum dynamics.

The equation (1) says exactly that the pair of maps (f_*, f^*) defines a morphism of Chu spaces

$$(f_*, f^*)\colon (P_A, M_A, d_A) \to (P_B, M_B, d_B).$$

Thus we see that in an entirely natural way, we can associate an operational theory with a sub-category of Chu spaces, more precisely of ***Chu***$(\textbf{\textit{Set}}, \mathcal{D})$ [54]. This sub-category will not in general be full, since not every Chu morphism will arise from a transformation in the theory.

However, this does not yet provide an account of compound systems. While Chu spaces have a standard monoidal structure, and indeed form $*$-autonomous categories [20], we should not in general expect that operational theories will give rise to *monoidal* sub-categories of Chu spaces. Rather, we should see the notion of compound system as an important degree of freedom, which is to be specified by the theory.

Thus given operational systems $A = (P_A, M_A, d_A)$ and $B = (P_B, M_B, d_B)$, we should be able to form a system $A \otimes B = (P_{A\otimes B}, M_{A\otimes B}, d_{A\otimes B})$.

What general properties should such a notion satisfy? One important requirement, which appears in one form or another in the various formulations of operational theories, is to have an *inclusion of pure tensors*. This is given by maps

$$\iota^{\mathsf{P}}_{A,B}\colon P_A \times P_B \to P_{A\otimes B}, \qquad \iota^{\mathsf{M}}_{A,B}\colon M_A \times M_B \to M_{A\otimes B}.$$

For readability, we shall write $p \otimes p'$ rather than $\iota^{\mathsf{P}}_{A,B}(p, p')$, and similarly for measurements.

The fundamental property which this inclusion must satisfy relates to the evaluation. For all $p \in P_A$, $p' \in P_B$, $m \in M_A$, $m' \in M_B$, we must have:

$$d_{A\otimes B}(p \otimes p', m \otimes m') = d_A(p, m) \cdot d_B(p', m'). \tag{2}$$

This expresses the probabilistic independence of pure tensors. Conceptually, pure tensors arise by preparing states or performing measurements independently on subsystems.

In addition, there are a number of coherence conditions which are needed to get a mathematically robust notion. Rather than writing these down in an ad hoc fashion, we shall now turn to a more systematic way of defining the categorical structure of operational theories, in which these conditions arise naturally from standard notions.

3.3. Operational representations: functorial formulation. We shall now take a different view, in which the structure of an operational theory arises from a symmetric monoidal category, which we think of as a *process category*. The operational theory will amount to a certain form of *representation* of this process category. The receiving category for the representation will be $(\boldsymbol{Set}, \times, \boldsymbol{1})$, viewed as a symmetric monoidal category.

Given a symmetric monoidal category $\boldsymbol{C}$, an operational representation of $\boldsymbol{C}$ is specified by the following data:

- A symmetric monoidal sub-category $\boldsymbol{C}_{\mathsf{t}}$ of $\boldsymbol{C}$. This will usually have the same objects as $\boldsymbol{C}$, and only those morphisms which correspond to admissible transformations.
- A symmetric monoidal functor $\mathsf{P}\colon \boldsymbol{C}_{\mathsf{t}} \to \boldsymbol{Set}$ which represents, for each object A of $\boldsymbol{C}_{\mathsf{t}}$, viewed as a type of system, the corresponding set of preparations or states.
- A contravariant symmetric monoidal functor $\mathsf{M}\colon \boldsymbol{C}_{\mathsf{t}}^{\mathrm{op}} \to \boldsymbol{Set}$ which for each A represents the measurements on A. Note that $\boldsymbol{C}_{\mathsf{t}}^{\mathrm{op}}$ is a symmetric monoidal category.
- A dinatural symmetric monoidal transformation

$$\mathsf{d}\colon \mathsf{P} \times \mathsf{M} \xrightarrow{..} \mathrm{K}_{\mathcal{D}}$$

  which gives the evaluation rule of the theory. Here $\mathrm{K}_{\mathcal{D}}$ is the constant functor valued at $\mathcal{D}$. Note that a constant symmetric monoidal functor valued at a set M is just a commutative monoid $(M, \cdot, 1)$ in $\boldsymbol{Set}$. We take $\mathcal{D}$ to be a commutative monoid under pointwise multiplication.

We shall assume that the functors P, M are embeddings, i.e. injective on objects and faithful.

Let us now unpack this definition.

- The general point of view is that the structure of the operational theory is controlled by the 'abstract' category $\boldsymbol{C}$. The types of the theory are the objects of $\boldsymbol{C}_{\mathsf{t}}$.
- Rather than a single set of preparations, we have a *variable set* P, which for each type A gives us a set P_A. Moreover, this acts functorially on the admissible transformations $f\colon A \to B$ in $\boldsymbol{C}_{\mathsf{t}}$ to produce functions $f_*\colon \mathsf{P}_A \to \mathsf{P}_B$, where $f_* := \mathsf{P}(f)$. Thus these functions take preparations on A to preparations on B, as already discussed.
- Similarly, the functor M specifies a variable set M_A of measurements for each system type A. The contravariant action of this functor is again as expected from our previous discussion.

The first new ingredient which picks up the issue of monoidal structure is that P and M are required to be *monoidal* functors. The fact that P and M are monoidal means that there are natural transformations

$$\iota^{\mathsf{P}}_{A,B} \colon \mathsf{P}_A \times \mathsf{P}_B \to \mathsf{P}_{A\otimes B}, \qquad \iota^{\mathsf{M}}_{A,B} \colon \mathsf{M}_A \times \mathsf{M}_B \to \mathsf{M}_{A\otimes B}.$$

i.e. inclusions of pure tensors. Naturality means that the diagrams

$$\begin{array}{ccc} \mathsf{P}_A \times \mathsf{P}_B & \xrightarrow{\iota^{\mathsf{P}}_{A,B}} & \mathsf{P}_{A\otimes B} \\ {\scriptstyle f_*\times g_*}\downarrow & & \downarrow{\scriptstyle (f\otimes g)_*} \\ \mathsf{P}_{A'} \times \mathsf{P}_{B'} & \xrightarrow[\iota^{\mathsf{P}}_{A',B'}]{} & \mathsf{P}_{A'\otimes B'} \end{array} \qquad \begin{array}{ccc} \mathsf{M}_A \times \mathsf{M}_B & \xrightarrow{\iota^{\mathsf{M}}_{A,B}} & \mathsf{M}_{A\otimes B} \\ {\scriptstyle f^*\times g^*}\uparrow & & \uparrow{\scriptstyle (f\otimes g)^*} \\ \mathsf{M}_{A'} \times \mathsf{M}_{B'} & \xrightarrow[\iota^{\mathsf{M}}_{A',B'}]{} & \mathsf{M}_{A'\otimes B'} \end{array}$$

commute. The coherence conditions for monoidal natural transformations complete the required properties of pure tensors.

The dinatural transformation $\mathsf{d}_A \colon \mathsf{P}_A \times \mathsf{M}_A \to \mathcal{D}$ represents the evaluation function. Dinaturality says that for each admissible transformation $f \colon A \to B$:

$$\begin{array}{ccccc} & & \mathsf{P}_B \times \mathsf{M}_B & & \\ & \nearrow^{f_*\times 1_B} & & \searrow^{\mathsf{d}_B} & \\ \mathsf{P}_A \times \mathsf{M}_B & & & & \mathcal{D} \\ & \searrow_{1_A\times f^*} & & \nearrow_{\mathsf{d}_A} & \\ & & \mathsf{P}_A \times \mathsf{M}_A & & \end{array}$$

Thus we see that dinaturality is exactly the Chu morphism condition (1). Monoidality of d is the equation (2).

3.4. Operational categories. If we are given an operational representation $(\boldsymbol{C}, \boldsymbol{C}_{\mathrm{t}}, \mathsf{P}, \mathsf{M}, \mathsf{d})$ we can construct from this a single category, recovering the picture given in Section 3.2.

For each object A of $\boldsymbol{C}$, we have the Chu space $(\mathsf{P}_A, \mathsf{M}_A, \mathsf{d}_A)$. By dinaturality of d, each morphism $f \colon A \to B$ gives rise to a Chu morphism

$$(f_*, f^*) \colon (\mathsf{P}_A, \mathsf{M}_A, \mathsf{d}_A) \to (\mathsf{P}_B, \mathsf{M}_B, \mathsf{d}_B).$$

By functoriality of P and M, we obtain a sub-category of Chu spaces.

Moreover, since P and M are embeddings, we can push the symmetric monoidal structure on $\boldsymbol{C}$ forward to this sub-category:

$$\mathsf{P}_A \otimes \mathsf{P}_B := \mathsf{P}_{A\otimes B}, \qquad f_* \otimes f'_* := (f \otimes f')_*,$$
$$\mathsf{M}_A \otimes \mathsf{M}_B := \mathsf{M}_{A\otimes B}, \qquad f^* \otimes f'^* := (f \otimes f')^*.$$

Thus we obtain a symmetric monoidal category, whose underlying category is a sub-category of Chu spaces. We call this the *operational category* arising from the operational representation.

3.5. Generalized representations. The structural properties of operational representations and categories are independent of the particular choice of the monoid $\mathcal{D}$ used in specifying the dinatural transformation d.

We shall define a *generalized operational representation with weights* $\mathcal{W}$, where $(\mathcal{W}, \cdot, 1)$ is a commutative monoid with a zero element, to be a tuple $(\boldsymbol{C}, \boldsymbol{C}_{\mathrm{t}}, \mathsf{P}, \mathsf{M}, \mathsf{d})$, where d now has the form

$$\mathsf{d}\colon \mathsf{P} \times \mathsf{M} \stackrel{..}{\to} \boldsymbol{K}_{\mathcal{W}}$$

and $\boldsymbol{K}_{\mathcal{W}}$ is the constant symmetric monoidal functor valued at $\mathcal{W}$. This yields the definition of operational representation given previously when $\mathcal{W} = \mathcal{D}$.

We now have a general scheme for representing symmetric monoidal categories as operational categories. So far, however, we have no examples. We shall now show how monoidal dagger categories give rise to operational representations in a canonical fashion, following the ideas of categorical quantum mechanics [7].

§4. Monoidal dagger categories. Monoidal dagger categories are the basic structures used in categorical quantum mechanics [7]. We shall briefly review the definitions, and give a number of examples.

A *dagger category* is a category $\boldsymbol{C}$ equipped with an identity-on-objects, contravariant, strictly involutive functor. Concretely, for each arrow $f : A \to B$, there is an arrow $f^{\dagger} : B \to A$, and this assignment satisfies:

$$1^{\dagger} = 1, \qquad (g \circ f)^{\dagger} = f^{\dagger} \circ g^{\dagger}, \qquad f^{\dagger\dagger} = f\,.$$

We define an arrow $f : A \to B$ in a dagger category to be a *dagger-isomorphism* if:

$$f^{\dagger} \circ f = 1_A, \qquad f \circ f^{\dagger} = 1_B.$$

A *symmetric monoidal dagger category* is a dagger category with a symmetric monoidal structure $(\boldsymbol{C}, \otimes, I, \lambda, \rho, \alpha, \sigma)$ such that

$$(f \otimes g)^{\dagger} = f^{\dagger} \otimes g^{\dagger}$$

and moreover the natural isomorphisms λ, ρ, α, σ are componentwise dagger-isos.

Examples.

- The category ***Hilb*** of Hilbert spaces and bounded linear maps, and its (full) sub-category ***FHilb*** of finite-dimensional Hilbert spaces. Here the dagger is the adjoint, and the tensor product has its standard interpretation for Hilbert spaces. More generally, any symmetric monoidal C*-category is an example [33, 28]. This includes categories of (right) Hilbert C*-modules, which are Hilbert spaces whose inner product takes values in an arbitrary C*-algebra instead of $\mathbb{C}$.

- The category ***Rel*** of sets and relations. Here the dagger is relational converse, while the monoidal structure is given by the cartesian product. This generalizes to relations valued in a commutative quantale [55], and to the category of relations for any regular category [18]. Small categories as objects and profunctors as morphisms behave very similarly to ***Rel***, even though they only form a bicategory [16].
- A common generalization of ***FHilb*** and ***FRel***, the category of finite sets and relations, is obtained by forming the category $\boldsymbol{FMat}(S)$, where S is a commutative semiring with involution. $\boldsymbol{FMat}(S)$ has finite sets as objects, and maps $X \times Y \to S$ as morphisms, which we think of as 'X times Y matrices'. Composition is by matrix multiplication, while the dagger is conjugate transpose, where conjugation of a matrix means elementwise application of the involution on S. The tensor product of X and Y is given by $X \times Y$, with the action on matrices given by componentwise multiplication. (This corresponds to the 'Kronecker product' of matrices). If we take $S = \mathbb{C}$, this yields a category equivalent to ***FHilb***, while if we take S to be the Boolean semiring $\{0, 1\}$ (with trivial involution), we get ***FRel***.
- An infinitary generalization of $\boldsymbol{FMat}(\mathbb{C})$ is given by ***LMat***. This category has arbitrary sets as objects, and as morphisms matrices $M : X \times Y \to \mathbb{C}$ such that for each $x \in X$, the family $\{M(x, y)\}_{y \in Y}$ is ℓ_2-summable; and for each $y \in Y$, the family $\{M(x, y)\}_{x \in X}$ is ℓ_2-summable. ***Hilb*** is equivalent to a (non-full) sub-category of ***LMat***.
- If $\boldsymbol{C}$ and $\boldsymbol{D}$ are symmetric monoidal dagger categories, then so is the category $[\boldsymbol{C}, \boldsymbol{D}]$ of functors $F : \boldsymbol{C} \to \boldsymbol{D}$ that preserve the dagger. Morphisms are natural transformations. This accounts for several interesting models. For example, setting $\boldsymbol{D} = \boldsymbol{FHilb}$ and letting $\boldsymbol{C}$ be a group, we obtain the category of unitary representations. Any topological or conformal quantum field theory is a sub-category of the case where $\boldsymbol{D} = \boldsymbol{FHilb}$ and $\boldsymbol{C}$ is the category of cobordisms [45, 8, 56]. Letting $\boldsymbol{C}$ be the discrete category $\mathbb{N}$, and letting $\boldsymbol{D}$ be either ***FHilb*** or ***FRel***, we recover $\boldsymbol{FMat}(\boldsymbol{D}(I, I))$.

The doubling construction. All of the above examples are variations on the theme of matrix categories. Indeed, it seems hard to find natural examples which are not of this form. However, there is a construction which produces a symmetric monoidal dagger category from *any* symmetric monoidal category. Although the construction is formal, it is interesting in our context since it can be seen as a form of *quantization*; it converts classical process categories into a form in which quantum constructions are meaningful.

Given a category $\boldsymbol{C}$, we define a dagger category $\boldsymbol{C}_{\leftrightarrows}$ as follows. The objects are the same as those of $\boldsymbol{C}$, and a morphism $(f, g) : A \to B$ is a pair of $\boldsymbol{C}$-morphisms $f : A \to B$, $g : B \to A$. Composition is defined componentwise; while $(f, g)^\dagger = (g, f)$. This is in fact the object part of the right adjoint to the evident forgetful functor ***DagCat*** $\to$ ***Cat***; see [38, 3.1.17]. Thus for each dagger category $\boldsymbol{C}$, there is a dagger functor $\eta_{\boldsymbol{C}}\colon \boldsymbol{C} \to \boldsymbol{C}_{\leftrightarrows}$ which is the identity on objects, and sends f to $(f, f^{\dagger})$. This has the universal property with respect to dagger functors $\boldsymbol{C} \to \boldsymbol{D}_{\leftrightarrows}$ for categories $\boldsymbol{D}$.

This cofree construction of a dagger category lifts to the level of symmetric monoidal categories. If $\boldsymbol{C}$ is a symmetric monoidal category, then $\boldsymbol{C}_{\leftrightarrows}$ is a symmetric monoidal dagger category, with the monoidal structure defined componentwise: thus $(f, g) \otimes (h, k) := (f \otimes h, g \otimes k)$. Note in particular that the structural isos in $\boldsymbol{C}$ turn into *dagger* isos in $\boldsymbol{C}_{\leftrightarrows}$.

4.1. Additional structure. We shall require two further structural ingredients. The first is *zero morphisms*: for each pair of objects A, B, a morphism $0_{A,B}\colon A \to B$ such that, for all $f\colon C \to A$ and $g\colon B \to D$,

$$0_{A,B} \circ f = 0_{C,B}, \qquad g \circ 0_{A,B} = 0_{A,D}.$$

Note that if zero morphisms exist, they are unique.

In the context of symmetric monoidal dagger categories, we further require that

$$f \otimes 0 = 0 = 0 \otimes g, \qquad 0^{\dagger} = 0.$$

Examples. All the examples of symmetric monoidal dagger categories given above have zero morphisms in an evident fashion. Functor categories have componentwise zero morphisms. Zero morphisms in $\boldsymbol{C}_{\leftrightarrows}$ are pairs of zero morphisms in $\boldsymbol{C}$. For more examples, see [39].

The final ingredient we shall require is a *trace ideal* in the sense of [4].[1] Firstly, we recall that in any monoidal category, the *scalars*, i.e. the endomorphisms of the tensor unit I, form a commutative monoid [44].

An *endomorphism ideal* in a symmetric monoidal category $\boldsymbol{C}$ is specified by a set $\mathcal{I}(A) \subseteq \mathrm{End}(A)$ for each object A, where $\mathrm{End}(A) = \boldsymbol{C}(A, A)$ is the set of endomorphisms on A. This is subject to the following closure conditions:

$$g\colon A \to B, f \in \mathcal{I}(A), h\colon B \to A \;\Rightarrow\; g \circ f \circ h \in \mathcal{I}(B)$$

$$f \in \mathcal{I}(A), g \in \mathcal{I}(B) \;\Rightarrow\; f \otimes g \in \mathcal{I}(A \otimes B), \qquad \mathcal{I}(I) = \mathrm{End}(I)$$

$$0 \in \mathcal{I}(A).$$

If $\boldsymbol{C}$ is a dagger category, $\mathcal{I}$ is a *dagger endomorphism ideal* when additionally

$$f \in \mathcal{I}(A) \;\Rightarrow\; f^{\dagger} \in \mathcal{I}(A),$$

but we will also call these endomorphism ideals for short. A *trace ideal* is an endomorphism ideal $\mathcal{I}$, together with a function

$$\mathrm{Tr}_A\colon \mathcal{I}(A) \to \mathrm{End}(I)$$

[1] Strictly speaking, we are defining the more restricted notion of *global trace* of an endomorphism, rather than a *parameterized trace* as in [4]. This restricted notion is all we shall need.

for each object A, subject to the following axioms:

$$\mathrm{Tr}_A(g \circ f) = \mathrm{Tr}_B(f \circ g) \quad (f : A \to B, g : B \to A, \\ g \circ f \in \mathcal{I}(A), f \circ g \in \mathcal{I}(B))$$

$$\mathrm{Tr}_{A \otimes B}(f \otimes g) = \mathrm{Tr}_A(f)\mathrm{Tr}_B(g), \qquad \mathrm{Tr}_I(s) = s.$$

A *dagger trace ideal* additionally satisfies

$$\mathrm{Tr}_A(f^\dagger) = \mathrm{Tr}_A(f)^\dagger,$$

but we will also call these trace ideal for short. We call a morphism $f \in \mathcal{I}(A)$ *trace class*.

Examples. All of the examples given above have trace ideals. In the case of finite matrices, the usual matrix trace is a total operation. In the case of ***Hilb***, we interpret trace class in the standard sense for Hilbert spaces, and similarly for ***LMat***. Through the GNS-embedding [33, Proposition 1.14], this also provides a trace ideal for any C*-category.

In the case of relations, the summation over the diagonal becomes a supremum in a complete semilattice, which is always defined.

Any symmetric monoidal dagger sub-category of $[\boldsymbol{C}, \boldsymbol{D}]$ inherits endomorphism ideals and zero morphisms from $\boldsymbol{D}$ componentwise, and has a trace function $\mathrm{Tr}(\alpha) = \bigvee_A \mathrm{Tr}(\alpha_A)$ as soon as $\boldsymbol{D}(I, I)$ has an operation $\bigvee$ satisfying $\bigvee_A s_A^\dagger = (\bigvee_A s_A)^\dagger$, $\bigvee_A s = s$, and $(\bigvee_A s_A t_A) = (\bigvee_A s_A)(\bigvee_A t_A)$, where A ranges over the objects of $\boldsymbol{C}$. This is the case when $\boldsymbol{C}$ is a finite group, as well as for topological quantum field theories.

The doubling construction turns trace ideals into dagger trace ideals. For $(f, g) : A \to A$, define $(f, g) \in \mathcal{I}(A)$ if and only if $f \in \mathcal{I}(A)$ and $g \in \mathcal{I}(A)$, and $\mathrm{Tr}_A(f, g) = (\mathrm{Tr}_A(f), \mathrm{Tr}_A(g))$. Thus if $\boldsymbol{C}$ is a symmetric monoidal category with zero morphisms and a trace ideal, $\boldsymbol{C}_{\leftrightarrows}$ is a dagger category with the same structure.

In Appendix B, we prove a number of results about trace ideals:

- We characterize when trace ideals exist, and to what extent they are unique.
- We show that we really need to restrict to ideals to consider traces: the category of Hilbert spaces does not support a trace on all morphisms.
- As a corollary, we derive that dual objects in the category of Hilbert spaces are necessarily finite-dimensional.
- Finally, we prove in some detail that the category of Hilbert spaces indeed has a trace ideal; the details turn out to be quite subtle.

This material would have unduly interrupted the main flow of the paper, but is of mathematical interest in its own right.

§5. From categorical quantum mechanics to operational categories. Let $\boldsymbol{C}$ be a symmetric monoidal dagger category with zero morphisms and a trace ideal. We shall show that $\boldsymbol{C}$ gives rise to an operational representation and operational category in a canonical fashion, directly inspired by quantum mechanics.

5.1. Transformations. We take $\boldsymbol{C}_{\mathrm{t}}$ to be the sub-category with the same objects as $\boldsymbol{C}$, and with dagger-isomorphisms as arrows. This is a groupoid, i.e. all morphisms are invertible.

It is easily seen to be a monoidal dagger sub-category of $\boldsymbol{C}$.

5.2. States. A morphism $f \in \mathrm{End}(A)$ in a dagger category is *positive* if for some $g : A \to B$, $f = g^\dagger \circ g$. We define a *state* on A to be a positive morphism $f \in \mathrm{End}(A)$ which is trace class, and such that $\mathrm{Tr}_A(f) = 1$. We write P_A for the set of states on A.

In ***Hilb***, this definition yields exactly the standard notion of *density operator* as used in quantum mechanics.

Pure states can also be defined in this setting. An arrow $\psi : I \to A$ has *unit norm* if $\psi^\dagger \circ \psi = 1$. Given such an arrow, $\psi \circ \psi^\dagger \in \mathsf{P}_A$. Indeed, this arrow is clearly positive, and

$$\mathrm{Tr}_A(\psi \circ \psi^\dagger) = \mathrm{Tr}_I(\psi^\dagger \circ \psi) = \mathrm{Tr}_I(1) = 1$$

using our assumption on ψ and the axioms for the trace.

Given a dagger isomorphism $f : A \to B$ in $\boldsymbol{C}$, the function $f_* : \mathsf{P}_A \to \mathsf{P}_B$ is defined by

$$f_* : s \mapsto f \circ s \circ f^\dagger.$$

Functoriality holds, since

$$g_* \circ f_*(s) = g \circ (f \circ s \circ f^\dagger) \circ g^\dagger = (g \circ f) \circ s \circ (g \circ f)^\dagger = (g \circ f)_*(s).$$

Inclusion of pure tensors is given by

$$\iota^{\mathsf{P}}_{A,B} : (s, t) \mapsto s \otimes t.$$

It is straightforward to check the coherence conditions.

5.3. Measurements. A *dagger idempotent*, or *projector*, on A is an arrow $P \in \mathrm{End}(A)$ such that

$$P^2 = P, \qquad P = P^\dagger.$$

A family $\{f_i\}_{i \in I}$ of endomorphisms on A is:

- *Pairwise disjoint* if $f_i \circ f_j = 0$, $i \neq j$;
- *Jointly monic* if for all $g, h : B \to A$:

$$[\forall i \in I.\, f_i \circ g = f_i \circ h] \Rightarrow g = h.$$

A *projective measurement* on A with finite set of outcomes $O' \subseteq O$ is a family of dagger idempotents $\{P_o\}_{o \in O'}$ on A which is pairwise disjoint and jointly monic. We take M_A to be the set of projective measurements on A.

The functorial action of the measurement functor on dagger isomorphisms $f : A \to B$ in $\mathbf{C}$ is defined by

$$f^*(P_o) = f^\dagger \circ P_o \circ f.$$

It is easily verified that f^* preserves disjointness and joint monicity of families of projectors, and hence carries projective measurements to projective measurements. Functoriality is also easily verified.

Inclusion of tensors is defined pointwise on projectors:

$$\iota^{\mathsf{P}}_{A,B} : (P_o, P_{o'}) \mapsto P_o \otimes P_{o'}.$$

Note that the combined measurement will have a finite set of outcomes which, perhaps with some relabelling, can be regarded as a subset of O.

5.4. Evaluation. The transformation d is defined as follows, where $s \in \mathsf{P}_A$, and $m = \{P_o\}_{o \in O'} \in \mathsf{M}_A$:

$$\mathsf{d}_A(s, m)(o) := \begin{cases} \mathrm{Tr}_A(s \circ P_o), & o \in O' \\ 0, & \text{otherwise.} \end{cases}$$

Note that d is valued in the commutative monoid of scalars $\mathcal{W} := \mathrm{End}(I)^O$. By the assumption of zero morphisms, this monoid has a zero element.

The dinaturality of this transformation, i.e. the Chu morphism condition, is just:

$$\mathrm{Tr}_B(f \circ s \circ f^\dagger \circ P_o) = \mathrm{Tr}_A(s \circ f^\dagger \circ P_o \circ f).$$

The monoidality of d is verified as follows:

$$\begin{aligned} \mathsf{d}_{A \otimes B}(s \otimes s', m \otimes m')(o, o') &= \mathrm{Tr}_{A \otimes B}(s \otimes s' \circ P_o \otimes P_{o'}) \\ &= \mathrm{Tr}_{A \otimes B}(s \circ P_o \otimes s' \circ P_{o'}) \\ &= \mathrm{Tr}_A(s \circ P_o)\mathrm{Tr}_B(s' \circ P_{o'}) \\ &= \mathsf{d}_A(s, m)(o) \cdot \mathsf{d}_B(s', m')(o'). \end{aligned}$$

5.5. The canonical operational representation. We collect the constructions described in this section together. Given a symmetric monoidal dagger category $\mathbf{C}$ with zero morphisms and a trace ideal, we have defined a sub-category $\mathbf{C}_{\mathrm{t}}$, monoidal functors P and M, and a dinatural transformation d.

PROPOSITION 3. *The tuple* $(\mathbf{C}, \mathbf{C}_{\mathrm{t}}, \mathsf{P}, \mathsf{M}, \mathsf{d})$ *is an operational representation with weights* $\mathcal{W}$. *We call this the canonical operational representation of* $\mathbf{C}$. *The corresponding operational category is the canonical operational category for* $\mathbf{C}$. ⊣

We say that the canonical representation is *distributional* if the monoid of scalars $\mathrm{End}(I)$ has an addition making it a commutative semiring, and for each state $s \in \mathsf{P}_A$ and measurement $m \in \mathsf{M}_A$:

$$\sum_{o \in O} \mathsf{d}_A(s, m)(o) = 1. \tag{4}$$

We say that it is *probabilistic* if moreover the image of d embeds into the semiring of non-negative reals.

§6. Examples of operational categories. We shall now examine the operational categories arising from various examples of symmetric monoidal dagger categories.

6.1. Hilbert spaces. The definitions of states, measurements and evaluation are directly inspired by those used in the standard Hilbert-space formulation of quantum mechanics. Thus it is immediate that the states in the canonical representation for ***Hilb*** are the density matrices, while the dagger-isomorphisms are the unitary transformations.

For measurements, we have the following result.

PROPOSITION 5. *Measurements in* ***Hilb*** *have exactly their standard meaning. More precisely, observables with finite discrete spectra correspond exactly to the interpretation in* ***Hilb*** *of the abstract notion of measurements as defined in Section* 5.3 *for dagger categories.*

PROOF. We think of the outcomes as labelling the eigenvalues of the observable; then the family $\{P_o\}_{o \in O'}$ should correspond to the *spectral decomposition* of the observable. Clearly, dagger idempotents correspond exactly to projectors in ***Hilb***, and so does the notion of a pairwise disjoint family of projectors. It remains to show that the joint monicity condition captures the fact that a pairwise disjoint family of projectors $\{P_i\}_{i \in I}$ yields a *resolution of the identity*, i.e.

$$\sum_{i \in I} P_i = 1_A.$$

Indeed, if $\sum_{i \in I} P_i = 1_A$ and $P_i \circ g = P_i \circ h$ for all i, then

$$\begin{aligned} g = 1_A \circ g = \left(\sum_{i \in I} P_i\right) \circ g = \sum_{i \in I} P_i \circ g \\ = \sum_{i \in I} P_i \circ h = \left(\sum_{i \in I} P_i\right) \circ h = 1_A \circ h = h. \end{aligned}$$

For the converse, suppose that $\sum_{i \in I} P_i \neq 1_A$. This implies that for some non-zero vector ψ, $P_i(\psi) = 0$ for all i. Then for $f : \mathbb{C} \to A$ given by $1 \mapsto \psi$, we have $P_i \circ f = P_i \circ 0$ for all i, so the family is not jointly monic. ⊣

Finally, the definition of d matches the standard statistical algorithm of quantum mechanics. Thus we obtain the standard interpretations of states, transformations, (projective) measurements, and probabilities of measurement outcomes.

The operational category arising from ***Hilb*** is of course probabilistic.

The same analysis holds for C*-categories through their GNS-construction, and for subcategories of $[\mathbf{C}, \mathbf{Hilb}]$ such as topological quantum field theories. states and measurements in such categories are just natural transformations whose components are states or measurements respectively. Because the tensor unit in such categories is the constant functor K_I, they have the same scalars as ***Hilb***. Therefore the induced operational categories are probabilistic.

6.2. Relations. We shall now give a general analysis of the operational representation for locale-valued relations. This level of generality will be useful when we go on look at non-locality in operational categories.

We recall that a *locale* [42] (also known as a *frame* or *complete Heyting algebra*) is a complete lattice Ω such that the following distributive law holds:

$$a \wedge \bigvee_{i \in I} b_i = \bigvee_{i \in I} a \wedge b_i.$$

The category $\mathbf{Rel}(\Omega)$ has sets as objects, while the morphisms $R : X \to Y$ are Ω-valued relations (or matrices) $R : X \times Y \to \Omega$. We write $[\![xRy]\!] = \omega$ for $R(x, y) = \omega$. Composition is relational composition (or matrix multiplication) evaluated in Ω. If $R \colon X \to Y$ and $S \colon Y \to Z$, then:

$$[\![x(S \circ R)z]\!] := \bigvee_{y \in Y} [\![xRy]\!] \wedge [\![ySz]\!].$$

Clearly, ***Rel*** is the special case that Ω is the Boolean semiring $\{\bot, \top\}$, where we identify $\bot$, the bottom element of the lattice, with 0, and $\top$, the top element, with 1. Note that the full sub-category $\mathbf{FRel}(\Omega)$ of finite sets is identical to $\mathbf{FMat}(\Omega)$, where we regard Ω as a semiring with idempotent addition and multiplication. Indeed, in the finite case, completeness of Ω need not be assumed, and we are simply in the case of matrices over idempotent semirings.

We shall take the tensor unit in $\mathbf{Rel}(\Omega)$ to be $I = \{\bullet\}$.

By an Ω-subset of a set X, we mean a function $X \to \Omega$. Any family $\{S_i\}$ of Ω-subsets of X has a 'union' $\bigvee_i S_i$ given by $x \mapsto \bigvee_i S_i(x)$, and an 'intersection' $\bigwedge_i S_i$ given by $x \mapsto \bigwedge_i S_i(x)$. In particular, we write $\top_X$ for the Ω-subset of X given by $x \mapsto \top$, and $\bot_X$ for the Ω-subset of X given by $x \mapsto \bot$. Given a set X, we say that a family $\{S_i\}_{i \in I}$ of Ω-subsets of X is a *disjoint cover* of X if:

$$S_i \wedge S_j = \bot_X \ \ (i \neq j), \qquad \bigvee_{i \in I} S_i = \top_X.$$

Given a Ω-subset S of X, we define a Ω-relation $\Delta_S \colon X \to X$ by

$$[\![x\Delta_S y]\!] = \begin{cases} S(x) & \text{if } x = y, \\ \bot & \text{if } x \neq y. \end{cases}$$

Note that

$$\Delta_S \circ \Delta_T = \bot_{X \times X} \iff S \wedge T = \bot_X, \quad \bigvee_{i \in I} \Delta_{S_i} = 1_A \iff \bigvee_{i \in I} S_i = \top_X. \tag{6}$$

PROPOSITION 7. *Projective measurements on X in $\textbf{Rel}(\Omega)$ consist of families of relations $\{\Delta_{S_i}\}_{i\in I}$, where $\{S_i\}_{i\in I}$ is a disjoint cover of X.*

PROOF. Clearly any family of relations of this form is a projective measurement. For the converse, suppose we have a projective measurement $\{P_i\}_{i\in I}$ on X. The fact that P_i is a projector in $\textbf{Rel}(\Omega)$ means that $[\![xP_iy]\!] = [\![yP_ix]\!]$ and $[\![xPz]\!] = \bigvee_y [\![xPy]\!] \wedge [\![yPz]\!]$, which implies that $[\![xP_ix]\!] \geq [\![xP_iy]\!]$. Suppose for a contradiction that $[\![xP_iy]\!] = \omega > \bot$ where $x \neq y$. Define $R, S\colon I \to X$ by $[\![\bullet Rx]\!] = \omega = [\![\bullet Sy]\!]$, and $[\![\bullet Rz]\!] = \bot = [\![\bullet Sz]\!]$ for other z. Then $[\![\bullet(P_i \circ R)x]\!] = \bigvee_z [\![\bullet Rz]\!] \wedge [\![zP_ix]\!] = \omega \wedge [\![xP_ix]\!] = \omega$ since $[\![xP_ix]\!] \geq [\![yP_ix]\!] = \omega$, and also $[\![\bullet(P_i \circ S)x]\!] = \omega$. Similarly $[\![\bullet(P_i \circ R)y]\!] = \omega = [\![\bullet(P_i \circ S)y]\!]$, and $[\![\bullet(P_i \circ R)z]\!] = \bot = [\![\bullet(P_i \circ S)z]\!]$ for other z. Hence $P_i \circ R = P_i \circ S$. Moreover, $P_j \circ R = P_j \circ S = \bot$ for any $j \neq i$, by disjointness of the family, since e.g. $\bot < [\![xP_jz]\!] \leq [\![xP_jx]\!]$ implies $P_i \circ P_j \neq \bot$. Thus $P_k \circ R = P_k \circ S$ for all $k \in I$, contradicting joint monicity. Hence P_i must have the form $P_i = \Delta_{S_i}$ for some $S_i \subseteq X$. The fact that the family $\{S_i\}_{i\in I}$ is a disjoint cover of X now follows from (6). ⊣

Next we analyze states in $\textbf{Rel}(\Omega)$. Firstly, we give an explicit description of the trace. If $R\colon X \to X$ is an Ω-valued relation,

$$[\![\bullet \mathrm{Tr}_X(R)\bullet]\!] = \bigvee_x [\![xRx]\!].$$

Thus the trace can be viewed as a predicate on endo-relations, which is satisfied to the extent that the relation has a 'fixpoint', i.e. a reflexive element.

Note that Ω-valued relations $R\colon I \to X$ of unit norm correspond to Ω-subsets S of X satisfying

$$\bigvee_x S(x) = \top.$$

The corresponding pure state is P_S, defined by $[\![xP_Sy]\!] = S(x) \wedge S(y)$.

We say that states s, t on X are *equivalent* if for all Ω-subsets S of X:

$$\mathrm{Tr}_X(s \circ \Delta_S) = \mathrm{Tr}_X(t \circ \Delta_S).$$

PROPOSITION 8. *Every state in $\textbf{Rel}(\Omega)$ is equivalent to a pure state.*

PROOF. If s is a state on X, then it satisfies $\top = \bigvee_x [\![xsx]\!]$, and for some relation R,

$$[\![xsy]\!] = \bigvee_z [\![xRz]\!] \wedge [\![yRz]\!].$$

Define an Ω-subset $S = \mathrm{dom}(s)$ of X by $x \mapsto [\![xsx]\!]$. We claim that s is equivalent to P_S. Indeed, for any Ω-subset T of X,

$$\mathrm{Tr}_X(s \circ \Delta_T) = \bigvee_x [x(s \circ \Delta_T)x]$$

$$
\begin{aligned}
&= \bigvee_{x,y} [\![x \Delta_T y]\!] \wedge [\![ysx]\!] \\
&= \bigvee_{x} T(x) \wedge [\![xsx]\!] \\
&= \bigvee_{x} T(x) \wedge S(x) \\
&= \bigvee_{x,y} [\![y P_S x]\!] \wedge [\![x \Delta_T y]\!] \\
&= \mathrm{Tr}_X(P_S \circ \Delta_T). \qquad \dashv
\end{aligned}
$$

Finally, we consider evaluation. The scalars in ***Rel***(Ω) can be identified with the locale Ω. Because states correspond to Ω-subsets S satisfying $\bigvee_x S(x) = \top$, and measurements to disjoint covers, we see that equation (4) is satisfied. Thus we have the following result.

PROPOSITION 9. *The operational category arising from **Rel**(Ω) is distributional.*

PROOF. Let Δ_S be a state, and m be a measurement given by a disjoint cover $\{S_o\}$ of X. Then

$$
\begin{aligned}
\sum_o \mathsf{d}_A(\Delta_S, m)(o) &= \bigvee_o \mathrm{Tr}_X(\Delta_S \circ \Delta_{S_o}) \\
&= \bigvee_{o,x,y} [\![x \Delta_S y]\!] \wedge [\![y \Delta_{S_o} x]\!] \\
&= \bigvee_{o,x} S(x) \wedge S_o(x) \\
&= \bigvee_{x} S(x) \wedge \left(\bigvee_o S_o(x) \right) \\
&= \bigvee_{x} S(x) \\
&= \top. \qquad \dashv
\end{aligned}
$$

Discussion. These results highlight two important differences between ***Rel***(Ω) and ***Hilb*** as operational categories. In ***Hilb***, every projector can appear as part of a projective measurement, while in ***Rel***(Ω) the collective conditions of disjointness and joint monicity impose the constraint that projectors have to be sub-identities Δ_S. Moreover, in ***Rel***(Ω) the distinction between *superpositions* of pure states, and *convex combinations* to form mixed states, is lost, so that every state is equivalent to a pure one. The relevance of this will become apparent when we discuss non-locality in ***Rel***(Ω) in Section 9.2.

§7. **Classical operational categories.** The construction of operational representations on monoidal dagger categories is directly inspired by quantum mechanics. However, operational theories should also include classical physics—or its discrete operational residue. Our notion of operational representation is indeed broad enough for this, as we shall now show.

The basic classical setting we shall consider is the category ***Stoch***. The objects are finite sets, and the morphisms $M : X \to Y$ are the $X \times Y$-matrices valued in $[0, 1]$ which are row-stochastic. Thus for each $x \in X$, we have a probability distribution on Y.

An alternative description of ***Stoch*** is as the Kleisli category for the monad of discrete probability distributions; see [40].

The monoidal structure is defined as for ***FMat***(S). Note that ***Stoch*** is not closed under matrix transposition. Indeed, we have the following result.

PROPOSITION 10. *There is no dagger structure on* ***Stoch***.

PROOF. Note that if a category $\boldsymbol{C}$ has a dagger structure, it is in particular self-dual, i.e. equivalent to $\boldsymbol{C}^{\mathrm{op}}$. However, the one-element set is terminal but not initial in ***Stoch***, which is thus not self-dual. ⊣

It follows that we cannot directly apply the construction of Section 5. One might consider using the formal doubling construction on ***Stoch*** to obtain a dagger symmetric monoidal category with a dagger trace ideal. But this would not yield the expected result; for example, the dagger would not be given by transpose of (bi-stochastic) matrices. However, it is easy to give a direct definition of an operational representation, as follows.

- The sub-category ***Stoch***$_{\mathrm{t}}$ is defined by restricting to the functions (deterministic transformations), represented as matrices by their characteristic maps. Thus if $f : X \to Y$ is a function, for each $x \in X$ the corresponding probability distribution is $\delta_{f(x)}$.
- A state on X is a morphism $I \to X$ in ***Stoch***, or equivalently a probability distribution on X. This is the classical notion of mixed state. The functorial action of states is described as follows. Given $f : X \to Y$, we define

$$f_*(s)(y) = \sum_{f(x)=y} s(x).$$

- A measurement on X is a function $m : X \to O$ with finite image $O' \subseteq O$. This is just a discrete random variable. The functorial action on $f : X \to Y$ is just

$$m \mapsto m \circ f.$$

- The evaluation is defined by:

$$\mathsf{d}_X(s, m)(o) = \sum_{m(x)=o} s(x).$$

The following result is easily verified.

PROPOSITION 11. *The above data specifies a probabilistic operational representation of* ***Stoch***. ⊣

Various generalizations of this construction are possible:

- We can generalize to 'distributions' over an arbitrary commutative semiring, as in [40]. This will still yield a distributional operational representation.
- We can generalize to probability measures over general measure spaces. This amounts to using the Kleisli category of the Giry monad [34].

§8. Non-locality in operational categories. Having set up a general framework for operational categories, we shall now investigate an important foundational notion in this general setting; namely *non-locality*.

Throughout this section, we fix a distributional operational representation $(\boldsymbol{C}, \boldsymbol{C}_{\mathrm{t}}, \mathsf{P}, \mathsf{M}, \mathsf{d})$ on a monoidal category $\boldsymbol{C}$.

8.1. Empirical models. We shall begin by showing how probability models of the form commonly studied in quantum information and quantum foundations can be interpreted in the corresponding operational category. In these models, there are n agents or sites, each of which has the choice of one of several measurement settings; and each measurement has a number of distinct outcomes. For each choice of a measurement setting by each of the agents, we have a probability distribution on the joint outcomes of the measurements.

We shall associate objects $A_1, \ldots, A_n$ with the n sites. We define $A := A_1 \otimes \cdots \otimes A_n$. We fix a state $s \in \mathsf{P}_A$. For each combination of measurements $(m_1, \ldots, m_n)$, where $m_i \in \mathsf{M}_{A_i}$ for $i = 1, \ldots, n$, we obtain the measurement $m := m_1 \otimes \cdots \otimes m_n$ by inclusion of pure tensors. Now the probability of obtaining a joint outcome $o := (o_1, \ldots, o_n)$ for m is given by

$$p(o|m) \;:=\; \mathsf{d}_A(s, m)(o).$$

We can regard these models as observational 'windows' on the operational theory. They represent the directly accessible information predicted by the theory, and provide the empirical yardstick by which it is judged.

8.2. Non-locality. We now define what it means for an empirical model of the kind described in the previous sub-section to exhibit non-locality. We shall follow the traditional route of using hidden variables explicitly, although we could equivalently, and perhaps more elegantly, formulate non-locality in terms of the (non-)existence of a joint distribution [31, 5].

We are assuming a fixed distributional model, with a semiring of weights $\mathcal{W}$. A *$\mathcal{W}$-distribution* on a set X is a function $d : X \to \mathcal{W}$ of finite support, such that

$$\sum_{x \in X} d(x) = 1.$$

A hidden-variable model for an empirical model is defined using a set Λ of hidden variables, with a fixed distribution d.[2] For each $\lambda \in \Lambda$, the model specifies a distribution $q^\lambda(o|m)$ on outcomes o for each choice of measurements m. The required condition for the hidden variable model to realize the empirical model p is that, for all m and o:

$$p(o|m) = \sum_{\lambda \in \Lambda} q^\lambda(o|m) \cdot d(\lambda).$$

That is, we recover the empirical probabilities by averaging over the hidden variables.

We say that the hidden-variable model is *local* if, for all $\lambda \in \Lambda$, $m = (m_1, \ldots, m_n)$, and $o := (o_1, \ldots, o_n)$:

$$q^\lambda(o|m) = \prod_{i=1}^{n} q^\lambda(o_i|m_i).$$

Here $q^\lambda(o_i|m_i)$ is the marginal:

$$q^\lambda(o_i|m_i) = \sum_{o'_i = o_i, m'_i = m_i} q^\lambda(o'|m').$$

We say that the empirical model p is *local* if it is realized by some local hidden-variable model; and *non-local* otherwise.

Note that the definition of non-locality makes sense for any distributional operational category. Thus we can lift these ideas to the general level of operational categories. We say that an operational category exhibits non-locality if it gives rise to a non-local empirical model. Ultimately, we have a criterion for ascribing non-locality to monoidal process categories themselves, relative to a given distributional operational representation.

§9. Examples of non-locality. We shall now investigate non-locality in a number of examples.

9.1. Hilbert spaces. As expected, the operational category arising from ***Hilb***, which is essentially the finite-dimensional part of standard quantum mechanics, does exhibit non-locality.

As a standard example—essentially the one used by Bell in his original proof of Bell's theorem—consider the following table.

	$(0,0)$	$(1,0)$	$(0,1)$	$(1,1)$
(a,b)	$1/2$	0	0	$1/2$
(a,b')	$3/8$	$1/8$	$1/8$	$3/8$
(a',b)	$3/8$	$1/8$	$1/8$	$3/8$
(a',b')	$1/8$	$3/8$	$3/8$	$1/8$

[2]The assumption of a fixed distribution d is technically the condition of 'λ-independence' [26].

It lists the probabilities that one of two outcomes (0 or 1) occurs when simultaneously measured with one of two measurements at two sites (a or a' at the first site, and b or b' at the second). This table can be realized in quantum mechanics, e.g. by a Bell state, written in the Z basis as

$$\frac{|\uparrow\uparrow\rangle + |\downarrow\downarrow\rangle}{\sqrt{2}},$$

subjected to spin measurements in the XY-plane of the Bloch sphere, at a relative angle of $\pi/3$.

A standard argument (see e.g. [15, 5]) shows that this table cannot be realized by a local hidden-variable model.

The same reasoning applies to C*-categories and subcategories of $[\boldsymbol{C}, \boldsymbol{Hilb}]$, taking the hidden variables componentwise. The constant functor valued e.g. at the model described above then still shows that such operational categories are non-local.

9.2. Relations. Suppose we are given an empirical model in the distributional operational category obtained from $\boldsymbol{Rel}(\Omega)$. The types are sets $X_1, \dots, X_n$, there is a state $s = \Delta_S$ for a Ω-subset S of $X := \prod_i X_i$ satisfying $\bigvee_x S(x) = \top$, and measurements $m_i = \{\Delta_{S_o^i}\}_{o \in O'}$, where $\{S_o^i\}$ is a disjoint cover of X_i. For each combination of measurements m and outcomes o, we have:

$$p(o|m) = \begin{cases} \bigvee_x S(x) \wedge S_o^i(x) & \text{if } o \in O', \\ 0 & \text{otherwise.} \end{cases}$$

We shall now construct a local hidden-variable model which realizes this empirical model, using the elements of X as the hidden variables. We define the distribution d_s on X as $x \mapsto S(x)$. Note that we are working over Ω (the locale of scalars in $\boldsymbol{Rel}(\Omega)$), so this is a well-defined distribution, which sums to 1.

We define $p^x(o|m) \equiv \bigwedge_i S_{o_i}^i(x_i)$, so this hidden-variable model is local by construction.

We must verify that this model agrees with the empirical model. This comes down to the following calculation for $o \in O'$:

$$p(o|m) = \bigvee_x S(x) \wedge S_o^i(x) = \bigvee_x S(x) \wedge \bigwedge_i S_{o_i}^i(x) = \bigvee_x p^x(o|m) \wedge d_s(x).$$

We conclude from this that $\boldsymbol{Rel}(\Omega)$, despite being a 'quantum-like' monoidal dagger-category, does *not* admit non-local behaviour. This stands in interesting counter-point to the fact that, as shown extensively in [3], relational models can be used to give 'logical' proofs of non-locality and contextuality, in the style of 'Bell's theorem without inequalities' [35]. The key point is that these logical proofs are based on showing the non-existence of global sections compatible with a given empirical model; while here we are looking at empirical models generated by *states* in $\boldsymbol{Rel}(\Omega)$, which are exactly sets of global elements.

The key feature of quantum mechanics, by contrast, is that quantum states under suitable measurements *are* able to realize families of probability distributions which have no global sections.

Bearing in mind that finite-dimensional quantum mechanics corresponds to the operational category arising from $\boldsymbol{FMat}(\mathbb{C})$, while $\boldsymbol{FRel}(\Omega)$ is $\boldsymbol{FMat}(\Omega)$, this shows that *idempotence of the scalars implies that only local behaviour can be realized*; thus non-locality can only arise in non-idempotent situations.

9.3. Classical stochastic maps. We now consider the case of classical stochastic maps, as discussed in Section 7. This is in fact quite similar to the case for $\boldsymbol{Rel}$. Given an empirical model realized by sets $X_1, \ldots, X_n$, a state s which is a probability distribution on $X := \prod_i X_i$, and measurements $m_i : X_i \to O$, we again take the hidden variables to be the elements of X. We can write s as a convex combination

$$s = \sum_{x \in X} \mu_x \delta_x .$$

Note that μ is a probability distribution on X. We can define $p^x(o|m) := \delta_o(m(x))$. Clearly $p^x(o|m) = \prod_i p^x(o_i|m_i)$, so this hidden-variable model is local.

It is straightforward to verify that the probabilities $p(o|m)$ are recovered by averaging over the deterministic hidden variables.

Thus we conclude, as expected, that $\boldsymbol{Stoch}$ does not exhibit non-locality.

In fact, we can say more than this. We can calibrate the expressiveness of an operational theory in terms of which empirical models it realizes. We shall now show that $\boldsymbol{Stoch}$ realizes *exactly* those models which have local hidden-variable realizations.

To see this, suppose we are given sets of measurements $M_1, \ldots, M_n$. We define $M := \bigsqcup_i M_i$, the disjoint union of these sets of measurements, and $X := O^M$. Thus elements of X simultaneously assign outcomes to all measurements. For each $m = (m_1, \ldots, m_n) \in \prod_i M_i$, we define a map $\hat{m} : X \to O$ by

$$\hat{m} : x \mapsto (x(m_1), \ldots, x(m_n)).$$

For each m, we get the probability distribution on outcomes given by

$$d_m : o \mapsto \sum_{\hat{m}(x) = o} s(x).$$

This is the empirical model realized by the state x, viewed as a probability distribution on the hidden variables X; and as shown e.g. in [5], all local models are of this form.

9.4. Signed stochastic maps. We shall now consider a variant of $\boldsymbol{Stoch}$ which has much greater expressive power in terms of the empirical models it realizes. This is the category $\boldsymbol{SStoch}$ of *signed* stochastic maps; real matrices such that each row sums to 1. Thus for each input, there is a 'signed probability measure' on outputs, which may include 'negative probabilities' [61, 27, 50, 30]. An operational representation can be defined for ***SStoch*** in the same fashion as for ***Stoch***; it is still distributional.

The following result can be extracted from [5, Theorem 5.9], using the same encoding of empirical models which we employed in the previous sub-section. The reader should refer to [5, Theorem 5.9] for the details, which are non-trivial.

PROPOSITION 12. *The class of empirical models which are realized by the operational category obtained from* **SStoch** *are exactly the no-signalling models; thus they properly contain the quantum models.*

This says that the operational category obtained from ***SStoch*** is *more* expressive, in terms of the empirical models it realizes, than the canonical operational category derived from ***Hilb***, which corresponds to quantum mechanics.

Example. We consider the bipartite system with two measurements at each site, each with outcomes $\{0, 1\}$. Thus the disjoint union M of the two sets of measurements has four elements, and $X = \{0, 1\}^M$ has 16 elements. Now consider the following state:

$$x := [1/2, 0, 0, 0, -1/2, 0, 1/2, 0, -1/2, 1/2, 0, 0, 1/2, 0, 0, 0].$$

The distributions it generates for the various measurement combinations can be listed in the following table.

	$(0,0)$	$(1,0)$	$(0,1)$	$(1,1)$
(a,b)	$1/2$	0	0	$1/2$
(a',b)	$1/2$	0	0	$1/2$
(a,b')	$1/2$	0	0	$1/2$
(a',b')	0	$1/2$	$1/2$	0

This can be recognized as the *Popescu-Rohrlich box* [53], which achieves super-quantum correlations.

The state x can be obtained from the PR-box specification by solving a system of linear equations; see [5] for details.

§**10. Final remarks.** This paper makes a first precise connection between monoidal categories, and the categorical quantum mechanics framework, on the one hand, and operational theories on the other. Clearly, this can be taken much further. We note a number of directions which it would be interesting to pursue.

- We have used our framework of operational categories to study non-locality in a general setting. In particular, we have a clear definition of whether a model of categorical quantum mechanics exhibits non-locality or not, as explained at the end of Section 8. As we saw, while Hilbert-space quantum mechanics does, the category of sets and relations, which forms a very useful 'foil' model for quantum mechanics in many respects [60, 22], does not. An important further direction is to apply a similar analysis to *contextuality*, which can be seen as a broader phenomenon, of which non-locality is a special case. In [5], a general setting is developed allowing a unified treatment of contextuality and non-locality. We would like to extend the present account to this setting, in which *compatibility* of measurements is explicitly represented, leading to a natural sheaf-theoretic structure.

- Such a development would also lead to a more satisfactory treatment of outcomes, in place of the somewhat clumsy device used in the present paper.
- It would also be interesting to interpret some of the general results which have been proved for operational theories, relating e.g. to no-broadcasting [11], teleportation [12], and information causality [10], in our categorical framework, and ultimately to obtain such results for classes of monoidal process categories.
- We would also like to examine the issue of axiomatization or 'reconstruction' of quantum mechanics from the categorical point of view.
- There are various constructions for turning a monoidal category of 'pure' states into one of 'mixed' states [57, 23]. It would be interesting to relate these constructions to our canonical operational categories. Similarly, there is a category embodying Spekkens' toy theory [60, 21]. It would be of interest to study the associated operational category.

Regarding related work, we note that in [13], the structure of the concrete category of convex operational theories is investigated.

Acknowledgements. Financial support from EPSRC Senior Research Fellowship EP/E052819/1 and the U.S. Office of Naval Research Grant Number N000141010357 is gratefully acknowledged. We thank Shane Mansfield for a number of useful comments, which in particular led to an improved formulation of Proposition 8.

Appendix A. First notions from category theory. We shall review some basic notions from category theory. For more detailed background, see [9].

A *category* $\mathbf{C}$ has a collection of objects $A, B, C, \ldots$, and arrows $f, g, h, \ldots$. Each arrow has specified *domain* and *codomain* objects: notation is $f : A \to B$ for an arrow f with domain A and codomain B. The collection of all arrows with domain A and codomain B is denoted as $\mathbf{C}(A, B)$. Given arrows $f : A \to B$ and $g : B \to C$, we can form the *composition* $g \circ f : A \to C$. Composition is associative, and there are identity arrows $1_A : A \to A$ for each object A, with $f \circ 1_A = f$, $1_A \circ g = g$, for every $f : A \to B$ and $g : C \to A$. An arrow $f : A \to B$ is called an *iso(morphism)* when $f \circ f^{-1} = 1_B$ and $f^{-1} \circ f = 1_A$ for some arrow $f^{-1} : B \to A$. An arrow $f : A \to B$ is *split monic* when $g \circ f = 1_A$ for some $g : B \to A$, and it is *split epic* when $f \circ g = 1_A$ for some $g : B \to A$; by abuse of notation, we will write $g = f^{-1}$ in both cases.

If $\boldsymbol{C}$ is a category, we write $\boldsymbol{C}^{\mathrm{op}}$ for the opposite category, with the same objects as $\boldsymbol{C}$, and arrows $A \to B$ corresponding to arrows $B \to A$ in $\boldsymbol{C}$.

If $\boldsymbol{C}$ and $\boldsymbol{D}$ are categories, a *functor* $F : \boldsymbol{C} \to \boldsymbol{D}$ assigns an object FA of $\boldsymbol{D}$ to each object A of $\boldsymbol{C}$; and an arrow $Ff : FA \to FB$ of $\boldsymbol{D}$ to every arrow $f : A \to B$ of $\boldsymbol{C}$. These assignments must preserve composition and identities: $F(g \circ f) = F(g) \circ F(f)$, and $F(1_A) = 1_{FA}$.

Given functors $F, G : \boldsymbol{C} \to \boldsymbol{D}$, a *natural transformation* $t : F \dot{\to} G$ is a family of arrows $\{t_A : FA \to GA\}$ indexed by the objects of $\boldsymbol{C}$, such that, for every $f : A \to B$ in $\boldsymbol{C}$, the following naturality diagram commutes:

$$\begin{array}{ccc} FA & \xrightarrow{t_A} & GA \\ {\scriptstyle Ff}\downarrow & & \downarrow{\scriptstyle Gf} \\ FB & \xrightarrow[t_B]{} & GB \end{array}$$

A *natural isomorphism* is a natural transformation whose components are isomorphisms. An *equivalence of categories* is a pair of functors $F : \boldsymbol{C} \to \boldsymbol{D}$ and $G : \boldsymbol{D} \to \boldsymbol{C}$ such that there are natural isomorphisms $F \circ G \cong 1_{\boldsymbol{D}}$ and $G \circ F \cong 1_{\boldsymbol{C}}$.

A *symmetric monoidal category* is a structure $(\boldsymbol{C}, \otimes, I, \alpha, \lambda, \rho, \sigma)$ where:

- $\boldsymbol{C}$ is a category;
- $\otimes : \boldsymbol{C} \times \boldsymbol{C} \to \boldsymbol{C}$ is a functor (*tensor*);
- I is a distinguished object of $\boldsymbol{C}$ (*unit*);
- $\alpha, \lambda, \rho, \sigma$ are natural isomorphisms (*structural isos*) with components

$$\alpha_{A,B,C} : A \otimes (B \otimes C) \to (A \otimes B) \otimes C$$
$$\lambda_A : I \otimes A \to A \qquad \rho_A : A \otimes I \to A$$
$$\sigma_{A,B} : A \otimes B \to B \otimes A$$

such that certain coherence diagrams commute.

Products are a classical example of symmetric monoidal structure; the category is then called *Cartesian*. The symmetric monoidal structure can also support entanglement; the category is then called *compact* [7].

Let $\boldsymbol{C}$ and $\boldsymbol{D}$ be symmetric monoidal categories. A symmetric monoidal functor

$$(F, e, m) : \boldsymbol{C} \to \boldsymbol{D}$$

comprises

- a functor $F : \boldsymbol{C} \to \boldsymbol{D}$,
- an arrow $e : I_{\boldsymbol{D}} \to FI_{\boldsymbol{C}}$,
- a natural transformation $m_{A,B} : FA \otimes FB \to F(A \otimes B)$,

subject to coherence conditions with the structural isomorphisms. The symmetric monoidal functor is called *strong* when m is a natural isomorphism.

Let $(F, e, m), (G, e', m') \colon \boldsymbol{C} \to \boldsymbol{D}$ be symmetric monoidal functors. A *monoidal natural transformation* between them is a natural transformation $t \colon F \stackrel{.}{\to} G$ such that the following diagrams commute.

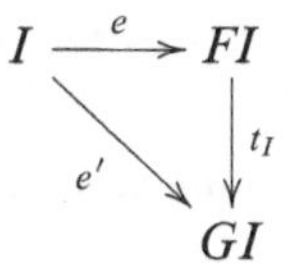

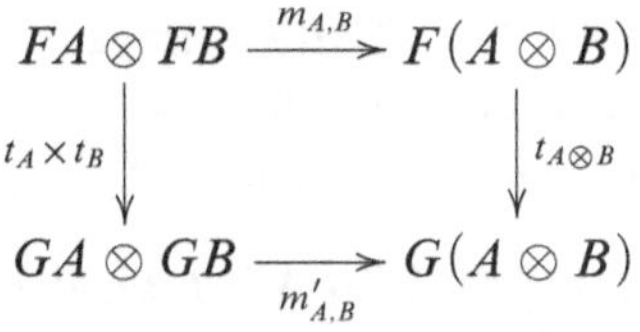

Appendix B. Trace ideals. This appendix further studies the notion of trace ideal, introduced in Section 4.1. It presents several technical results that are mathematically interesting, but would break up the flow of the main text. For example, we characterize when trace ideals exist, and to what extent they are unique. Also, we show that we really need to restrict to ideals to consider traces: the category of Hilbert spaces does not support a trace on all morphisms. As a conceptually satisfying corollary, we derive that dual objects in the category of Hilbert spaces are necessarily finite-dimensional. Finally, we prove in some detail that the category of Hilbert spaces indeed has a trace ideal; this was claimed in Section 4.1, but the details are quite subtle.

B.1. Existence. The question whether a category allows a trace ideal at all can be answered as follows.

A sub-category $\boldsymbol{D}$ of $\boldsymbol{C}$ is called *tracial* when endomorphisms in $\boldsymbol{C}$ factoring through $\boldsymbol{D}$ can only do so in a way unique up to isomorphism. More precisely: if $f_1 \colon X \to Y$, $f_2 \colon Y \to X$, $f_1' \colon X \to Y'$, $f_2' \colon Y' \to X$ are morphisms of $\boldsymbol{C}$, and Y and Y' are objects of $\boldsymbol{D}$, and $f_2 \circ f_1 = f_2' \circ f_1'$, then there is a morphism $i \colon Y \to Y'$ in $\boldsymbol{D}$ that is either split monic or split epic, such that $f_1' = i \circ f_1$ and $f_2' = f_2 \circ i^{-1}$.

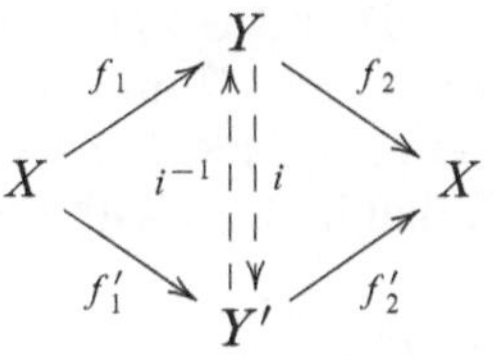

The category $\boldsymbol{C}$ is called *traceable* when the full sub-category consisting of the monoidal unit I is tracial. Notice that traceability generalizes the fact, holding in any monoidal category, that the scalars are commutative.

Proposition 13. *Any dagger monoidal tracial sub-category $\boldsymbol{D}$ of $\boldsymbol{C}$ with a trace ideal induces a trace ideal*

$$\mathcal{I}(X) = \{f \in \mathbf{C}(X, X) \mid f = f_2 \circ f_1 \textit{ with } f_1 \colon X \to Y, f_2 \colon Y \to X \textit{ and } Y \textit{ in } \boldsymbol{D}, f_1 \circ f_2 \in \mathcal{I}_{\boldsymbol{D}}(Y)\}$$

$$\mathrm{Tr}(f) = \mathrm{Tr}_{\boldsymbol{D}}(f_1 \circ f_2)$$

on $\boldsymbol{C}$.

PROOF. One directly checks that $\mathcal{I}(X)$ is an endomorphism ideal; in particular $\mathcal{I}(I) = \boldsymbol{D}(I, I) = \boldsymbol{C}(I, I)$. Because $\boldsymbol{D}$ is tracial, Tr is well-defined. The axioms for the trace function are also readily verified. ⊣

THEOREM 14. *A dagger monoidal category has a unique minimal trace ideal*

$$\mathcal{I}(X) = \{f : X \to X \mid f \text{ factors through } I\}$$
$$\mathrm{Tr}(f) = b \circ a, \text{ when } f = a \circ b \text{ with } a : I \to X \text{ and } b : X \to I$$

and hence has any trace ideal whatsoever, if and only if it is traceable.

PROOF. That the given data form a trace ideal follows from the previous proposition, because the full sub-category consisting of just the monoidal unit I is certainly (totally) traced. To see that this trace ideal is minimal, i.e. that any trace ideal must contain this one, follows from the first and third axioms of endomorphism ideal. ⊣

As a consequence of the previous theorem, the evaluation of measurements on pure states is completely determined by the structure of the category, independent of the trace ideal. If $s = \psi \circ \psi^\dagger$ is a pure state on X, and $\{P_o\}$ a measurement, then for every outcome o:

$$\mathrm{Tr}(s \circ P_o) = \mathrm{Tr}(\psi \circ \psi^\dagger \circ P_o) = \mathrm{Tr}(\psi^\dagger \circ P_o \circ \psi) = \psi^\dagger \circ P_o \circ \psi.$$

Therefore, the only possible freedom the choice of a trace ideal brings comes out in behaviour on mixed states.

B.2. Uniqueness. We now consider uniqueness of trace ideals. The following proposition proves that trace ideals are a categorical invariant, in the sense that they are preserved under equivalence. A dagger monoidal equivalence is a pair of functors $F : \boldsymbol{C} \to \boldsymbol{D}$ and $G : \boldsymbol{D} \to \boldsymbol{C}$ that form an equivalence of categories, such that $F(f^\dagger) = F(f)^\dagger$ and $G(f^\dagger) = G(f)^\dagger$, and there are natural isomorphisms $F(I) \cong I$, $G(I) \cong I$, $F(X \otimes Y) \cong F(X) \otimes F(Y)$ and $G(X \otimes Y) \cong G(X) \otimes G(Y)$ that interact with the coherence isomorphisms in the appropriate way.

PROPOSITION 15. *Trace ideals are preserved under dagger monoidal equivalence: if* $F : \boldsymbol{C} \to \boldsymbol{D}$ *and* $G : \boldsymbol{D} \to \boldsymbol{C}$ *are strong monoidal functors that preserve daggers and form an equivalence of categories, and* $(\mathcal{I}, \mathrm{Tr}^{\mathcal{I}})$ *is a trace ideal in* $\boldsymbol{C}$, *then*

$$\mathcal{J}(X) = G^{-1}(\mathcal{I}(G(X))) = \{g \in \boldsymbol{D}(X, X) \mid G(g) \in \mathcal{I}(G(X))\},$$
$$\mathrm{Tr}_X^{\mathcal{J}}(g) = F(\mathrm{Tr}_{G(X)}^{\mathcal{I}}(G(g))),$$

form a trace ideal in $\boldsymbol{D}$.

PROOF. First, observe that if $f \in \mathcal{I}(X)$, and $g: X \to Y$ is an isomorphism with inverse h, then $\mathrm{Tr}(f) = \mathrm{Tr}(gfh)$. Then, to verify that $\mathcal{J}$ is an endomorphism ideal, the first requirement follows from functoriality of G; the second from the fact that G is monoidal; and the third from fullness of G together with monoidality of G. It is a dagger endomorphism ideal because G preserves daggers. Verifying that $\mathrm{Tr}^{\mathcal{J}}$ satisfies the requirements is completely analogous, except that the last condition additionally uses $F(G(s)) \cong s$. ⊣

However, trace ideals need not be unique. In fact, there may even be more than one trace function making a fixed endomorphism ideal into a trace ideal, as the following example shows.

EXAMPLE 16. A *tracial state* on a C*-algebra A is a linear map $\tau: A \to \mathbb{C}$ satisfying $\tau(a^*a) \geq 0$, $\tau(1) = 1$, and $\tau(ab) = \tau(ba)$. There exists a unital C*-algebra A with distinct tracial states $\tau \neq \tau': A \to \mathbb{C}$ [46].

Make a category $\boldsymbol{C}$ as follows. Objects are natural numbers. There are only endomorphisms. Morphisms $0 \to 0$ are complex numbers; the identity is 0, and composition is addition. For $n \geq 1$, morphisms $n \to n$ are elements of the n-fold direct sum $A \oplus A \oplus \cdots \oplus A$; the identity is $(1, 1, \ldots, 1)$, and composition is pointwise multiplication.

We give this category a monoidal structure by letting the tensor product of objects n and m be $n + m$. If one of n or m is 0, the action on morphisms is by scalar multiplication. For $n, m \geq 1$, the action on morphisms is clear. The monoidal unit is the object 0.

Taking $\mathcal{I}(X)$ to be all endomorphisms on X certainly gives an endomorphism ideal. Define $\mathrm{Tr}_0(z) = z$, and $\mathrm{Tr}_n(a_1, \ldots, a_n) = \sum_{i=1}^{n} \tau(a_i)$ for $n \geq 1$. This satisfies all the conditions needed to make $\mathcal{I}$ into a trace ideal. But the very same construction with τ' gives a different trace function.

The previous example is in stark contrast to Cartesian categories or compact categories, where traces are unique; see [59] and [37], respectively. The counterexample above is somewhat artificial, because all morphisms are endomorphisms. It remains unclear whether trace ideals on, for example, compact categories, are unique.

B.3. The need for trace ideals. We will now show that in the category ***Hilb***, there exists no trace ideal consisting of all morphisms. More precisely, we will show that ***Hilb*** is not an instance of the established notion of *traced monoidal category* [43]. This notion asks not just for traces of all endomorphisms, but requires a 'partial trace' of morphisms $f: X \otimes U \to Y \otimes U$, resulting in a morphism $\mathrm{Tr}^U(f): X \to Y$. There are then several additional axioms, such as the following naturality:

$$\mathrm{Tr}^U(f) \circ g = \mathrm{Tr}^U(f \circ (g \otimes 1_U)) \quad \text{for } f: X \otimes U \to Y \otimes U, g: X' \to X.$$

We will now show that the monoidal category (***Hilb***, $\otimes$) cannot be traced monoidal. Subsequently, we will show that it *does* have a trace ideal. This justifies working with trace ideals in monoidal categories instead of traced monoidal categories. We are indebted to Peter Selinger for the following proof.

LEMMA 17. *Suppose* $(\textbf{Hilb}, \otimes)$ *is traced monoidal. Then* $\mathrm{Tr}(f+g) = \mathrm{Tr}(f) + \mathrm{Tr}(g)$ *for all endomorphisms* $f, g \colon H \to H$.

PROOF. Choose an orthonormal basis $\{|0\rangle, |1\rangle\}$ for $\mathbb{C}^2$, and write $|+\rangle = |0\rangle + |1\rangle$. Recall that $\mathbb{C}^2 \otimes H \cong H \oplus H$. Define $F \colon \mathbb{C}^2 \otimes H \to H$ via the block matrix $\begin{pmatrix} f & g \end{pmatrix}$. Hence $F \circ (|0\rangle \otimes 1_H) = f$ and $F \circ (|1\rangle \otimes 1_H) = g$. Now:

$$\begin{aligned}
\mathrm{Tr}(f+g) &= \mathrm{Tr}(F \circ (|+\rangle \otimes 1_H)) \\
&= \mathrm{Tr}(F) \circ |+\rangle && \text{(by naturality)} \\
&= (\mathrm{Tr}(F) \circ |0\rangle) + (\mathrm{Tr}(F) \circ |1\rangle) \\
&= \mathrm{Tr}(F \circ (|0\rangle \otimes 1_H)) + \mathrm{Tr}(F \circ (|1\rangle \otimes 1_H)) && \text{(by naturality)} \\
&= \mathrm{Tr}(f) + \mathrm{Tr}(g).
\end{aligned}$$

The third equality uses that composition is bilinear. ⊣

THEOREM 18. *The monoidal category* $(\textbf{Hilb}, \otimes)$ *is not traced monoidal.*

PROOF. Suppose $(\textbf{Hilb}, \otimes)$ was traced monoidal. Let H be an infinite-dimensional Hilbert space. Then there exist isomorphisms $\varphi \colon H \oplus \mathbb{C} \xrightarrow{\cong} H$ and $\psi \colon H \xrightarrow{\cong} \mathbb{C} \oplus H$. Write them in block matrix form as $\varphi = \begin{pmatrix} \varphi_1 & \varphi_2 \end{pmatrix}$ and $\psi = \begin{pmatrix} \psi_1 \\ \psi_2 \end{pmatrix}$. Consider the morphisms $f_1, f_2, f_3 \colon H \oplus \mathbb{C} \oplus H \to H \oplus \mathbb{C} \oplus H$ given by the following block matrices.

$$f_1 = \begin{pmatrix} 1 & 0 & 0 \\ 0 & 0 & 0 \\ 0 & 0 & 0 \end{pmatrix} \qquad f_2 = \begin{pmatrix} 1 & 0 & 0 \\ 0 & 1 & 0 \\ 0 & 0 & 0 \end{pmatrix} \qquad f_3 = \begin{pmatrix} 0 & 0 & 0 \\ 0 & 1 & 0 \\ 0 & 0 & 0 \end{pmatrix}$$

Let $g = \varphi \oplus \psi \colon (H \oplus \mathbb{C}) \oplus H \to H \oplus (\mathbb{C} \oplus H)$. Then

$$\begin{aligned}
g \circ f_2 &= \begin{pmatrix} \varphi_1 & \varphi_2 & 0 \\ 0 & 0 & \psi_1 \\ 0 & 0 & \psi_2 \end{pmatrix} \circ \begin{pmatrix} 1 & 0 & 0 \\ 0 & 1 & 0 \\ 0 & 0 & 0 \end{pmatrix} = \begin{pmatrix} \varphi_1 & \varphi_2 & 0 \\ 0 & 0 & 0 \\ 0 & 0 & 0 \end{pmatrix} \\
&= \begin{pmatrix} 1 & 0 & 0 \\ 0 & 0 & 0 \\ 0 & 0 & 0 \end{pmatrix} \circ \begin{pmatrix} \varphi_1 & \varphi_2 & 0 \\ 0 & 0 & \psi_1 \\ 0 & 0 & \psi_2 \end{pmatrix} = f_1 \circ g.
\end{aligned}$$

Hence

$$\mathrm{Tr}(f_1) = \mathrm{Tr}(f_1 \circ g \circ g^{-1}) = \mathrm{Tr}(g \circ f_2 \circ g^{-1}) = \mathrm{Tr}(f_2 \circ g^{-1} \circ g) = \mathrm{Tr}(f_2).$$

But $\mathrm{Tr}(f_2) = \mathrm{Tr}(f_1 + f_3) = \mathrm{Tr}(f_1) + \mathrm{Tr}(f_3)$ by Lemma 17. And because f_3 has finite rank, we know that $\mathrm{Tr}(f_3) = \mathrm{Tr}(1_{\mathbb{C}}) = 1$. Thus $\mathrm{Tr}(f_2) = \mathrm{Tr}(f_2) + 1$, which is a contradiction. ⊣

B.4. Dual objects in *Hilb* are finite-dimensional. The previous theorem allows an interesting corollary. Recall that the main characteristic of compact categories is that objects have duals: objects L, R in a monoidal category are called *dual* when there are maps $\eta: I \to R \otimes L$ and $\varepsilon: L \otimes R \to I$ making the following two composites identities.

$$L \cong L \otimes I \xrightarrow{1 \otimes \eta} L \otimes (R \otimes L) \cong (L \otimes R) \otimes L \xrightarrow{\varepsilon \otimes 1} I \otimes L \cong L$$

$$R \cong I \otimes R \xrightarrow{\eta \otimes 1} (R \otimes L) \otimes R \cong R \otimes (L \otimes R) \xrightarrow{1 \otimes \varepsilon} R \otimes I \cong R$$

It is well-known that if $H \in$ ***Hilb*** is finite-dimensional, then H and H^* are dual objects by $\eta(1) = \sum_{i=1}^{n} |i\rangle \otimes \langle i|$ and $\varepsilon(|i\rangle) = 1$, for any choice of orthonormal basis $\{|i\rangle\}_{i=1,\ldots,n}$ for H; see [44, 43, 7]. This recipe does not work when H is infinite-dimensional, because $\sum_i |i\rangle$ does not converge in that case. However, this does not exclude the possibility that there might be other H^*, η, ε making H into a dual object. No rigorous proof that infinite-dimensional Hilbert spaces cannot have duals has been published, as far as we know.

COROLLARY 19. *Objects in* $(\textbf{\textit{Hilb}}, \otimes)$ *with duals are precisely finite-dimensional Hilbert spaces.*

PROOF. Let H be an infinite-dimensional Hilbert space. Suppose H has a dual object H^*. For $f: H \to H$, define $\mathrm{Tr}^H(f)$ as the following composite.

$$I \xrightarrow{\eta} H^* \otimes H \cong H \otimes H^* \xrightarrow{f \otimes 1_{H^*}} H \otimes H^* \xrightarrow{\varepsilon} I$$

This satisfies all equations for a trace function, as far as these make sense 'locally', for just one object H. In ***Hilb***, the object $\mathbb{C}$ always has a dual, and if H and K have duals, then so does $H \oplus K$. Now, notice that the proof of Theorem 18 only uses the trace properties 'locally', i.e. for the objects $\mathbb{C}, H, \mathbb{C}^2 \otimes H \cong H \oplus H, H \oplus \mathbb{C}, H \oplus \mathbb{C} \oplus H$. Hence the contradiction it results in holds here, too. ⊣

In fact, in any monoidal category with biproducts, one can show that if $A \cong A \oplus I$, then $\mathrm{Tr}^A(1_A) = \mathrm{Tr}^A(1_A) + 1$. We thank Jamie Vicary for this observation.

B.5. Trace class maps form a trace ideal in *Hilb*. To show that the usual trace of continuous linear maps between Hilbert spaces does in fact give a trace ideal requires some work, as virtually all textbooks only consider endomorphisms, whereas the defining conditions of trace ideals also involve morphisms between different objects.

We need to recall some terminology; for any unexplained terms, we refer to [17]. Other good references are [32, 58]. A linear map $f: H \to K$ between Hilbert spaces is *Hilbert-Schmidt* when $\sum_n \|f(e_n)\|_K^2 < \infty$ for an orthonormal basis (e_n) of H. A positive continuous linear map $f: H \to H$ is *trace class* when $\sum_n |\langle e_n \mid f(e_n)\rangle| < \infty$ for an orthonormal basis (e_n) of H. An arbitrary continuous linear map $f: H \to H$ is trace class when its absolute value $|f|: H \to H$ is trace class. Both definitions are independent of the choice of basis (e_n). If f is trace class, then $\langle e_n \mid f(e_n)\rangle$ is absolutely summable, and hence the following *trace property* holds:

$$\mathrm{Tr}(f) = \sum_n \langle e_n \mid f(e_n)\rangle$$

is a well-defined complex number. The *Cauchy-Schwarz inequality* states that

$$|\langle x \mid y\rangle| \leq \left(\|x\|^2 \cdot \|y\|^2\right)^{1/2}$$

for any two elements x, y of a Hilbert space. The *Hölder inequality* states that

$$\sum_n |x_n \cdot y_n| \leq \left(\sum_n |x_n|^2\right)^{1/2} \cdot \left(\sum_n |y_n|^2\right)^{1/2}$$

for any two sequences (x_n) and (y_n) of complex numbers with $\sum_n |x_n|^2 < \infty$ and $\sum_n |y_n|^2 < \infty$.

LEMMA 20. *Let* $H \underset{g}{\overset{f}{\rightleftarrows}} K$ *be morphisms in* ***Hilb***. *Then* $g \circ f$ *is trace class if and only if* f *and* g *are Hilbert-Schmidt.*

PROOF. By polar decomposition, there is a unique partial isometry $w \colon H \to K$ satisfying $g \circ f = w \circ |g \circ f|$ and $\ker(w) = \ker(g \circ f)$. It follows that $|g \circ f| = w^\dagger \circ g \circ f$. Hence, for an orthonormal basis (e_n) of H,

$$\begin{aligned}
\sum_n |\langle e_n \mid |g \circ f|(e_n)\rangle| &= \sum_n |\langle e_n \mid w^\dagger \circ g \circ f(e_n)\rangle| \\
&= \sum_n |\langle g^\dagger \circ w(e_n) \mid f(e_n)\rangle| \\
&\leq \sum_n \left(\|g^\dagger \circ w(e_n)\|^2 \cdot \|f(e_n)\|^2\right)^{1/2} \quad \text{(by Cauchy-Schwarz)} \\
&= \sum_n \big| \|g^\dagger \circ w(e_n)\| \cdot \|f(e_n)\| \big| \\
&\leq \left(\sum_n \|g^\dagger \circ w(e_n)\|^2\right)^{1/2} \cdot \left(\sum_n \|f(e_n)\|^2\right)^{1/2}. \\
&\qquad \text{(by Hölder)}
\end{aligned}$$

Therefore gf is trace class if and only if $\sum_n \|f(e_n)\|^2 < \infty$ and $\sum_n \|g^\dagger \circ w(e_n)\|^2 < \infty$. Because w is a partial isometry, the latter inequality holds if and only if $\sum_n \|g(e_n)\|^2 < \infty$. That is, $g \circ f$ is trace class if and only if f and g are Hilbert-Schmidt. ⊣

PROPOSITION 21. *The category* ***Hilb*** *has a dagger trace ideal consisting of the usual trace class maps and the usual trace function.*

PROOF. That the trace class maps on a Hilbert space H are closed under adjoint and tensor products is easily seen. Also, any morphism $\mathbb{C} \to \mathbb{C}$ is trivially trace class. Now suppose that $f: H \to H$ is trace class. By the previous lemma, we can write $f = f_2 \circ f_1$ for Hilbert-Schmidt maps f_i. If $g: H \to K$ and $h: K \to H$ are arbitrary morphisms, then $g \circ f_2$ and $f_1 \circ h$ are again Hilbert-Schmidt. Therefore, by the previous lemma again, $g \circ f \circ h = (g \circ f_2) \circ (f_1 \circ h)$ is trace class. Thus trace class maps indeed form an endomorphism ideal.

One easily sees from the trace property that trace is the identity on scalars, is multiplicative on tensor products, and preserves daggers. To prove that $\mathrm{Tr}(g \circ f) = \mathrm{Tr}(f \circ g)$ for $f: H \to K$ and $g: K \to H$ with both $f \circ g$ and $g \circ f$ trace class, we rely on *Lidskii's trace formula* for separable H: if h is trace class, then ($\sum_n \lambda_n(h)$ is absolutely convergent and) $\mathrm{Tr}(h) = \sum_n \lambda_n(h)$, where $\lambda_n(h)$ are the eigenvalues counted up to algebraic multiplicity [58, Theorem 3.7]. But $g \circ f$ and $f \circ g$ have precisely the same spectrum, so that $\mathrm{Tr}(g \circ f) = \sum_n \lambda_n(g \circ f) = \sum_n \lambda_n(f \circ g) = \mathrm{Tr}(f \circ g)$.

Finally, we claim that for positive, trace class functions $h: H \to H$ on any (possibly nonseparable) Hilbert space H, Lidskii's formula still holds, which finishes the proof that trace class operators form a trace ideal, because we may then replace $g \circ f$ and $f \circ g$ above by their absolute value. Pick an orthonormal basis $\{e_i\}$ for H. Since h is trace class, $\sum_i \langle e_i \mid h(e_i) \rangle = \sum_i \|\sqrt{h}(e_i)\|$ is summable. Hence $\ker(h)^\perp = \ker(\sqrt{h})^\perp$ can only contain countably many e_i. Because h is positive, its range is $\ker(h^\dagger)^\perp = \ker(h)^\perp$. Thus $h: H \to H$ restricts to a function $h\colon \ker(h)^\perp \to \ker(h)^\perp$ on a separable space. ⊣

We have written the above example out in more detail than the reader might have thought necessary, because it is easy to overlook subtleties. For example, it is not true that if $f: H \to K$ and $g: K \to H$ are morphisms such that $g \circ f$ is trace class, then $f \circ g$ is trace class, too. For a counterexample, let $H = K = \ell^2(\mathbb{N})$, and define $f(x, y) = (0, x)$ and $g(x, y) = (x, 0)$. Then certainly $g \circ f = 0$ is trace class. But it is easy to see that $f^\dagger(x, y) = (y, 0)$, that $g = g^\dagger = g^\dagger \circ g$, and hence that $g = g^\dagger \circ g = (f \circ g)^\dagger \circ (f \circ g) \geq 0$. Therefore $|f \circ g| = g$, and

$$\begin{aligned}\mathrm{Tr}(f \circ g) &= \sum_{m,n} \langle |f \circ g|(e_m, e_n) \mid (e_m, e_n) \rangle \\ &= \sum_{m,n} \langle e_m \mid e_m \rangle + \langle 0 \mid e_n \rangle = \dim(H) = \infty,\end{aligned}$$

so that $f \circ g$ is not trace class.

REFERENCES

[1] S. Abramsky, *No-Cloning in Categorical Quantum Mechanics*, ***Semantic Techniques in Quantum Computation*** (S. Gay and I. Mackie, editors), Cambridge University Press, 2010, pp. 1–28.

[2] S. Abramsky, *Big toy models: Representing physical systems as Chu spaces*, **Synthese**, vol. 186(3) (2012), pp. 697–718.

[3] ———, *Relational Hidden Variables and Non-Locality*, **Studia Logica**, vol. 101 (2013), no. 2, pp. 411–452, available as arXiv:1007.2754.

[4] S. Abramsky, R. Blute, and P. Panangaden, *Nuclear and trace ideals in tensored*-categories*, **Journal of Pure and Applied Algebra**, vol. 143 (1999), pp. 3–47.

[5] S. Abramsky and A. Brandenburger, *The sheaf-theoretic structure of non-locality and contextuality*, **New Journal of Physics**, vol. 13 (2011), p. 113036.

[6] S. Abramsky and B. Coecke, *A categorical semantics of quantum protocols*, **Proceedings of the 19th Annual IEEE Symposium on Logic in Computer Science**, IEEE, 2004, pp. 415–425.

[7] ———, *Categorical quantum mechanics*, **Handbook of Quantum Logic and Quantum Structures: Quantum Logic**, Elsevier Science, 2008, pp. 261–324.

[8] M. F. Atiyah, *Topological quantum field theories*, **Publications Mathematiques de l'I.H.É,S.**, vol. 68 (1988), pp. 175–186.

[9] S. Awodey, ***Category Theory***, Oxford University Press, 2010.

[10] H. Barnum, J. Barrett, L. O. Clark, M. Leifer, R. Spekkens, N. Stepanik, A. Wilce, and R. Wilke, *Entropy and information causality in general probabilistic theories*, **New Journal of Physics**, vol. 12 (2010), p. 033024.

[11] H. Barnum, J. Barrett, M. Leifer, and A. Wilce, *Generalized no-broadcasting theorem*, **Physical review letters**, vol. 99 (2007), no. 24, p. 240501.

[12] ———, *Teleportation in general probabilistic theories*, **Arxiv preprint arXiv:0805.3553**, (2008).

[13] H. Barnum, R. Duncan, and A. Wilce, *Symmetry, compact closure and dagger compactness for categories of convex operational models*, **Journal of Philosophical Logic**, vol. 42 (2013), pp. 501–523, DOI 10.1007/s10992-013-9280-8.

[14] M. Barr, ***-autonomous categories***, Lecture Notes in Mathematics, vol. 752, Springer, 1979.

[15] J. S. Bell, *On the Einstein-Podolsky-Rosen paradox*, **Physics**, vol. 1 (1964), no. 3, pp. 195–200.

[16] J. Bénabou, *Distributors at work*, 2000, available at http://www.mathematik.tu-darmstadt.de/~streicher/FIBR/DiWo.pdf.

[17] J. Blank, P. Exner, and M. Havlíček, ***Hilbert space operators in quantum physics***, second ed., Springer, 2008.

[18] C. Butz, *Regular categories and regular logic*, Technical Report LS-98-2, BRICS, October 1998.

[19] G. Chiribella, G. M. D'Ariano, and P. Perinotti, *Informational derivation of quantum theory*, **Physical Review A**, vol. 84 (2011), no. 1, p. 012311.

[20] P.-H. Chu, *Constructing $*$-autonomous categories*, In *$*$-Autonomous Categories*? [14], pp. 103–137.

[21] B. Coecke and B. Edwards, *Spekkens' toy theory as a category of processes*, **Mathematical Foundations of Information Flow** (S. Abramsky and M. Mislove, editors), American Mathematical Society, 2012, arXiv:1108.1978.

[22] B. Coecke, B. Edwards, and R. W. Spekkens, *Phase groups and the origin of non-locality for qubits*, **Electronic Notes in Theoretical Computer Science**, vol. 270 (2011), no. 2, pp. 15–36.

[23] B. Coecke and C. Heunen, *Pictures of complete positivity in arbitrary dimension*, **Quantum Physics and Logic 2012**, Electronic Proceedings in Theoretical Computer Science, vol. 158, 2014, pp. 1–14.

[24] B. Coecke and A. Kissinger, *The compositional structure of multipartite quantum entanglement*, **Automata, Languages and Programming**, (2010), pp. 297–308.

[25] B. Dakic and C. Brukner, *Quantum Theory and Beyond: Is Entanglement Special?*, ***Deep Beauty: Understanding the Quantum World through Mathematical Innovation***, Cambridge University Press, 2011, pp. 365–392.

[26] W. M. Dickson, ***Quantum Chance and Non-Locality***, Cambridge University Press, 1999.

[27] P. A. M. Dirac, *The physical interpretation of quantum mechanics*, ***Proceedings of the Royal Society of London. Series A, Mathematical and Physical Sciences***, vol. 180 (1942), no. 980, pp. 1–40.

[28] S. Doplicher and J. E. Roberts, *A new duality theory for compact groups*, ***Inventiones mathematicae***, vol. 98 (1989), pp. 157–218.

[29] R. Duncan and S. Perdrix, *Rewriting measurement-based quantum computations with generalised flow*, ***Automata, Languages and Programming***, (2010), pp. 285–296.

[30] R. P. Feynman, *Negative probability*, ***Quantum Implications: Essays in Honour of David Bohm*** (B. J. Hiley and F. D. Peat, editors), Routledge and Kegan Paul, 1987, pp. 235–248.

[31] A. Fine, *Joint distributions, quantum correlations and commuting observables*, ***Journal of Mathematical Physics***, vol. 23 (1982), p. 1306.

[32] D. J. H. Garling, ***Inequalities***, Cambridge University Press, 2007.

[33] P. Ghez, R. Lima, and J. E. Roberts, *W^*-categories*, ***Pacific Journal of Mathematics***, vol. 120 (1985), pp. 79–109.

[34] M. Giry, *A categorical approach to probability theory*, ***Categorical Aspects of Topology and Analysis***, Springer, 1982, pp. 68–85.

[35] D. M. Greenberger, M. A. Horne, A. Shimony, and A. Zeilinger, *Bell's theorem without inequalities*, ***American Journal of Physics***, vol. 58 (1990), p. 1131.

[36] L. Hardy, *Quantum theory from five reasonable axioms*, ***Arxiv preprint quant-ph/0101012***, (2001).

[37] M. Hasegawa, *On traced monoidal closed categories*, ***Mathematical Structures in Computer Science***, vol. 19 (2008), pp. 217–244.

[38] C. Heunen, ***Categorical Quantum Models and Logics***, Ph.D. thesis, Radboud University Nijmegen, 2009.

[39] C. Heunen and B. Jacobs, *Quantum logic in dagger kernel categories*, ***Order***, vol. 27 (2010), no. 2, pp. 177–212.

[40] B. Jacobs, *Convexity, duality and effects*, ***Theoretical Computer Science*** (Cristian S. Calude and Vladimiro Sassone, editors), IFIP Advances in Information and Communication Technology, vol. 323, Springer, Berlin, Heidelberg, 2010, pp. 1–19.

[41] J. M. Jauch, ***Foundations of Quantum Mechanics***, Addison-Wesley, 1968.

[42] P. T. Johnstone, ***Stone Spaces***, Studies in Advanced Mathematics, vol. 3, Cambridge University Press, 1986.

[43] A. Joyal, R. Street, and D. Verity, *Traced monoidal categories*, ***Mathematical Proceedings of the Cambridge Philosophical Society***, vol. 3 (1996), pp. 447–468.

[44] G. M. Kelly and M. L. Laplaza, *Coherence for compact closed categories*, ***Journal of Pure and Applied Algebra***, vol. 19 (1980), pp. 193–213.

[45] J. Kock, ***Frobenius Algebras and 2D Topological Quantum Field Theories***, London Mathematical Society Student Texts, no. 59, Cambridge University Press, 2003.

[46] R. Longo, *A remark on crossed product of C^*-algebras*, ***Journal of the London Mathematical Society (2)***, vol. 23 (1981), pp. 531–533.

[47] G. Ludwig, ***Foundations of Quantum Mechanics***, vol. 1, Springer-Verlag, 1983.

[48] G. W. Mackey, ***Mathematical Foundations of Quantum Mechanics***, Benjamin, 1963.

[49] L. Masanes and M. P. Müller, *A derivation of quantum theory from physical requirements*, ***New Journal of Physics***, vol. 13 (2011), p. 063001.

[50] J. E. Moyal, *Quantum mechanics as a statistical theory*, ***Mathematical Proceedings of the Cambridge Philosophical Society***, vol. 45 (1949), no. 1, pp. 99–124.

[51] A. Peres, ***Quantum Theory: Concepts and Methods***, vol. 57, Kluwer, 1993.

[52] C. Piron, ***Foundations of Quantum Physics***, WA Benjamin, Inc., Reading, MA, 1976.

[53] S. Popescu and D. Rohrlich, *Quantum nonlocality as an axiom*, ***Foundations of Physics***, vol. 24 (1994), no. 3, pp. 379–385.

[54] V. R. Pratt, *Chu spaces from the representational viewpoint*, ***Annals of Pure and Applied Logic***, vol. 96 (1999), no. 1-3, pp. 319–333.

[55] K. I. Rosenthal, ***Quantales and their applications***, Pitman Research Notes in Mathematics, Longman Scientific & Technical, 1990.

[56] G. Segal, *The definition of conformal field theory*, ***Topology, Geometry and Quantum Field Theory***, London Mathematical Society Lecture Note Series, vol. 308, Cambridge University Press, 2004, pp. 421–577.

[57] P. Selinger, *Dagger compact closed categories and completely positive maps*, ***Quantum Programming Languages***, Electronic Notes in Theoretical Computer Science, vol. 170, Elsevier, 2007, pp. 139–163.

[58] B. Simon, ***Trace Ideals and Their Applications***, Mathematical surveys and monographs, no. 120, American Mathematical Society, 1979.

[59] A. Simpson and G. Plotkin, *Complete axioms for categorical fixed-point operators*, ***Logic in Computer Science***, 2000, pp. 30–41.

[60] R. W. Spekkens, *Evidence for the epistemic view of quantum states: A toy theory*, ***Physical Review A***, vol. 75 (2007), no. 3, p. 032110.

[61] E. Wigner, *On the quantum correction for thermodynamic equilibrium*, ***Physical Review***, vol. 40 (1932), no. 5, p. 749.

DEPARTMENT OF COMPUTER SCIENCE
UNIVERSITY OF OXFORD
WOLFSON BUILDING, PARKS ROAD
OXFORD OX1 3QD
E-mail: samson.abramsky@cs.ox.ac.uk
E-mail: heunen@cs.ox.ac.uk

RELATING OPERATOR SPACES VIA ADJUNCTIONS

BART JACOBS AND JORIK MANDEMAKER

Abstract. This chapter uses categorical techniques to describe relations between various sets of operators on a Hilbert space, such as self-adjoint, positive, density, effect and projection operators. These relations, including various Hilbert-Schmidt isomorphisms of the form $\mathrm{tr}(A-)$, are expressed in terms of dual adjunctions, and maps between them. Of particular interest is the connection with quantum structures, via a dual adjunction between convex sets and effect modules. The approach systematically uses categories of modules, via their description as Eilenberg-Moore algebras of a monad.

§1. **Introduction.** There is a recent exciting line of work connecting research in the semantics of programming languages and logic, and research in the foundations of quantum physics, including quantum computation and logic, see [9] for an overview. This paper fits in that line of work. It concentrates on operators (on Hilbert spaces) and organises and relates these operators according to their algebraic structure. This is to a large extent not more than a systematic presentation of known results and connections in the (modern) language of category theory. However, the approach leads to clarifying results, like Theorem 14 that relates density operators and effects via a dual adjunction between convex sets and effect modules (extending earlier work [25]). It is in line with many other dual adjunctions and dualities that are relevant in programming logics [31, 1, 30]. Indeed, via this dual adjunction we can put the work [11] on quantum weakest preconditions in perspective (see especially Remark 15).

The article begins by describing the familiar sets of operators (bounded, self-adjoint, positive) on a (finite-dimensional) Hilbert space in terms of functors to categories of modules. The dual adjunctions involved are made explicit, basically via dual operation $V \mapsto V^*$, see Section 2. Since the algebraic structure of these sets of operators is described in terms of modules over various semirings, namely over complex numbers $\mathbb{C}$ (for bounded operators), over real numbers $\mathbb{R}$ (for self-adjoint operators), and over non-negative real numbers $\mathbb{R}_{\geq 0}$ (for positive operators), it is useful to have a uniform description of such modules. It is provided in Section 3, via the notion of algebra of

Logic and Algebraic Structures in Quantum Computing
Edited by J. Chubb, A. Eskandarian and V. Harizanov
Lecture Notes in Logic, 45
© 2016, Association for Symbolic Logic

a monad (namely the multiset monad). This abstract description provides (co)limits and the monoidal closed structure of such algebras (from [33]) for free. We then use that convex sets can also be described as such algebras of a monad (namely the distribution monad), and elaborate the connection with effect modules (also known as convex effect algebras, see [36]). In this setting we discuss various 'Gleason-style' correspondences, between projections, effects, and density matrices. We borrow the probabilistic Gelfand duality between (Banach) effect modules and (compact) convex sets from [29] for the final steps in our analysis. This duality formalises the difference between the approaches of Heisenberg (focusing on observables/effects) and Schrödinger (focusing on states), see e.g. [22]. It allows us to reconstruct all sets of operators on a Hilbert space from its projections, see Table 1 for an overview. The main contribution of the paper thus lies in a systematic description.

We should emphasise that the investigations in this paper concentrate on finite-dimensional Hilbert and vector spaces.

1.1. Operator overview. For a (finite-dimensional) Hilbert space H we shall study the following sets of operators $H \to H$.

$$\mathcal{B}(H) \hookleftarrow \mathcal{SA}(H) \hookleftarrow \mathcal{P}os(H) \hookleftarrow \mathcal{E}f(H) \;\; \begin{array}{l} \hookleftarrow \mathcal{P}r(H) \\ \hookleftarrow \mathcal{DM}(H) \end{array} \tag{1}$$

where:

Notation	Description	Structure
$\mathcal{B}(H)$	bounded/continuous linear	vector space over $\mathbb{C}$
$\mathcal{SA}(H)$	self-adjoint $A^\dagger = A$	vector space over $\mathbb{R}$
$\mathcal{P}os(H)$	positive: $A \geq 0$	module over $\mathbb{R}_{\geq 0}$
$\mathcal{E}f(H)$	effect: $0 \leq A \leq I$	effect module over $[0, 1]$
$\mathcal{P}r(H)$	projection: $A^2 = A = A^\dagger$	orthomodular lattice
$\mathcal{DM}(H)$	density: $A \geq 0$ and $\mathrm{tr}(A) = 1$	convex set

The emphasis lies on the 'structure' column. It describes the algebraic structure of the sets of operators that will be relevant here. It is not meant to capture all the structure that is present. For instance, the set $\mathcal{B}(H)$ of endomaps is not only a vector space over the complex numbers, but actually a C^*-algebra.

As is well-known, operators on Hilbert spaces behave in a certain sense as numbers. For instance, by taking H to be the trivial space $\mathbb{C}$ of complex numbers, the diagram (1) becomes:

$$\mathbb{C} \hookleftarrow \mathbb{R} \hookleftarrow \mathbb{R}_{\geq 0} \hookleftarrow [0,1] \;\begin{matrix} \hookleftarrow \{0,1\} \\ \hookleftarrow \{1\} \end{matrix}$$

§2. **Operators and duality.** This section concentrates on the first three sets of operators in (1), namely on $\mathcal{B}(H) \hookleftarrow \mathcal{SA}(H) \hookleftarrow \mathcal{P}os(H)$. It will focus on isomorphisms $V \cong V^*$, for $V = \mathcal{B}(H), \mathcal{SA}(H), \mathcal{P}os(H)$. These isomorphisms turn out to be natural in H, in categories of modules (or vector spaces). This serves as motivation for further investigation of the structures involved, in subsequent sections. Only later will we study the density and effect operators $\mathcal{DM}(H)$ and $\mathcal{E}f(H)$, capturing 'states and statements' in quantum logic. The material in this section thus serves as preparation. It is not new, except possibly for the presentation in terms of maps of adjunctions.

We start by recalling that the category $\textbf{Vect}_{\mathbb{C}}$ of vector spaces over the complex numbers $\mathbb{C}$ carries an involution given by conjugation: for a vector space V we write $\overline{V}$ for the conjugate space, with the same vectors as V, but with scalar multiplication given by $z \bullet_{\overline{V}} x = \overline{z} \bullet_V x$, where the complex number $z \in \mathbb{C}$ has conjugate $\overline{z} \in \mathbb{C}$. This yields an 'involution' endofunctor $\overline{(-)}\colon \textbf{Vect}_{\mathbb{C}} \to \textbf{Vect}_{\mathbb{C}}$ which is the identity on morphisms. A linear map $f : \overline{V} \to W$ is sometimes called *conjugate* linear, because it satisfies $f(z \bullet v) = \overline{z} \bullet f(v)$. Complex conjugation $z \mapsto \overline{z}$ is an example of a conjugate linear (isomorphism) $\overline{\mathbb{C}} \xrightarrow{\cong} \mathbb{C}$ in $\textbf{Vect}_{\mathbb{C}}$. We refer to [6, 16, 27] for more information on involutions in a categorical setting.

We shall write $V \multimap W$ for the 'exponent' vector space of linear maps $V \to W$ between vector spaces V and W. There is the standard correspondence between linear functions $U \to (V \multimap W)$ and $U \otimes V \to W$.

One uses this exponent $\multimap$ to form the dual space $V^* = \overline{V} \multimap \mathbb{C}$. If V is finite-dimensional, say with a basis $e_1, \ldots, e_n$, written in 'ket' notation as $|\,j\,\rangle = e_j$, there is the familiar isomorphism of V with its dual space $V^* = \overline{V} \multimap \mathbb{C}$ given as:

$$\begin{array}{ccc} V & \xrightarrow{\quad\cong\quad} & V^* \cong \overline{V} \multimap \mathbb{C} \\ \left(\sum_j z_j |\,j\,\rangle\right) & \longmapsto & \left(\sum_j z_j \langle\, j\,|\right), \end{array} \tag{2}$$

where the 'bra' $\langle\, j\,|\colon \overline{V} \to \mathbb{C}$ sends a vector $w = \left(\sum_k w_k |\,k\,\rangle\right)$ to its j-th coordinate $\langle j \mid w\rangle = \sum_k \overline{w_k} \langle j \mid k\rangle = \overline{w_j}$. Clearly, this yields an isomorphism $V \xrightarrow{\cong} V^*$, because these functions $\langle\, j\,|$ form a 'dual' basis for V^*. This isomorphism (2) is a famous example of a non-natural mapping, depending on a choice of basis. It will play a crucial role below, where V is a vector space of operators on a Hilbert space.

The mapping $V \mapsto V^* = (\overline{V} \multimap \mathbb{C})$ yields a functor $\mathbf{\mathit{Vect}}_{\mathbb{C}} \to (\mathbf{\mathit{Vect}}_{\mathbb{C}})^{\mathrm{op}}$; for a map $C: H \to K$ we have $C^*: K^* \to H^*$ given by $f \mapsto f \circ \overline{C}$. This functor $(-)^*$ is adjoint to itself, in the sense that there is a bijective correspondence (suggested by the double lines) as on the left below, forming an adjunction as on the right.

$$
\begin{array}{c}
V \longrightarrow (\overline{W} \multimap \mathbb{C}) \\
\hline\hline
V \otimes \overline{W} \longrightarrow \mathbb{C} \cong \overline{\mathbb{C}} \\
\hline\hline
\overline{V} \otimes W \cong \overline{V \otimes \overline{W}} \longrightarrow \mathbb{C} \\
\hline\hline
W \longrightarrow (\overline{V} \multimap \mathbb{C})
\end{array}
\qquad\qquad
\mathbf{\mathit{Vect}}_{\mathbb{C}} \underset{(-)^* = \overline{(-)} \multimap \mathbb{C}}{\overset{(-)^* = \overline{(-)} \multimap \mathbb{C}}{\rightleftarrows}} (\mathbf{\mathit{Vect}}_{\mathbb{C}})^{\mathrm{op}} \quad (\bot)
$$

In the next step, let ***FdHilb*** be the category of finite-dimensional Hilbert spaces with bounded linear maps between them. One can drop the boundedness requirement, because a linear map between finite-dimensional spaces is automatically bounded (i.e. continuous). As is usual, we write $\mathcal{B}(H)$ for the homset of endomaps $H \to H$ in ***FdHilb***. This set $\mathcal{B}(H)$ of "operators on H" is a vector space over $\mathbb{C}$, of dimension n^2 with the outer products $|\, j\,\rangle\langle\, k\,|$, for $j, k \leq n$, as basis—assuming a basis $|\,1\,\rangle, \ldots, |\,n\,\rangle$ for H. Such outer product projections $|\, j\,\rangle\langle\, k\,|$ may be understood as the matrix with only 0s except for a single 1 in the j-th row of the k-th column. In general, an operator $A: H \to H$ can be written as matrix $A = \sum_{j,k} A_{jk} |\, j\,\rangle\langle\, k\,|$, where the matrix entries A_{jk} may be described as $\langle\, j\,|A|\,k\,\rangle$.

The mapping $H \mapsto \mathcal{B}(H)$ is functorial, and will be used here as functor $\mathcal{B}$: ***FdHilb*** $\to$ $\mathbf{\mathit{Vect}}_{\mathbb{C}}$. On a map $C: H \to K$ it yields a linear function $\mathcal{B}(H) \to \mathcal{B}(K)$, written as $\mathcal{B}(C)$, and given by:

$$
\mathcal{B}(C)\Big(H \xrightarrow{A} H\Big) = \Big(K \xrightarrow{C^\dagger} H \xrightarrow{A} H \xrightarrow{C} K\Big). \tag{3}
$$

The operator $C^\dagger = \overline{C}^T$ is the conjugate transpose of C, satisfying $\langle Cv \mid w\rangle = \langle v \mid C^\dagger w\rangle$. It makes ***FdHilb*** into a dagger category, see e.g. [2]. This dagger forms an involution on the vector space $\mathcal{B}(H)$. Also, it is adjoint to itself, as in:

$$
\begin{array}{c}
V \xrightarrow{\ f\ } W \\
\hline\hline
W \underset{f^\dagger}{\longrightarrow} V
\end{array}
\qquad\qquad
\mathbf{\mathit{FdHilb}} \underset{(-)^\dagger}{\overset{(-)^\dagger}{\rightleftarrows}} \mathbf{\mathit{FdHilb}}^{\mathrm{op}} \quad (\bot)
$$

In the next result we apply the duality isomorphism $V \cong V^*$ in (2) for $V = \mathcal{B}(H)$. As we shall see, it involves the trace operation $\mathrm{tr}: \mathcal{B}(H) \to \mathbb{C}$ of which we first recall some basic facts. For $A \in \mathcal{B}(H)$ the trace $\mathrm{tr}(A)$ can be defined as the sum $\sum_j A_{jj}$ of the diagonal matrix values. This definition is independent of the choice of matrix/basis. This trace tr satisfies the following basic properties.

$$
\begin{array}{ll}
\operatorname{tr}(A+B) = \operatorname{tr}(A) + \operatorname{tr}(B) & \\
\operatorname{tr}(zA) = z\operatorname{tr}(A) & \text{where } z \in \mathbb{C} \\
\operatorname{tr}(AB) = \operatorname{tr}(BA) & \text{the so-called cyclic property} \\
\operatorname{tr}(A^T) = \operatorname{tr}(A) & \text{where } (-)^T \text{ is the transpose operation} \\
\operatorname{tr}(A^\dagger) = \overline{\operatorname{tr}(A)} & \text{which results from previous points} \\
\operatorname{tr}(A) \geq 0 & \text{when } A \text{ is positive: } A \geq 0, \text{ i.e. } \langle v \mid Av \rangle \geq 0.
\end{array}
$$

PROPOSITION 1. *For a finite-dimensional Hilbert space H the duality isomorphism* (2) *applied to the vector space $\mathcal{B}(H)$ of endomaps boils down to a trace calculation, namely*:

$$\mathcal{B}(H) \xrightarrow[\cong]{\mathfrak{hs}_{\mathcal{B}}} \mathcal{B}(H)^* = \overline{\mathcal{B}(H)} \multimap \mathbb{C} \quad \text{is} \quad \mathfrak{hs}_{\mathcal{B}}(A) = \lambda B.\, tr(AB^\dagger),$$

where the λ-notation is borrowed from the λ-calculus, and used for function abstraction: $\lambda B.\cdots$ *describes the function* $B \mapsto \cdots$.

This map $\mathcal{B}(H) \xrightarrow{\cong} \mathcal{B}(H)^*$ is independent of the choice of basis. More categorically, it yields a natural isomorphism involving adjoint $(-)^\dagger$ and dual $(-^* = \overline{(-)} \multimap \mathbb{C}$ in:

$$\mathcal{B} \circ (-)^\dagger \xRightarrow[\cong]{\mathfrak{hs}_{\mathcal{B}}} (-)^* \circ \mathcal{B},$$

Pictorially, this $\mathfrak{hs}_{\mathcal{B}}$ is a natural transformation $\mathbf{FdHilb} \overset{\longrightarrow}{\underset{\longrightarrow}{\Downarrow\cong}} \mathbf{Vect}_{\mathbb{C}}^{\mathrm{op}}$ *between the two functors* $\mathbf{FdHilb} \rightrightarrows \mathbf{Vect}_{\mathbb{C}}^{\mathrm{op}}$ *given in*:

$$
\begin{array}{ccccc}
 & & \mathbf{FdHilb}^{\mathrm{op}} & & \\
 & \overset{(-)^\dagger}{\nearrow} & & \overset{\mathcal{B}}{\searrow} & \\
\mathbf{FdHilb} & & & & \mathbf{Vect}_{\mathbb{C}}{}^{\mathrm{op}} \\
 & \underset{\mathcal{B}}{\searrow} & & \underset{(-)^*}{\nearrow} & \\
 & & \mathbf{Vect}_{\mathbb{C}} & &
\end{array}
$$

Moreover, this $\mathfrak{hs}_{\mathcal{B}}$ is part of a map of adjunctions (*see* [34, *IV*,7]) *in the following situation*.

$$
\begin{array}{ccc}
\mathbf{FdHilb} & \underset{(-)^\dagger}{\overset{(-)^\dagger}{\underset{\longleftarrow}{\overset{\longrightarrow}{\perp}}}} & \mathbf{FdHilb}^{\mathrm{op}} \\
{\scriptstyle \mathcal{B}}\downarrow & & \downarrow{\scriptstyle \mathcal{B}} \\
\mathbf{Vect}_{\mathbb{C}} & \underset{(-)^*}{\overset{(-)^*}{\underset{\longleftarrow}{\overset{\longrightarrow}{\perp}}}} & \mathbf{Vect}_{\mathbb{C}}{}^{\mathrm{op}}
\end{array}
$$

The letters 'h' and 's' in the map $\mathfrak{hs}_{\mathcal{B}}$ stand for Hilbert and Schmidt, since the inner product $(A, B) \mapsto \operatorname{tr}(AB^\dagger) = \mathfrak{hs}_{\mathcal{B}}(A)(B)$ is commonly named after them. The subscript $\mathcal{B}$ is added because we shall encounter analogues of this isomorphism for other operators. We drop the subscript when confusion is unlikely.

Proof. If $|\,1\,\rangle, \dots, |\,n\,\rangle$ is a basis for H, then the map $\mathfrak{hs}\colon \mathcal{B}(H) \to \mathcal{B}(H)^*$ becomes, according to (2),

$$\begin{aligned} A = \Big(\sum_{j,k} A_{jk}|\,j\,\rangle\langle\,k\,|\Big) \longmapsto \lambda B.\, \textstyle\sum_{j,k} A_{jk}\overline{B_{jk}} &= \lambda B.\, \textstyle\sum_{j,k} A_{jk}(B^\dagger)_{kj} \\ &= \lambda B.\, \textstyle\sum_{j} (AB^\dagger)_{jj} \\ &= \lambda B.\, \mathrm{tr}(AB^\dagger). \end{aligned}$$

Since the trace of a matrix is basis-independent, so is this isomorphism $\mathfrak{hs}$. Naturality amounts to commutation of the following diagram, for each map $C\colon H \to K$ in ***FdHilb***.

$$\begin{array}{ccc} \mathcal{B}(H) & \xrightarrow[\cong]{\mathfrak{hs}_H} & \mathcal{B}(H)^* = \overline{\mathcal{B}(H)} \multimap \mathbb{C} \\ {\scriptstyle \mathcal{B}(C^\dagger)}\uparrow & & \uparrow{\scriptstyle \mathcal{B}(C)^*} \\ \mathcal{B}(K) & \xrightarrow[\cong]{\mathfrak{hs}_K} & \mathcal{B}(K)^* = \overline{\mathcal{B}(K)} \multimap \mathbb{C} \end{array}$$

This diagram commutes because:

$$\begin{aligned} (\mathcal{B}(C)^* \circ \mathfrak{hs}_K)(A)(B) &= (\mathfrak{hs}_K(A) \circ \mathcal{B}(C))(B) \\ &= \mathfrak{hs}_K(A)(\mathcal{B}(C)(B)) \\ &= \mathfrak{hs}_K(A)(CBC^\dagger) \\ &= \mathrm{tr}(A(CBC^\dagger)^\dagger) \\ &= \mathrm{tr}(ACB^\dagger C^\dagger) \\ &= \mathrm{tr}(C^\dagger ACB^\dagger) \qquad \text{by the cyclic property} \\ &= \mathfrak{hs}_H(C^\dagger AC)(B) \\ &= \mathfrak{hs}_H(\mathcal{B}(C^\dagger)(A))(B) \\ &= (\mathfrak{hs}_H \circ \mathcal{B}(C^\dagger))(A)(B). \end{aligned}$$

Finally, we use the basic fact that, because these $\mathfrak{hs}_H$'s are natural in H and componentwise isomorphisms, the inverses $\mathfrak{hs}_H^{-1}$ are also natural in H, see e.g. [4, Lemma 7.11]. The details of the map of adjunctions in the above diagram are left to the interested reader. ⊣

The remarkable thing about this result is that whereas the maps $V \xrightarrow{\cong} V^*$ in (2) are not natural, the instantiations $\mathfrak{hs}\colon \mathcal{B}(H) \xrightarrow{\cong} \mathcal{B}(H)^*$ are, because they involve a trace calculation that is base-independent. We briefly describe the inverse of $\mathfrak{hs} = \lambda A.\, \mathrm{tr}(A-)\colon \mathcal{B}(H) \xrightarrow{\cong} \mathcal{B}(H)^* = (\overline{\mathcal{B}(H)} \multimap \mathbb{C})$, via a choice of basis $|\,1\,\rangle, \dots, |\,n\,\rangle$ for H. So suppose we have a linear map $f\colon \overline{\mathcal{B}(H)} \to \mathbb{C}$. Define an operator $\mathfrak{hs}^{-1}(f) \in \mathcal{B}(H)$ with matrix entries:

$$\big(\mathfrak{hs}^{-1}(f)\big)_{jk} = f(|\,j\,\rangle\langle\,k\,|). \tag{4}$$

Then we recover f via the trace calculation:

$$\begin{aligned}
\mathfrak{hs}\big(\mathfrak{hs}^{-1}(f)\big)(B) &= \mathrm{tr}\big(\mathfrak{hs}^{-1}(f)B^{\dagger}\big) \\
&= \mathrm{tr}\Big(\sum_{j,k} \mathfrak{hs}^{-1}(f)_{jk}|\, j\,\rangle\langle\, k\,|)B^{\dagger}\Big) \\
&= \sum_{j,k} \mathfrak{hs}^{-1}(f)_{jk}\, \mathrm{tr}\left(|\, j\,\rangle\langle\, k\,| B^{\dagger}\right) \\
&= \sum_{j,k} f(|\, j\,\rangle\langle\, k\,|)\, \mathrm{tr}\left(\langle\, k\,| B^{\dagger} |\, j\,\rangle\right) \\
&= \sum_{j,k} f(|\, j\,\rangle\langle\, k\,|)\, \mathrm{tr}\left((B^{\dagger})_{kj}\right) \\
&= \sum_{j,k} f(|\, j\,\rangle\langle\, k\,|)\overline{B_{jk}} \\
&= \sum_{j,k} f(B_{jk}|\, j\,\rangle\langle\, k\,|), \qquad \text{because } f \text{ is conjugate linear} \\
&= f\Big(\sum_{j,k} B_{jk}|\, j\,\rangle\langle\, k\,|\Big) \\
&= f(B).
\end{aligned}$$

Again, this mapping $f \mapsto A_f$ is independent of a choice of basis, because its inverse $A \mapsto \mathrm{tr}(A-)$ does not depend on such a choice.

Self-adjoint operators. We now restrict ourselves to self-adjoint operators $\mathcal{SA}(H) \hookrightarrow \mathcal{B}(H)$. We recall that an operator $A\colon H \to H$ is called self-adjoint (or Hermitian) if $A^{\dagger} = A$. In terms of matrices this means that $A_{jk} = \overline{A_{kj}}$. In particular, all entries A_{jj} on the diagonal are real numbers, and so is the trace (as sum of these A_{jj}). The set of self-adjoint operators $\mathcal{SA}(H)$ forms a vector space over $\mathbb{R}$. The mapping $H \mapsto \mathcal{SA}(H)$ can be extended to a functor $\mathcal{SA}\colon$ ***Hilb*** $\to$ $\boldsymbol{Vect}_{\mathbb{R}}$, by:

$$\mathcal{SA}(C)\Big(H \xrightarrow{A} H\Big) = \Big(K \xrightarrow{C^{\dagger}} H \xrightarrow{A} H \xrightarrow{C} K\Big).$$

like for $\mathcal{B}$ in (3). This is well-defined since if A is self-adjoint then so is $\mathcal{SA}(C)(A)$, since:

$$\big(\mathcal{SA}(C)(A)\big)^{\dagger} = \big(CAC^{\dagger}\big)^{\dagger} = C^{\dagger\dagger}A^{\dagger}C^{\dagger} = CAC^{\dagger} = \mathcal{SA}(C)(A).$$

There are serveral ways to turn a linear operator into a self-adjoint one. For instance, for each complex number $z \in \mathbb{C}$ and $B \in \mathcal{B}(H)$ we have self-adjoint operators:

$$zB + \overline{z}B^{\dagger} \qquad \text{and} \qquad izB - i\overline{z}B^{\dagger}. \tag{5}$$

In this way we obtain mappings $\mathcal{B}(H) \to \mathcal{SA}(H)$ in $\boldsymbol{Vect}_{\mathbb{R}}$. If the real part $Re(z)$ is non-zero, the mapping:

$$B \longmapsto \frac{1}{2Re(z)}\left(zB + \overline{z}B^{\dagger}\right)$$

is a left-inverse of the inclusion a $\mathcal{SA}(H) \hookrightarrow \mathcal{B}(H)$, making it a split mono.

By moving from $\mathcal{B}$ to $\mathcal{SA}$ we get the following analogue of Proposition 1.

PROPOSITION 2. *For $H \in$ **FdHilb**, the subset $\mathcal{SA}(H) \hookrightarrow \mathcal{B}(H)$ of self-adjoint operators on H is a vector space over $\mathbb{R}$, for which one obtains a natural isomorphism in $\boldsymbol{Vect}_{\mathbb{R}}$:*

$$\mathcal{SA}(H) \xrightarrow[\cong]{\mathfrak{hs}_{\mathcal{SA}}} \mathcal{SA}(H)^{*} = \mathcal{SA}(H) \multimap \mathbb{R} \quad by \quad \mathfrak{hs}_{\mathcal{SA}}(A)(B) = tr(AB). \tag{6}$$

It gives rise to a map of adjunctions:

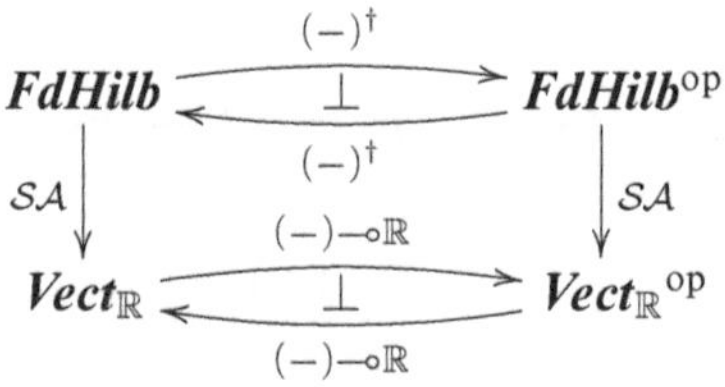

PROOF. If $A, B \colon H \to H$ are self adjoint operators, then $\mathrm{tr}(AB^{\dagger}) = \mathrm{tr}(AB)$ is a real number, since:

$$\overline{\mathrm{tr}(AB)} = \mathrm{tr}\left((AB)^{\dagger}\right) = \mathrm{tr}\left(B^{\dagger}A^{\dagger}\right) = \mathrm{tr}\left(BA\right) = \mathrm{tr}(AB).$$

Conversely, suppose we have a (linear) map $f \colon \mathcal{SA}(H) \to \mathbb{R}$ in $\boldsymbol{Vect}_{\mathbb{R}}$. It can be extended to a function $f' \colon \overline{\mathcal{B}(H)} \to \mathbb{C}$ via

$$f'(B) = \frac{1}{2}\left(f(B + B^{\dagger}) + if(iB - iB^{\dagger})\right)$$

using, as described in (5), that $B + B^{\dagger}$ and $iB - iB^{\dagger}$ are self-adjoint. It is not hard to see that f' preserves sums of operators and satisfies $f'(zB) = \overline{z}f'(B)$. This f' really extends f since in the special case when B is self-adjoint we get $f'(B) = \frac{1}{2}(f(2B) + if(0)) = f(B)$ by linearity.

By Proposition 1 there is a unique $A \in \mathcal{B}(H)$ with:

$$f' = \mathfrak{hs}_{\mathcal{B}}(A) = \mathrm{tr}(A(-)^{\dagger}) \colon \overline{\mathcal{B}(H)} \longrightarrow \mathbb{C}.$$

We now put $\mathfrak{hs}_{\mathcal{SA}}{}^{-1}(f) = \frac{1}{2}(A + A^{\dagger}) \in \mathcal{SA}(H)$, and check for $B \in \mathcal{SA}(H)$:

$$\begin{aligned}
&\mathfrak{hs}_{\mathcal{SA}}\left(\mathfrak{hs}_{\mathcal{SA}}{}^{-1}(f)\right)(B) \\
&\quad = \mathrm{tr}\left(\mathfrak{hs}_{\mathcal{SA}}{}^{-1}(f)B\right) \\
&\quad = \frac{1}{2}\left(\mathrm{tr}(AB) + \mathrm{tr}(A^{\dagger}B)\right)
\end{aligned}$$

$$
\begin{aligned}
&= \frac{1}{2}(\mathrm{tr}(AB^{\dagger}) + \mathrm{tr}((BA)^{\dagger})) && \text{since } B \text{ is self-adjoint}\\
&= \frac{1}{2}(f'(B) + \overline{\mathrm{tr}(BA)})\\
&= \frac{1}{2}(f(B) + \overline{\mathrm{tr}(AB)}) && \text{since } f(B) = f'(B) \text{ when } B \in \mathcal{SA}(H)\\
&= \frac{1}{2}(f(B) + \overline{f(B)})\\
&= \frac{1}{2}(f(B) + f(B)) && \text{because } f(B) \text{ is real valued}\\
&= f(B) && \text{since } f \text{ is linear.}
\end{aligned}
$$

In the other direction, one obtains $\mathfrak{hs}_{\mathcal{SA}}{}^{-1}(\mathfrak{hs}_{\mathcal{SA}}(A)) = A$ by uniqueness.

We prove uniqueness in the self-adjoint case too. Assume a self-adjoint operator $C \in \mathcal{SA}(H)$ also satisfies $f = \mathfrak{hs}_{\mathcal{SA}}(C)\colon \mathcal{SA}(H) \to \mathbb{R}$. We need to prove $C = A = \frac{1}{2}(A' + A'^{\dagger})$. We plan to show $A_{jk} = C_{jk}$ wrt. an arbitrary basis, and thus $A = C$. We prove the equality $A_{jk} = C_{jk}$ in two steps, by proving that both their real and imaginary parts are the same.

$$
\begin{aligned}
Re(C_{jk}) &= \frac{1}{2}\left(C_{jk} + \overline{C_{jk}}\right)\\
&= \frac{1}{2}\left(C_{jk} + (C^{\dagger})_{kj}\right)\\
&= \frac{1}{2}\left(C_{jk} + C_{kj}\right)\\
&= \frac{1}{2}\left(\langle j\,|C|\,k\rangle + \langle k\,|C|\,j\rangle\right)\\
&= \frac{1}{2}(\mathrm{tr}(\langle j\,|C|\,k\rangle) + \mathrm{tr}(\langle k\,|C|\,j\rangle))\\
&= \frac{1}{2}(\mathrm{tr}(C(|\,k\rangle\langle j\,| + |\,j\rangle\langle k\,|)))\\
&= \frac{1}{2}(\mathrm{tr}(A(|\,k\rangle\langle j\,| + |\,j\rangle\langle k\,|)))\\
&\quad \text{by assumption, using that } |\,k\rangle\langle j\,| + |\,j\rangle\langle k\,| \text{ is self-adjoint}\\
&= \ldots \text{ (as before)}\\
&= Re(A_{jk}).
\end{aligned}
$$

Similarly, $Im(C_{jk}) = Im(A_{jk})$, by writing $Im(C_{jk}) = \frac{1}{2}\left(-iC_{jk} + i\overline{C_{jk}}\right)$ and using the self-adjoint operator $-i|\,j\rangle\langle k\,| + i|\,k\rangle\langle j\,|$. ⊣

Implicitly, the proof gives a formula for the inverse $\mathfrak{hs}_{\mathcal{SA}}{}^{-1}$ of the Hilbert-Schmidt map for self-adjoint operators.

Positive operators. An operator $A\colon H \to H$ is called positive if the inner product $\langle Ax \mid x\rangle$ is a non-negative real number, for each $x \in H$. In that case one writes $A \geq 0$. This is equivalent to: $A = BB^\dagger$, for some operator B, and also to: all eigenvalues are non-negative reals. In a spectral decomposition $A = \sum_j \lambda_j |\, j\,\rangle\langle\, j\,|$ a positive operator A has eigenvalues $\lambda_j \in \mathbb{R}_{\geq 0}$ for all j. Hence the trace $\mathrm{tr}(A)$ is a non-negative real number. The set of positive operators on H is written here as $\mathcal{P}os(H)$. It forms a module over the semiring $\mathbb{R}_{\geq 0}$ of non-negative reals since positive operators are closed under addition and under scalar multiplication with $r \in \mathbb{R}_{\geq 0}$. A positive operator is clearly self-adjoint, since $A^\dagger = (BB^\dagger)^\dagger = B^{\dagger\dagger}B^\dagger = BB^\dagger = A$. Thus there are inclusion maps $\mathcal{P}os(H) \hookrightarrow \mathcal{SA}(H) \hookrightarrow \mathcal{B}(H)$. We can describe taking positive operators as a functor $\mathcal{P}os\colon \boldsymbol{Hilb} \to \boldsymbol{Mod}_{\mathbb{R}_{\geq 0}}$ from Hilbert spaces to modules over the non-negative real numbers. The action of $\mathcal{P}os$ on maps is like for $\mathcal{SA}$ and $\mathcal{B}$ in (3), and is well-defined, since if $C\colon H \to K$ in $\boldsymbol{Hilb}$ and $A \geq 0$, then $\mathcal{P}os(C)(A) = CAC^\dagger \geq 0$ since for each $x \in K$,

$$\langle CAC^\dagger x \,|\, x\rangle \;=\langle AC^\dagger x \,|\, C^\dagger x\rangle \;\geq\; 0.$$

As an aside we recall that via positivity one obtains the Löwner order on arbitrary operators A, B, defined as: $A \leq B$ iff $B - A \geq 0$. Thus: $A \leq B$ iff $\exists P \in \mathcal{P}os(H).\, A + P = B$. Hence the spaces $\mathcal{P}os(H) \hookrightarrow \mathcal{SA}(H) \hookrightarrow \mathcal{B}(H)$ are actually ordered (see also [37, 15]).

PROPOSITION 3. *For $H \in \boldsymbol{FdHilb}$, the subset $\mathcal{P}os(H) \hookrightarrow \mathcal{SA}(H)$ of positive operators is a module over the non-negative reals $\mathbb{R}_{\geq 0}$, for which there is a natural isomorphism in $\boldsymbol{Mod}_{\mathbb{R}_{\geq 0}}$:*

$$\mathcal{P}os(H) \xrightarrow[\cong]{\mathfrak{hs}_{\mathcal{P}os}} \mathcal{P}os(H)^* = \mathcal{P}os(H) \multimap \mathbb{R}_{\geq 0} \;\; by \;\; \mathfrak{hs}_{\mathcal{P}os}(A)(B) = tr(AB). \qquad (7)$$

This isomorphism gives rise to a map of adjunctions:

$$\begin{array}{ccc}
\boldsymbol{FdHilb} & \underset{(-)^\dagger}{\overset{(-)^\dagger}{\rightleftarrows}}\;\bot & \boldsymbol{FdHilb}^{\mathrm{op}} \\
\downarrow{\scriptstyle \mathcal{P}os} & & \downarrow{\scriptstyle \mathcal{P}os} \\
\boldsymbol{Mod}_{\mathbb{R}_{\geq 0}} & \underset{(-)\multimap\mathbb{R}_{\geq 0}}{\overset{(-)\multimap\mathbb{R}_{\geq 0}}{\rightleftarrows}}\;\bot & \boldsymbol{Mod}_{\mathbb{R}_{\geq 0}}{}^{\mathrm{op}}
\end{array}$$

PROOF. We first have to check that $\mathrm{tr}(AB) \geq 0$, for $A, B \in \mathcal{P}os(H)$, so that indeed $\mathrm{tr}(A-)$ has type $\mathcal{P}os(H) \to \mathbb{R}_{\geq 0}$. We do so by first writing the spectral decomposition as $A = \sum_j \lambda_j |\, j\,\rangle\langle\, j\,|$, with $\lambda_j \geq 0$. Then:

$$\begin{aligned}
\mathrm{tr}(AB) = \sum_j \lambda_j \mathrm{tr}(|\, j\,\rangle\langle\, j\,|B) = \sum_j \lambda_j \mathrm{tr}(\langle\, j\,|B|\, j\,\rangle) &= \sum_j \lambda_j \mathrm{tr}(\langle Bj \,|\, j\rangle) \\
&= \sum_j \lambda_j \langle Bj \,|\, j\rangle \\
&\geq 0, \quad \text{since } \langle Bj \,|\, j\rangle \geq 0.
\end{aligned}$$

These maps $\mathfrak{hs}_{\mathcal{P}os} = \mathrm{tr}(A-)$ clearly preserve the module structure: additions and scalar multiplication (with a non-negative real number). Next, assume we have a linear map $f : \mathcal{P}os(H) \to \mathbb{R}_{\geq 0}$ in $\boldsymbol{Mod}_{\mathbb{R}_{\geq 0}}$. Like before, we wish to extend it, this time to a map $f' \colon \mathcal{SA}(H) \multimap \mathbb{R}$. If we have an arbitrary self-adjoint operator $B \in \mathcal{SA}(H)$ we can write it as difference $B = B_p - B_n$ of its positive and negative parts $B_p, B_n \in \mathcal{P}os(H)$. One way to do it is to write $B = \sum_j \lambda_j | j \rangle\langle j |$ as spectral decomposition, and to separate the (real-valued) eigenvalues λ_j into negative and non-negative ones. Then take:

$$B_p = \sum_{\lambda_j \geq 0} \lambda_j | j \rangle\langle j | \quad \text{and} \quad B_n = \sum_{\lambda_j < 0} -\lambda_j | j \rangle\langle j |. \tag{8}$$

Now we can define $f'(B) = f(B_p) - f(B_n) \in \mathbb{R}$. This outcome is independent of the choice of B_p, B_n, since if $C, D \in \mathcal{P}os(H)$ also satisfy $B = C - D$, then $B_p + D = C + B_n$, so that by linearity:

$$f(B_p) + f(D) = f(B_p + D) = f(C + B_n) = f(C) + f(B_n),$$

and thus:

$$f'(B) = f(B_p) - f(B_n) = f(C) - f(D).$$

It is not hard to see that the resulting function $f' \colon \mathcal{SA}(H) \to \mathbb{R}$ is linear (in $\boldsymbol{Vect}_{\mathbb{R}}$). Hence by Proposition 2 there is a unique $A = \mathfrak{hs}_{\mathcal{SA}}{}^{-1}(f') \in \mathcal{SA}(H)$ with $f' = \mathfrak{hs}_{\mathcal{SA}}(A) = \mathrm{tr}(A-) \colon \mathcal{SA}(H) \to \mathbb{R}$. For a positive operator $B \in \mathcal{P}os(H)$ we then get $\mathrm{tr}(AB) = f'(B) = f(B) \geq 0$, since $B = B_p$ for such a positive B.

We now write $A = A_p - A_n$ as in (8), where $A_n = \sum_{\lambda_j < 0} -\lambda_j | j \rangle\langle j |$. Projection operators of the form $| j \rangle\langle j |$ are positive, so that we get for each j with $\lambda_j < 0$

$$\begin{aligned} 0 \;\leq\; \mathrm{tr}(A | j \rangle\langle j |) &= \mathrm{tr}(A_p | j \rangle\langle j |) - \mathrm{tr}(A_n | j \rangle\langle j |) \\ &= 0 - (-\lambda_j) \\ &= \lambda_j. \end{aligned}$$

But this is impossible, since we assumed $\lambda_j < 0$. Hence $A_n = 0$, and $A = A_p$ is a positive operator. Thus we have $f = \mathrm{tr}(A-) \colon \mathcal{P}os(H) \to \mathbb{R}_{\geq 0}$, as required, so that we can take $\mathfrak{hs}_{\mathcal{P}os}{}^{-1}(f) = \mathfrak{hs}_{\mathcal{SA}}{}^{-1}(f')$.

We briefly check uniqueness: if $C \in \mathcal{P}os(H)$ also satisfies $f = \mathrm{tr}(C-)$, then for an arbitrary $B \in \mathcal{SA}(H)$,

$$f'(B) = f(B_p) - f(B_n) = \mathrm{tr}(CB_p) - \mathrm{tr}(CB_n) = \mathrm{tr}(C(B_p - B_n)) = \mathrm{tr}(CB).$$

But then $C = A$ by the uniqueness from Proposition 2. ⊣

This concludes our description of the spaces of operators $\mathcal{B}(H) \hookleftarrow \mathcal{SA}(H) \hookleftarrow \mathcal{P}os(H)$ on a (finite-dimensional) Hilbert space H, as naturally self-dual modules. Before we proceed to density operators $\mathcal{DM}(H)$ and effects $\mathcal{E}f(H)$

on H we wish to explore and exploit the similarities between these modules (over $\mathbb{C}$, $\mathbb{R}$, and $\mathbb{R}_{\geq 0}$) in terms of algebras of a monad.

§3. Categories of modules as algebras. We recall that a semiring [20] is like a ring but without an additive inverse. Modules are vector spaces except that the scalars need only be a ring, and not a field. Here we generalise further and will also consider modules over a semiring. In fact we have already done in the previous section, when we talked about positive operators forming a module over the non-negative reals $\mathbb{R}_{\geq 0}$. As we now proceed more systematically, we shall see that such a module over a semiring consists of a commutative monoid of vectors, with scalar multiplication by elements of the semiring. It will be captured as algebra of the multiset monad.

In this section we thus start with the standard description of categories of modules, over a semiring S, as categories of algebras of a monad, namely of the multiset monad $\mathcal{M}_S$ associated with S. We shall be especially interested in the examples $S = \mathbb{R}_{\geq 0}, \mathbb{R}, \mathbb{C}$ giving us a uniform description of the categories of modules in which the spaces of operators $\mathcal{P}os(H)$, $\mathcal{SA}(H)$, $\mathcal{B}(H)$ on a Hilbert space H live. The general theory of monads—see e.g. [34, 5, 35, 7]—gives us certain structure for free, see Theorem 4 below.

The main result in this section, Theorem 6, relates the three spaces of operators $\mathcal{P}os(H)$, $\mathcal{SA}(H)$, $\mathcal{B}(H)$ via free constructions between categories of modules.

To start, let S be a semiring, consisting of a commutative additive monoid $(S, +, 0)$ and a multiplicative monoid $(S, \cdot, 1)$, where multiplication distributes over addition. One can define a "multiset" functor $\mathcal{M}_S \colon \textbf{\textit{Sets}} \to \textbf{\textit{Sets}}$ by:

$$\mathcal{M}_S(X) = \{\varphi \colon X \to S \mid \mathrm{supp}(\varphi) \text{ is finite}\},$$

where $\mathrm{supp}(\varphi) = \{x \in X \mid \varphi(x) \neq 0\}$ is the support of φ. For a function $f \colon X \to Y$ one defines $\mathcal{M}_S(f) \colon \mathcal{M}_S(X) \to \mathcal{M}_S(Y)$ by:

$$\mathcal{M}_S(f)(\varphi)(y) = \textstyle\sum_{x \in f^{-1}(y)} \varphi(x). \tag{9}$$

Such a (finite) multiset $\varphi \in \mathcal{M}_s(X)$ may be written as formal sum $s_1|\, x_1 \,\rangle + \cdots + s_k|\, x_k \,\rangle$ where $\mathrm{supp}(\varphi) = \{x_1, \ldots, x_k\}$ and $s_i = \varphi(x_i) \in S$ describes the "multiplicity" of the element x_i. The ket notation $|\, x_i \,\rangle$ is justified because these elements are vectors, and useful, because it distinguishes x as element of X and as vector in $\mathcal{M}_S(X)$. These formal sum are quotiented by the usual commutativity and associativity relations. Also, the same element $x \in X$ may be counted multiple times, so that $s_1|\, x \,\rangle + s_2|\, x \,\rangle$ is considered to be the same as $(s_1 + s_2)|\, x \,\rangle$. With this formal sum notation one can write the application of $\mathcal{M}_S$ on a map f as $\mathcal{M}_S(f)(\sum_i s_i|\, x_i \,\rangle) = \sum_i s_i|\, f(x_i) \,\rangle$.

This multiset functor is a monad, whose unit $\eta \colon X \to \mathcal{M}_S(X)$ is $\eta(x) = 1|\, x \,\rangle$, and multiplication $\mu \colon \mathcal{M}_S(\mathcal{M}_S(X)) \to \mathcal{M}_S(X)$ is $\mu(\sum_i s_i|\, \varphi_i \,\rangle)(x) = \sum_i s_i \cdot \varphi_i(x)$, where $\cdot$ is multiplication in S.

In order to emphasise that elements of $\mathcal{M}_S(X)$ are *finite* multisets, one may call $\mathcal{M}_S$ the *finitary* multiset monad. In order to include non-finite multisets, one has to assume that suitable infinite sums exist in the underlying semiring S. This is less natural.

For the semiring $S = \mathbb{N}$ one gets the free commutative monoid $\mathcal{M}_{\mathbb{N}}(X)$ on a set X. The monad $\mathcal{M}_{\mathbb{N}}$ is also known as the 'bag' monad, containing ordinary ($\mathbb{N}$-valued) multisets. If $S = \mathbb{Z}$ one obtains the free Abelian group $\mathcal{M}_{\mathbb{Z}}(X)$ on X. The Boolean semiring $2 = \{0, 1\}$ yields the finite powerset monad $\mathcal{P}_{fin} = \mathcal{M}_2$. Here we shall be mostly interested in the cases where S is $\mathbb{R}_{\geq 0}$, $\mathbb{R}$, or $\mathbb{C}$.

An (Eilenberg-Moore) algebra $\alpha\colon \mathcal{M}_S(X) \to X$ for the multiset monad corresponds to a monoid structure on X—given by $x + y = \alpha(1|\,x\,\rangle + 1|\,y\,\rangle)$—together with a scalar multiplication $\bullet\colon S \times X \to X$ given by $s \bullet x = \alpha(s|\,x\,\rangle)$. It preserves the additive structure (of S and of X) in each coordinate separately. This makes X a module, for the semiring S. Conversely, such an S-module structure on a commutative monoid M yields an algebra $\mathcal{M}_S(M) \to M$ by $\sum_i s_i|\,x_i\,\rangle \mapsto \sum_i s_i \bullet x_i$. Thus the category of algebras $\mathit{Alg}(\mathcal{M}_S)$ is equivalent to the category $\boldsymbol{Mod}_S$ of S-modules. When S happens to be a field, this category $\boldsymbol{Mod}_S$ is the category $\boldsymbol{Vect}_S$ of vector spaces over S. Thus we have a uniform description of the three categories of relevance in the previous section, namely:

$$\mathit{Alg}(\mathcal{M}_{\mathbb{R}_{\geq 0}}) = \boldsymbol{Mod}_{\mathbb{R}_{\geq 0}}$$

$$\mathit{Alg}(\mathcal{M}_{\mathbb{R}}) = \boldsymbol{Mod}_{\mathbb{R}} = \boldsymbol{Vect}_{\mathbb{R}} \qquad \mathit{Alg}(\mathcal{M}_{\mathbb{C}}) = \boldsymbol{Mod}_{\mathbb{C}} = \boldsymbol{Vect}_{\mathbb{C}}.$$

We continue this section with a basic result in the theory of monads, which is stated without proof, but with a few subsequent pointers.

THEOREM 4. *Let $\boldsymbol{A}$ be a symmetric monoidal category, which is both complete and cocomplete, and let $T\colon \boldsymbol{A} \to \boldsymbol{A}$ be a monad on $\boldsymbol{A}$. The category $\mathit{Alg}(T)$ of algebras is:*

(a) *also complete, with limits as in $\boldsymbol{A}$;*
(b) *cocomplete as soon as certain special colimits exist in $\mathit{Alg}(T)$, namely colimits of reflexive pairs;*
(c) *symmetric monoidal closed in case these colimits exist and the monad T is symmetric monoidal (commutative), where the free algebra functor $F\colon \boldsymbol{A} \to \mathit{Alg}(T)$ preserves the monoidal structure (i.e. is strong monoidal).* ⊣

A category of algebras is always "as complete" as its underlying category, see e.g. [35, 5]. Cocompleteness always holds for algebras over ***Sets*** and follows from a result of Linton's, see [5, § 9.3, Prop. 4] using the existence of coequalisers of reflexive pairs in ***Sets***. We shall mostly use this result for $\boldsymbol{A} = \boldsymbol{Sets}$, so that we don't have to worry about these special colimits; the monoidal structure on the underlying category ***Sets*** is thus cartesian. Monoidal structure $(I, \otimes)$ in categories of algebras goes back to [33] (see also [23]). The tensor unit

I is simply $F(1)$, for the free algebra functor $F\colon \mathbf{Sets} \to \mathit{Alg}(T)$ and the final (singleton) set 1. The tensor $\otimes$ is obtained as a suitable coequaliser of algebras. Algebra maps $X \otimes Y \to Z$ then correspond to bi-homomorphisms $UX \times UY \to UZ$. In particular, there is a universal bi-homomorphism $\otimes\colon UX \times UY \to U(X \otimes Y)$. The free functor preserves these tensors.

The multiset monad $\mathcal{M}_S$ is symmetric monoidal if S is a (multiplicatively) commutative semiring. In that case categories $\mathbf{Mod}_S$ are monoidal closed, with $S \cong \mathcal{M}_S(1)$ as tensor unit. Maps $M \otimes N \to K$ in $\mathbf{Mod}_S$ correspond to bilinear maps $M \times N \to K$ (linear in each argument separately). The associated exponent is written as $\multimap$, like before.

For modules $M, N \in \mathbf{Mod}_S$ there are obvious correspondences:

$$\frac{\dfrac{M \longrightarrow (N \multimap S)}{M \otimes N \longrightarrow S}}{\dfrac{N \otimes M \longrightarrow S}{N \longrightarrow (M \multimap S)}}$$

This means that there are adjunctions:

$$\mathbf{Mod}_S \underset{(-)\multimap S}{\overset{(-)\multimap S}{\rightleftarrows}} (\mathbf{Mod}_S)^{\mathrm{op}} \quad \bot \tag{10}$$

as used in the previous section.

In summary, we have a sequence of categories of algebras of monads:

$$\begin{array}{ccccc} \mathit{Alg}(\mathcal{M}_{\mathbb{R}_{\geq 0}}) & \longleftarrow & \mathit{Alg}(\mathcal{M}_{\mathbb{R}}) & \longleftarrow & \mathit{Alg}(\mathcal{M}_{\mathbb{C}}) \\ \| & & \| & & \| \\ \mathbf{Mod}_{\mathbb{R}_{\geq 0}} & & \mathbf{Vect}_{\mathbb{R}} & & \mathbf{Vect}_{\mathbb{C}} \end{array} \tag{11}$$

where the maps between them can be understood as arising from maps of monads in the other direction:

$$\mathcal{M}_{\mathbb{R}_{\geq 0}} \Rightarrow \mathcal{M}_{\mathbb{R}} \Rightarrow \mathcal{M}_{\mathbb{C}} \qquad \text{via semiring inclusions} \qquad \mathbb{R}_{\geq 0} \rightarrowtail \mathbb{R} \rightarrowtail \mathbb{C}.$$

This follows from the following general result.

PROPOSITION 5. *A homomorphism of semirings $f\colon S \to S'$, preserving both the additive and multiplicative monoid structures, gives rise to a map of monads $\mathcal{M}_S \Rightarrow \mathcal{M}_{S'}$, by $\left(\sum_j s_j|\, x_j\rangle\right) \mapsto \left(\sum_j f(s_j)|\, x_j\rangle\right)$, and thus to a functor $\mathit{Alg}(\mathcal{M}_{S'}) \to \mathit{Alg}(\mathcal{M}_S)$, by $(\mathcal{M}_{S'}(X) \to X) \longmapsto (\mathcal{M}_S(X) \to \mathcal{M}_{S'}(X) \to X)$. This functor always has a left adjoint.* $\dashv$

The left adjoint exists because categories of modules $\mathit{Alg}(\mathcal{M}_S) = \mathbf{Mod}_S$ are cocomplete; it can be constructed via a coequaliser, see e.g. [23, 29]. Thus, modules over their semirings have the structure of a bifibration [24].

The different spaces of operators $\mathcal{B}(H)$, $\mathcal{SA}(H)$, $\mathcal{P}os(H)$ on a Hilbert space H turn out to be related via free constructions. This was used implicitly in the proofs of Propositions 2 and 3 in the previous section, and also in [8].

THEOREM 6. *Write the left adjoints to the two forgetful functors in* (11) *as*:

$$\mathbf{Mod}_{\mathbb{R}_{\geq 0}} \xrightarrow{\mathcal{R}} \mathbf{Vect}_{\mathbb{R}} \xrightarrow{\mathcal{C}} \mathbf{Vect}_{\mathbb{C}}. \tag{12}$$

For a finite-dimensional Hilbert space H, the canonical inclusion morphisms $\mathcal{P}os(H) \hookrightarrow \mathcal{SA}(H)$ in $\mathbf{Mod}_{\mathbb{R}_{\geq 0}}$, and $\mathcal{SA}(H) \hookrightarrow \mathcal{B}(H)$ in $\mathbf{Vect}_{\mathbb{R}}$ yield via these adjunctions (transposed) maps that turn out to be isomorphisms:

$$\mathcal{R}(\mathcal{P}os(H)) \xrightarrow{\cong} \mathcal{SA}(H) \qquad \textit{and} \qquad \mathcal{C}(\mathcal{SA}(H)) \xrightarrow{\cong} \mathcal{B}(H).$$

Thus we have the following situation of triangles commuting up-to-isomorphism.

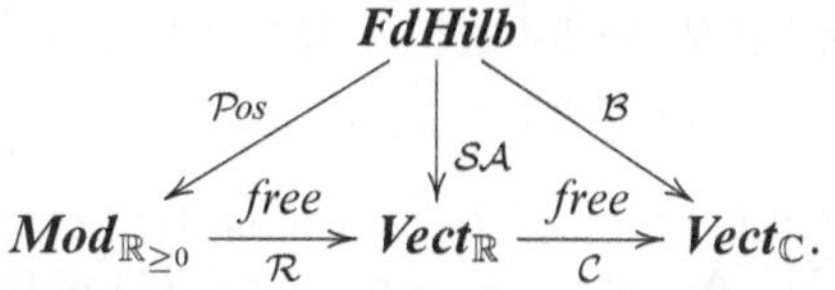

PROOF. The proof uses explicit constructions of the left adjoints $\mathcal{R}$ and $\mathcal{C}$ in (12). A module X over $\mathbb{R}_{\geq 0}$ can be turned into a vector space over $\mathbb{R}$ via the same construction that turns a commutative monoid into a commutative group

$$\mathcal{R}(X) = (X \times X)/\sim \qquad \text{where} \qquad (x_1, x_2) \sim (y_1, y_2)$$
$$\iff \exists z.\, x_1 + y_2 + z = y_1 + x_2 + z.$$

Addition is done componentwise: $[x_1, x_2]+[y_1, y_2] = [x_1+y_1, x_2+y_2]$, minus by reversal: $-[x_1, x_2] = [x_2, x_1]$, and scalar multiplication $\bullet\colon \mathbb{R} \times \mathcal{R}(X) \to \mathcal{R}(X)$ via:

$$r \bullet [x_1, x_2] = \begin{cases} [r \bullet x_1, r \bullet x_2] & \text{if } r \geq 0 \\ [(-r) \bullet x_2, (-r) \bullet x_1] & \text{if } r < 0 \end{cases}$$

(Notice the reversal of the x_i in the second case.)

A vector space X over $\mathbb{R}$ can be turned into a vector space over $\mathbb{C}$, simply via $\mathcal{C}(X) = X \times X$. The additive structure is obtained pointwise, and scalar multiplication $\bullet\colon \mathbb{C} \times \mathcal{C}(X) \to \mathcal{C}(X)$ is done as follows.

$$(a + ib) \bullet (x_1, x_2) = (a \bullet x_1 - b \bullet x_2, b \bullet x_1 + a \bullet x_2).$$

The inclusion morphism $\mathcal{P}os(H) \hookrightarrow \mathcal{SA}(H)$ in $\mathbf{Mod}_{\mathbb{R}_{\geq 0}}$ yields as transpose the map $\varphi\colon \mathcal{R}(\mathcal{P}os(H)) \to \mathcal{SA}(H)$ in $\mathbf{Vect}_{\mathbb{R}}$ given by $\varphi([B_1, B_2]) = B_1 - B_2$. It is surjective since each $A \in \mathcal{SA}(H)$ can be written as $A = A_p - A_n$ for $A_p, A_n \in \mathcal{P}os(H)$ as in (8). Thus $A = \varphi([A_p, A_n])$.

Similarly, the inclusion $\mathcal{SA}(H) \hookrightarrow \mathcal{B}(H)$ in $\mathbf{Vect}_{\mathbb{R}}$ gives rise to a transpose $\psi\colon \mathcal{C}(\mathcal{SA}(H)) \to \mathcal{B}(H)$ in $\mathbf{Vect}_{\mathbb{C}}$, given by $\psi(B_1, B_2) = B_1 + iB_2$. Also this map is surjective since each $A \in \mathcal{B}(H)$ can be written as $A = \frac{1}{2}(A + A^{\dagger}) + \frac{1}{2}i(-iA + iA^{\dagger})$, where $A + A^{\dagger}$ and $-iA + A^{\dagger}$ are self-adjoints. Hence $A = \psi(\frac{1}{2}(A + A^{\dagger}), \frac{1}{2}(-iA + iA^{\dagger}))$. ⊣

§4. Convex sets and effect modules. In the previous section we have seen how the spaces of operators $\mathcal{P}os(H)$, $\mathcal{SA}(H)$, $\mathcal{B}(H)$ fit in the context of modules. The spaces $\mathcal{DM}(H)$ of density operators (states) and $\mathcal{E}f(H)$ of effects (statements/predicates) require more subtle structures that will be introduced in this section, namely convex sets and effect modules. We show that they are related by a dual adjunction, and that there exists a map of adjunctions from Hilbert spaces, like in Section 2.

First we recall the definition of the two sets of operators that are relevant in this section.

$$\mathcal{DM}(H) = \{A \in \mathcal{P}os(H) \mid \mathrm{tr}(A) = 1\}$$
$$\mathcal{E}f(H) = \{A \in \mathcal{P}os(H) \mid A \leq I\},$$

where I is the identity map $H \to H$ and $\leq$ is the Löwner order (described before Proposition 3). A further subset of $\mathcal{E}f(H)$ is the set of projections, given as:

$$\mathcal{P}r(H) = \{A \in \mathcal{B}(H) \mid A^{\dagger} = A = AA\}.$$

For a projection $A \in \mathcal{P}r(H)$ there is an orthosupplement $A^{\perp} \in \mathcal{P}r(H)$ with $A + A^{\perp} = I$. This shows $A \leq I$, since $I - A = A^{\perp}$ is positive.

Before we investigate the algebraic structure of these sets of operators we briefly mention the following alternative formulation of effects. It is used for instance in [11], where these effects A are called predicates; they give a "quantum expectation value" $\mathrm{tr}(AB)$ for a density matrix B (see the map $\mathfrak{hs}_{\mathcal{E}f}$ in Theorem 14 below, elaborated in Remark 15).

LEMMA 7. *A positive operator $A \in \mathcal{P}os(H)$ is an effect if and only if all of its eigenvalues are in* $[0, 1]$.

PROOF. Suppose A is an effect with spectral decomposition $A = \sum_j \lambda_j |j\rangle\langle j|$, where we may assume that the eigenvectors $|\, j\,\rangle$ form an orthonormal basis. The eigenvalues λ_j are necessarily real and positive. They satisfy:

$$\lambda_j = \lambda_j \langle j \mid j\rangle = \langle\, j\,|\lambda_j|\, j\,\rangle = \langle\, j\,|A|\, j\,\rangle \leq \langle\, j\,|I|\, j\,\rangle = \langle j \mid j\rangle = 1.$$

Conversely, assume a positive operator A with spectral decomposition $A = \sum_j \lambda_j |\, j\,\rangle\langle\, j\,|$ where the $|\, j\,\rangle$ form an orthonormal basis and $\lambda_j \in [0, 1]$. Then: $A = \sum_j \lambda_j |\, j\,\rangle\langle\, j\,| \leq \sum_j |\, j\,\rangle\langle\, j\,| = I$. ⊣

4.1. Convex sets. We start with convex sets, and (conveniently) describe them via a monad, so that we can benefit from general results like in Theorem 4. Analogously to the multiset monad one defines the distribution monad

$\mathcal{D}$: ***Sets*** $\to$ ***Sets*** as:

$$\mathcal{D}(X) = \{\varphi\colon X \to [0,1] \mid \mathrm{supp}(\varphi) \text{ is finite and } \textstyle\sum_{x\in X}\varphi(x) = 1\}. \qquad (13)$$

Elements of $\mathcal{D}(X)$ are convex combinations $s_1|\,x_1\,\rangle + \cdots + s_k|\,x_k\,\rangle$, where the probabilities $s_i \in [0,1]$ satisfy $\sum_i s_i = 1$. Unit and multiplication making $\mathcal{D}$ a monad can be defined as for the multiset monad $\mathcal{M}_S$. This multiplication μ is well-defined since:

$$\sum_x \mu\Big(\sum_i s_i\varphi_i\Big)(x) = \sum_x\sum_i s_i\cdot\varphi_i(x) = \sum_i s_i\cdot\Big(\sum_x \varphi_i(x)\Big) = \sum_i s_i = 1.$$

The distribution monad $\mathcal{D}$ is always symmetric monoidal (commutative). Here it is defined for probabilities in the unit interval [0, 1], but the more general structure of an "effect monoid" may be used instead, see [26].

The following result goes back to [39], see also [32, 13, 25].

THEOREM 8. *The category $Alg(\mathcal{D})$ of algebras of the monad $\mathcal{D}$ is the category **Conv** of convex sets with affine maps between them.* ⊣

Here we shall identify such a convex set simply with an algebra $a\colon \mathcal{D}(X) \to X$ of the monad $\mathcal{D}$. It thus consists of a set X in which there is an interpretation $a(\sum_j s_j|\,x_j\,\rangle) \in X$ for each formal convex combination $\sum_j s_j|\,x_j\,\rangle \in \mathcal{D}(X)$. In particular, for each $r \in [0,1]$ and $x, y \in X$ there is an interpretation of the convex sum $rx + (r-1)y$, namely as $a(r|\,x\,\rangle + (1-r)|\,y\,\rangle) \in X$. The unit interval [0, 1] of real numbers is an obvious example of a convex set. Actually, it is a free one since $[0,1] \cong \mathcal{D}(\{0,1\})$. Affine maps preserve such interpretations of convex combinations. We recall that in the present context all such convex combinations involve only finitely many elements x_j.

LEMMA 9. *Let $\boldsymbol{FdHilb}_{\mathrm{Un}}$ be the category of finite-dimensional Hilbert spaces with unitary maps between them. Taking density operators yields a functor $\mathcal{DM}\colon \boldsymbol{FdHilb}_{\mathrm{Un}} \to$ **Conv** $= Alg(\mathcal{D})$.*

PROOF. As is well-known, the set $\mathcal{DM}(H)$ of density operators is convex: given finitely many $A_j \in \mathcal{DM}(H)$ and $r_j \in [0,1]$ with $\sum_j r_j = 1$, the operator $A = \sum_j r_j A_j$ is positive and has trace 1, since:

$$\mathrm{tr}(A) = \mathrm{tr}\Big(\sum_j r_j A_j\Big) = \sum_j r_j \mathrm{tr}(A_j) = \sum_j r_j = 1.$$

Moreover, if $U\colon H \to K$ is unitary—i.e. $UU^\dagger = I$ and (thus) $U^\dagger U = I$, so that $U^\dagger = U^{-1}$—then $\mathcal{DM}(U)(A) = UAU^\dagger\colon K \to K$ is in $\mathcal{DM}(K)$, if $A \in \mathcal{DM}(H)$, since:

$$\mathrm{tr}\big(\mathcal{DM}(U)(A)\big) = \mathrm{tr}(UAU^\dagger) = \mathrm{tr}(U^\dagger UA) = \mathrm{tr}(IA) = \mathrm{tr}(A) = 1. \qquad \dashv$$

The three example categories $Alg(\mathcal{M}_S)$ of interest here—that arise from multiset monads $\mathcal{M}_S$ for $S = \mathbb{R}_{\geq 0}, \mathbb{R}, \mathbb{C}$—are different from the category $Alg(\mathcal{D})$ of convex sets in at least three aspects:

- These categories $Alg(\mathcal{M}_S)$ are dually self-adjoint via the functor $(-) \multimap S$ as in (10).
- They have biproducts, because the monads $\mathcal{M}_S$ are 'additive', see [10].
- The tensor unit in $Alg(\mathcal{D}) = \boldsymbol{Conv}$ is the singleton set 1, since $\mathcal{D}(1) \cong 1$, so that tensors $\otimes$ in $\boldsymbol{Conv}$ have projections (see [23]).

The mapping $X \mapsto \boldsymbol{Conv}(X, [0,1])$, for X a convex set, does not yield an adjunction as in (10), but does lead to an interesting dual adjunction with a category of 'effect modules'. This will be the described in the next subsection.

But first we conclude this part on convex sets with an observation like in Theorem 6. There is an obvious map of monads $\mathcal{D} \Rightarrow \mathcal{M}_{\mathbb{R}_{\geq 0}}$, that gives rise to an inclusion functor $\boldsymbol{Mod}_{\mathbb{R}_{\geq 0}} = Alg(\mathcal{M}_{\mathbb{R}_{\geq 0}}) \to Alg(\mathcal{D}) = \boldsymbol{Conv}$, saying that modules over non-negative reals are convex sets—in a trivial manner. For general reasons, this functor has a left adjoint, that can be described explicitly in terms of a representation contruction that goes back to [38] (see also [25]). This left adjoint $\mathcal{S}\colon \boldsymbol{Conv} \to \boldsymbol{Mod}_{\mathbb{R}_{\geq 0}}$ is given on $X \in \boldsymbol{Conv}$ by:

$$\mathcal{S}(X) = \{0\} + \mathbb{R}_{>0} \times X,$$

with addition for $u, v \in \mathcal{S}(X)$, in trivial cases given by $u + 0 = u = 0 + u$ and:

$$(s, x) + (t, y) = (s + t,\ \frac{s}{s+t}x + \frac{t}{s+t}y).$$

A scalar multiplication $\bullet\colon \mathbb{R}_{\geq 0} \times \mathcal{S}(X) \to \mathcal{S}(X)$ is defined as:

$$s \bullet u = \begin{cases} 0 & \text{if } u = 0 \text{ or } s = 0 \\ (s \cdot t, x) & \text{if } u = (t, x) \text{ and } s \neq 0. \end{cases}$$

This makes $\mathcal{S}(X)$ a module over $\mathbb{R}_{\geq 0}$.

THEOREM 10. *For a finite-dimensional Hilbert space H, transposing the inclusion $\mathcal{DM}(H) \hookrightarrow \mathcal{P}os(H)$ in* $\boldsymbol{Conv}$ *gives an isomorphism $\mathcal{S}(\mathcal{DM}(H)) \xrightarrow{\cong} \mathcal{P}os(H)$ in $\boldsymbol{Mod}_{\mathbb{R}_{\geq 0}}$. In this way one obtains a triangle commuting up-to-isomorphism*:

$$\begin{array}{ccc} & \boldsymbol{FdHilb}_{\mathrm{Un}} & \\ {\scriptstyle \mathcal{DM}} \swarrow & & \searrow {\scriptstyle \mathcal{P}os} \\ Alg(\mathcal{D}) = \boldsymbol{Conv} & \xrightarrow[\mathcal{S}]{\textit{free}} & \boldsymbol{Mod}_{\mathbb{R}_{\geq 0}} = Alg(\mathcal{M}_{\mathbb{R}_{\geq 0}}) \end{array}$$

PROOF. The induced map $\mathcal{S}(\mathcal{DM}(H)) = \{0\} + \mathbb{R}_{\geq 0} \times \mathcal{DM}(H) \xrightarrow{\varphi} \mathcal{P}os(H)$ is given by $0 \mapsto 0$ and $(r, A) \mapsto rA$. It is injective, since if $rA = sB$ for $A, B \in \mathcal{DM}(H)$, then $r = r \cdot \mathrm{tr}(A) = \mathrm{tr}(rA) = \mathrm{tr}(sB) = s \cdot \mathrm{tr}(B) = s$, and thus $A = B$. It is also surjective: since each non-zero $B \in \mathcal{P}os(H)$ can be written as $B = \mathrm{tr}(B)(\frac{B}{\mathrm{tr}(B)}) = \varphi(\mathrm{tr}(B), \frac{B}{\mathrm{tr}(B)})$, where the operator $\frac{B}{\mathrm{tr}(B)}$ has trace 1 by construction. $\dashv$

By combining this result with Theorem 6 we see that each of the spaces of operators $\mathcal{B}(H)$, $\mathcal{SA}(H)$, $\mathcal{P}os(H)$ can be obtained from the space $\mathcal{DM}(H)$ of density operators via free constructions. As we will see in Theorem 14 below, density matrices and effects can be translated back and forth: $\mathcal{E}f(H) \cong$ $\boldsymbol{Conv}(\mathcal{DM}(H), [0, 1])$ and $\mathcal{DM}(H) \cong \boldsymbol{EMod}(\mathcal{E}f(H), [0, 1])$. Hence these density operators and effects are in a sense most fundamental among the operators on a Hilbert space.

4.2. Effect modules. Effect modules are structurally like modules over a semiring. But instead of a semiring of scalars one uses an effect monoid, such as the unit interval [0, 1]. Such an effect monoid is a monoid in the category of effect algebras, just like a semiring is a monoid in the category of commutative monoids. Thus, in order to define an effect module, we need the notion of effect algebra and of monoid in effect algebras. This will be introduced first.

But in order to define an effect algebra, we need the notion of partial commutative monoid (PCM). Before reading the definition of PCM, think of the unit interval [0, 1] with addition $+$. This $+$ is obviously only a partial operation, which is commutative and associative in a suitable sense. This will be formalised next.

A partial commutative monoid (PCM) consists of a set M with a zero element $0 \in M$ and a partial binary operation $\ovee\colon M \times M \to M$ satisfying the three requirements below. They involve the notation $x \perp y$ for: $x \ovee y$ is defined; in that case x, y are called orthogonal.

1. Commutativity: $x \perp y$ implies $y \perp x$ and $x \ovee y = y \ovee x$;
2. Associativity: $y \perp z$ and $x \perp (y \ovee z)$ implies $x \perp y$ and $(x \ovee y) \perp z$ and also $x \ovee (y \ovee z) = (x \ovee y) \ovee z$;
3. Zero: $0 \perp x$ and $0 \ovee x = x$;

For each set X the lift $\{0\} + X$ of X, obtained by adjoining a new element 0, is an example of a PCM, with $u \ovee 0 = u = 0 \ovee u$, and $\ovee$ undefined otherwise. Such structures are also studied under the name ‘partially additive monoid’, see [3].

The notion of effect algebra is due to [18], see also [14] for an overview.

DEFINITION 11. An effect algebra is a PCM $(E, 0, \ovee)$ with an orthosupplement. The latter is a unary operation $(-)^{\perp}\colon E \to E$ satisfying:

1. $x^{\perp} \in E$ is the unique element in E with $x \ovee x^{\perp} = 1$, where $1 = 0^{\perp}$;
2. $x \perp 1 \Rightarrow x = 0$.

A homomorphism $E \to D$ of effect algebras is given by a function $f : E \to D$ between the underlying sets satisfying $f(1) = 1$, and if $x \perp x'$ in E then both $f(x) \perp f(x')$ in D and $f(x \ovee x') = f(x) \ovee f(x')$.

Effect algebras and their homomorphisms form a category, called $\boldsymbol{EA}$.

The unit interval [0, 1] is a PCM with sum of $r, s \in [0, 1]$ defined if $r + s \leq 1$, and in that case $r \ovee s = r + s$. The unit interval is also an effect algebra with $r^{\perp} = 1 - r$. Each orthomodular lattice is an effect algebra, see [14, 17] for more information and examples. In particular, the projections $\mathcal{P}r(H)$ of a Hilbert space form an effect algebra, with $P \perp Q$ iff $P \leq Q^{\perp}$. In [26] a notion of 'convex category' is introduced in which homsets $Hom(X, 2)$ are effect algebras (where $2 = 1 + 1$ and 1 is final). Most importantly in the current setting, the set of effects $\mathcal{E}f(H)$, consisting of positive operators $A \leq I$ is an effect algebra, with $A \perp B$ iff $A + B \leq I$, and in that case $A \ovee B = A + B$; further, $A^{\perp} = I - A$. This yields a functor $\mathcal{E}f\colon \mathbf{FdHilb}_{\mathrm{Un}} \to \mathbf{EA}$.

In [28] it is shown that the category $\mathbf{EA}$ is symmetric monoidal, where morphisms $E \otimes D \to C$ in $\mathbf{EA}$ correspond to 'bimorphisms' $f\colon E \times D \to C$, satisfying $f(1,1) = 1$, and for all $x, x' \in E$ and $y, y' \in D$,

$$\begin{cases} x \perp x' \Longrightarrow f(x,y) \perp f(x',y) \text{ and } f(x \ovee x', y) = f(x,y) \ovee f(x',y) \\ y \perp y' \Longrightarrow f(x,y) \perp f(x,y') \text{ and } f(x, y \ovee y') = f(x,y) \ovee f(x,y'). \end{cases}$$

The tensor unit is the two-element effect algebra $2 = \{0, 1\}$. Since 2 is at the same time initial in $\mathbf{EA}$ we have a 'tensor with coprojections' (see [23] for 'tensors with projections'). One can think of elements of the tensor $E \otimes D$ as finite sums $\ovee_j x_j \otimes y_j$, where one identifies:

$$\begin{array}{cc} 0 \otimes y = 0 & x \otimes 0 = 0 \\ (x \ovee x') \otimes y = (x \otimes y) \ovee (x' \otimes y) & x \otimes (y \ovee y') = (x \otimes y) \ovee (x \otimes y'), \end{array}$$

when $x \perp x'$ and $y \perp y'$.

Example 12. For an arbitrary set X the powerset $\mathcal{P}(X)$ is a Boolean algebra, and so an orthomodular lattice, and thus an effect algebra. For $U, V \in \mathcal{P}(X)$ one has $U \perp V$ iff $U \cap V = \emptyset$ and in that case $U \ovee V = U \cup V$. The tensor product $[0,1] \otimes \mathcal{P}(X)$ of effect algebras is then given by the set of step functions $f\colon X \to [0,1]$; such functions have only finitely many output values. When X is a finite set, say with n elements, then $[0,1] \otimes \mathcal{P}(X) \cong [0,1]^n$, see [21].

As special case we have $[0,1] \otimes \{0,1\} \cong [0,1]$, since $\{0,1\}$ is the tensor unit. One writes $MO(n)$ for the orthomodular lattice with $2n + 2$ elements, namely $0, 1, i, i^{\perp}$, for $1 \leq i \leq n$, with only minimal equations. Thus $MO(0) = \{0,1\}$ and $MO(1) \cong \mathcal{P}(\{0,1\})$, so that $[0,1] \otimes MO(1) \cong [0,1]^2$. It can be shown that $[0,1] \otimes MO(2)$ is an octahedron.

Using this symmetric monoidal structure $(\otimes, 2)$ on $\mathbf{EA}$ we can consider, in a standard way, the category $Mon(\mathbf{EA})$ of monoids in the category $\mathbf{EA}$ of effect algebras. Such monoids are similar to semirings, which are monoids in the category of commutative monoids, i.e. objects of $Mon(\mathbf{CMon})$. A monoid $S \in Mon(\mathbf{EA})$ consists of a set S carrying effect algebra structure $(0, \ovee, (-)^{\perp})$ and a monoid structure, written multiplicatively, as in: $S \otimes S \xrightarrow{\cdot} S \leftarrow 2$. Since 2 is initial, the latter map $S \leftarrow 2$ does not add any structure. The monoid structure on S is thus determined by a bimorphism $\cdot\colon S \times S \to S$ that preserves $\ovee$ in each variable separately and satisfies $1 \cdot x = x = x \cdot 1$.

For such a monoid $S \in Mon(\mathbf{EA})$ we can consider the category $Act_S(\mathbf{EA}) = \mathbf{EMod}_S$ of S-monoid actions (scalar multiplications), or 'effect modules' over S (see [34, VII,§4]). Again this is similar to the situation in Section 3 where the category $\mathbf{Mod}_S$ of modules over a semiring S may be described as the category $Act_S(\mathbf{CMon})$ of commutative monoids with S-scalar multiplication. In this section an effect module $X \in Act_S(\mathbf{EA})$ thus consists of an effect algebra X together with an action (or scalar multiplication) $\bullet \colon S \otimes X \to X$, corresponding to a bimorphism $S \times X \to X$. A homomorphism of effect modules $X \to Y$ consists of a map of effect algebras $f : X \to Y$ preserving scalar multiplication $f(s \bullet x) = s \bullet f(x)$ for all $s \in S$ and $x \in X$.

By completely general reasoning the forgetful functor $\mathbf{EMod}_S \to \mathbf{EA}$ has a left adjoint, given by tensoring with S, as in:

$$\begin{array}{c} \mathbf{EMod}_S = Act_S(\mathbf{EA}) \\ {\scriptstyle S\otimes(-)}\uparrow \dashv \downarrow \\ \mathbf{EA} \end{array} \tag{14}$$

See [34, VII,§4] for details.

The main example of a (commutative) monoid in $\mathbf{EA}$ is the unit interval $[0, 1] \in \mathbf{EA}$ via ordinary multiplication. If $r_1 + r_2 \leq 1$, then we have the familiar distributivity in each variable, as in:

$$s \cdot (r_1 \ovee r_2) = s \cdot (r_1 + r_2) = (s \cdot r_1) + (s \cdot r_2) = (s \cdot r_1) \ovee (s \cdot r_2).$$

We shall be most interested in the associated category $\mathbf{EMod}_{[0,1]} = Act_{[0,1]}(\mathbf{EA})$. In the sequel 'effect module' will mean 'effect module over [0, 1]'. In particular, we shall write $\mathbf{EMod}$ for $\mathbf{EMod}_{[0,1]}$. These effect modules have been studied earlier under the name 'convex effect algebras', see [36]. We prefer the name 'effect module' to emphasise the similarity with ordinary modules.

The effects $\mathcal{E}f(H)$ of a Hilbert space form an example of an effect module, with the usual scalar multiplication $[0, 1] \times \mathcal{E}f(H) \to \mathcal{E}f(H)$. It is not hard to see that this mapping $H \mapsto \mathcal{E}f(H)$ yields a functor $\mathbf{FdHilb}_{\mathrm{Un}} \to \mathbf{EMod}$.

A (dual) adjunction between convex sets and effect algebras is described in [25]. Here it is strengthened to an adjunction between convex sets and effect modules.

Proposition 13. *By "homming into* [0, 1]*" one obtains an adjunction:*

$$\mathbf{Conv} \underset{\mathbf{EMod}(-,[0,1])}{\overset{\mathbf{Conv}(-,[0,1])}{\rightleftarrows}} \mathbf{EMod}^{\mathrm{op}} \quad (\bot)$$

Proof. Given a convex set, the homset $\mathbf{Conv}(X, [0, 1])$ of affine maps is an effect module, with $f \perp g$ iff $\forall x \in X.\, f(x) + g(x) \leq 1$. In that case one defines $f \ovee g = \lambda x \in X.\, f(x) + g(x)$. It is easy to see that this is again an affine function. Similarly, the pointwise scalar product $r \bullet f =$

$\lambda x \in X . r \cdot f(x)$ yields an affine function. This mapping $X \mapsto \boldsymbol{Conv}(X, [0,1])$ gives a contravariant functor since for $h\colon X \to X'$ in $\boldsymbol{Conv}$ pre-composition with h yields a map $(-) \circ h\colon \boldsymbol{Conv}(X', [0,1]) \to \boldsymbol{Conv}(X, [0,1])$ of effect modules.

In the other direction, given an effect module Y, the homset $\boldsymbol{EMod}(Y, [0,1])$ of effect module maps yields a convex set: for a formal convex sum $\sum_j r_j | f_j \rangle$, where $f_j\colon Y \to [0,1]$ in $\boldsymbol{EMod}$, we can define an actual sum $f\colon Y \to [0,1]$ by $f(y) = \sum_j r_j \cdot f_j(y)$. This f forms a map of effect modules. Again, functoriality is obtained via pre-composition.

The dual adjunction between $\boldsymbol{Conv}$ and $\boldsymbol{EMod}$ involves a bijective correspondence that is obtained by swapping arguments, like in (10). For $X \in \boldsymbol{Conv}$ and $Y \in \boldsymbol{EMod}$, we have:

$$\frac{X \xrightarrow{\;f\;} \boldsymbol{EMod}(Y, [0,1]) \qquad \text{in } \boldsymbol{Conv}}{Y \xrightarrow[\;g\;]{} \boldsymbol{Conv}(X, [0,1]) \qquad \text{in } \boldsymbol{EMod}}$$

What needs to be checked is that for a map f of convex sets as indicated, the swapped version $\widehat{f} = \lambda y \in Y . \lambda x \in X . f(x)(y)\colon Y \to \boldsymbol{Conv}(X, [0,1])$ is a map of effect modules—and similarly for g. This is straightforward. ⊣

With this adjunction in place we can give a clearer picture of density matrices and effects, forming a map of adjunctions (like in Section 2. The isomorphisms involved are well-known, see e.g. [8], but the framing of the relevant structure in terms of maps of adjunctions is new.

THEOREM 14. *There is a 'dual adjunction' between convex sets and effect modules as in the lower part of the diagram below. Further, there are natural isomorphisms*:

$$\begin{array}{ccc} \mathcal{E}f(H) \xrightarrow[\cong]{\mathfrak{hs}_{\mathcal{E}f}} \boldsymbol{Conv}\left(\mathcal{DM}(H), [0,1]\right) & \qquad & \mathcal{DM}(H) \xrightarrow[\cong]{\mathfrak{hs}_{\mathcal{DM}}} \boldsymbol{EMod}\left(\mathcal{E}f(H), [0,1]\right) \\ A \longmapsto tr(A-) & & B \longmapsto tr(B-) \end{array} \tag{15}$$

that give rise to a map of adjunctions given by states $\mathcal{DM}$ and statements (effects) $\mathcal{E}f$ in:

$$\begin{array}{ccc} \boldsymbol{FdHilb}_{\mathrm{Un}} & \underset{(-)^{\dagger}}{\overset{(-)^{\dagger}}{\rightleftarrows}} \; \bot & \boldsymbol{FdHilb}_{\mathrm{Un}}{}^{\mathrm{op}} \\ \mathcal{DM} \downarrow & & \downarrow \mathcal{E}f \\ \boldsymbol{Conv} & \underset{\boldsymbol{EMod}(-,[0,1])}{\overset{\boldsymbol{Conv}(-,[0,1])}{\rightleftarrows}} \; \bot & \boldsymbol{EMod}^{\mathrm{op}} \end{array}$$

PROOF. This map of adjunctions involves natural isomorphisms (15), in the categories $\boldsymbol{EMod}$ and $\boldsymbol{Conv}$. We start with the first one, labeled $\mathfrak{hs}_{\mathcal{E}f}$ in (15), and note that it is well-defined: for $A \in \mathcal{E}f(H)$ and $B \in \mathcal{DM}(H)$ one has:

$$\mathfrak{hs}_{\mathcal{E}f}(A)(B) = \mathrm{tr}(AB) \leq \mathrm{tr}(IB) = \mathrm{tr}(B) = 1.$$

Injectivity of $\mathfrak{hs}_{\mathcal{E}f}$ is obtained as follows. Assume $A_1, A_2 \in \mathcal{E}f(H)$ satisfy $\mathfrak{hs}_{\mathcal{E}f}(A_1) = \mathfrak{hs}_{\mathcal{E}f}(A_2)$, i.e. $\mathrm{tr}(A_1-) = \mathrm{tr}(A_2-)\colon \mathcal{DM}(H) \to [0,1]$. for an arbitrary non-zero element $x \in H$ there is a density matrix $B_x = \frac{|\,x\,\rangle\langle\,x\,|}{|x|^2}\colon H \to H$. Thus $\mathrm{tr}(A_1 B_x) = \mathrm{tr}(A_2 B_x)$. Then:

$$\begin{aligned}\langle (A_1 - A_2)x \mid x\rangle &= \langle x \mid A_1 x\rangle - \langle x \mid A_2 x\rangle \\ &= \mathrm{tr}(\langle\, x\,|A_1|\,x\,\rangle) - \mathrm{tr}(\langle\, x\,|A_2|\,x\,\rangle) \\ &= \mathrm{tr}(A_1|\,x\,\rangle\langle\,x\,|) - \mathrm{tr}(A_2|\,x\,\rangle\langle\,x\,|) \\ &= |x|^2(\mathrm{tr}(A_1 B_x) - \mathrm{tr}(A_2 B_x)) \\ &= 0.\end{aligned}$$

Since this equation holds for all $x \in H$, including $x = 0$, we get $A_1 - A_2 \geq 0$, and thus $A_2 \leq A_1$. Similarly $A_1 \leq A_2$, and thus $A_1 = A_2$.

For surjectivity of $\mathfrak{hs}_{\mathcal{E}f}$ assume a morphism of convex sets $h\colon \mathcal{DM}(H) \to [0,1]$. We turn it into a linear map $h'\colon \mathcal{P}os(H) \to \mathbb{R}_{\geq 0}$ in the category $\boldsymbol{Mod}_{\mathbb{R}_{\geq 0}}$ of modules over $\mathbb{R}_{\geq 0}$ via:

$$h'(B) = \begin{cases} 0 & \text{if } B = 0\text{, or equivalently, } \mathrm{tr}(B) = 0 \\ \mathrm{tr}(B) \cdot h\left(\frac{B}{\mathrm{tr}(B)}\right) & \text{otherwise.} \end{cases}$$

This is well-defined since $\mathrm{tr}(\frac{B}{\mathrm{tr}(B)}) = \frac{\mathrm{tr}(B)}{\mathrm{tr}(B)} = 1$. We check linearity of h'. It is easy to see that $h'(rB) = rh'(B)$, for $r \in \mathbb{R}_{\geq 0}$, and for non-zero $B, C \in \mathcal{P}os(H)$ we have:

$$\begin{aligned} & h'(B) + h'(C) \\ &\quad = \mathrm{tr}(B)\cdot h\left(\frac{B}{\mathrm{tr}(B)}\right) + \mathrm{tr}(C)\cdot h\left(\frac{C}{\mathrm{tr}(C)}\right) \\ &\quad = \mathrm{tr}(B+C)\cdot\left(\frac{\mathrm{tr}(B)}{\mathrm{tr}(B+C)}\cdot h\left(\frac{B}{\mathrm{tr}(B)}\right) + \frac{\mathrm{tr}(C)}{\mathrm{tr}(B+C)}\cdot h\left(\frac{C}{\mathrm{tr}(C)}\right)\right) \\ &\quad = \mathrm{tr}(B+C)\cdot\left(h\left(\frac{\mathrm{tr}(B)}{\mathrm{tr}(B+C)}\cdot\frac{B}{\mathrm{tr}(B)} + \frac{\mathrm{tr}(C)}{\mathrm{tr}(B+C)}\cdot\frac{C}{\mathrm{tr}(C)}\right)\right)\end{aligned}$$

since h preserves convex sums and:

$$\frac{\mathrm{tr}(B)}{\mathrm{tr}(B+C)} + \frac{\mathrm{tr}(C)}{\mathrm{tr}(B+C)} = \frac{\mathrm{tr}(B)}{\mathrm{tr}(B)+\mathrm{tr}(C)} + \frac{\mathrm{tr}(C)}{\mathrm{tr}(B)+\mathrm{tr}(C)} = 1$$

$$\begin{aligned} &\quad = \mathrm{tr}(B+C)\cdot h\left(\frac{B+C}{\mathrm{tr}(B+C)}\right) \\ &\quad = h'(B+C).\end{aligned}$$

By Proposition 3 there is a unique $A = \mathfrak{hs}_{\mathcal{P}os}{}^{-1}(h') \in \mathcal{P}os(H)$ with $h' = \mathrm{tr}(A-)\colon \mathcal{P}os(H) \to \mathbb{R}_{\geq 0}$. For a density operator $B \in \mathcal{DM}(H) \hookrightarrow \mathcal{P}os(H)$ we get $\mathrm{tr}(AB) = h'(B) = h(B) \in [0, 1]$. We claim that A is an effect, i.e. is in $\mathcal{E}f(H) \hookrightarrow \mathcal{P}os(H)$. Write $A = \sum_j \lambda_j |\, j\,\rangle\langle\, j\,|$ as spectral decomposition, where the $|\, j\,\rangle$ form an orthonormal basis. By Lemma 7 we need to prove $\lambda_j \leq 1$. Each operator $|\, j\,\rangle\langle\, j\,|$ is a density matrix, and thus $\lambda_j = \mathrm{tr}(A|\, j\,\rangle\langle\, j\,|) = h'(|\, j\,\rangle\langle\, j\,|) = h(|\, j\,\rangle\langle\, j\,|) \leq 1$.

We turn to the second map $\mathfrak{hs}_{\mathcal{DM}}$ in (15). Injectivity is obtained like for $\mathfrak{hs}_{\mathcal{E}f}$, using that each operator $|\, x\,\rangle\langle\, x\,|$ is a projection and thus an effect. For surjectivity assume a map of effect modules $g\colon \mathcal{E}f(H) \to [0, 1]$, we extend it to a linear map $g'\colon \mathcal{P}os(H) \to \mathbb{R}_{\geq 0}$ by:

$$g'(B) = n \cdot g\left(\frac{1}{n}B\right) \qquad \text{where } n \in \mathbb{N} \text{ is such that } \frac{1}{n}B \in \mathcal{E}f(H).$$

Such an n can be found in the following way. Take the spectral decomposition $B = \sum_j \lambda_j |\, j\,\rangle\langle\, j\,|$, where $\lambda_j \geq 0$, because B is positive, and the $|\, j\,\rangle$ form an orthonormal basis. We can find an $n \in \mathbb{N}$ with $\lambda_j \leq n$ for each j. Then $\frac{1}{n}B = \sum_j \frac{1}{n}\lambda_j |\, j\,\rangle\langle\, j\,|$ is an effect by Lemma 7. We also have to check that the definition of g' is independent of the choice of n: if also $\frac{1}{m}B \in \mathcal{E}f(H)$, assume, without loss of generality $m \leq n$; then we use that g is a map of $[0, 1]$-actions:

$$n \cdot g\left(\frac{1}{n}B\right) = n \cdot g\left(\frac{m}{n} \cdot \frac{1}{m}B\right) = n \cdot \frac{m}{n} \cdot g\left(\frac{1}{m}B\right) = m \cdot g\left(\frac{1}{m}B\right).$$

It is easy to see that the map g' is linear. Hence by Proposition 3 there is a (unique) $B = \mathfrak{hs}_{\mathcal{P}os}{}^{-1}(g') \in \mathcal{P}os(H)$ with $g' = \mathrm{tr}(B-)\colon \mathcal{P}os(H) \to \mathbb{R}_{\geq 0}$. Then for $A \in \mathcal{E}f(H)$ we have $g(A) = g'(A) = \mathrm{tr}(BA) \in [0, 1]$. In particular $1 = g(I) = \mathrm{tr}(BI) = \mathrm{tr}(B)$, so that $B \in \mathcal{DM}(H)$.

One of the equations that $\mathfrak{hs}_{\mathcal{E}f}$ and $\mathfrak{hs}_{\mathcal{DM}}$ should satisfy to ensure that we have a map of adjunctions is the following; the other one is similar and left to the reader.

$$\begin{array}{ccc}
\mathcal{DM}(H) & \xrightarrow{\eta = \lambda B.\, \lambda h.\, h(B)} & \textit{\textbf{EMod}}\left(\textit{\textbf{Conv}}(\mathcal{DM}(H), [0,1]), [0,1]\right) \\
 & \searrow\!\!\!= & \downarrow{\scriptstyle (-)\circ \mathfrak{hs}_{\mathcal{E}f}} \\
 & & \textit{\textbf{EMod}}(\mathcal{E}f(H), [0,1]) \\
 & & \downarrow{\scriptstyle \mathfrak{hs}_{\mathcal{DM}}{}^{-1}} \\
 & & \mathcal{DM}(H)
\end{array}$$

This triangle commutes since for $B \in \mathcal{DM}(H)$,

$$\begin{aligned}
\left(\mathfrak{hs}_{\mathcal{DM}}{}^{-1} \circ ((-) \circ \mathfrak{hs}_{\mathcal{E}f}) \circ \eta\right)(B) &= \mathfrak{hs}_{\mathcal{DM}}{}^{-1} \circ \eta(B) \circ \mathfrak{hs}_{\mathcal{E}f} \\
&= \mathfrak{hs}_{\mathcal{DM}}{}^{-1} \circ \lambda A.\, \eta(B)(\mathfrak{hs}_{\mathcal{E}f}(A)) \\
&= \mathfrak{hs}_{\mathcal{DM}}{}^{-1} \circ \lambda A.\, \mathfrak{hs}_{\mathcal{E}f}(A)(B)
\end{aligned}$$

$$
\begin{aligned}
&= \mathfrak{hs}_{\mathcal{DM}}{}^{-1} \circ \lambda A.\, \mathrm{tr}(AB) \\
&= \mathfrak{hs}_{\mathcal{DM}}{}^{-1} \circ \lambda A.\, \mathrm{tr}(BA) \\
&= \mathfrak{hs}_{\mathcal{DM}}{}^{-1} \circ \mathrm{tr}(B-) \\
&= B. \qquad \dashv
\end{aligned}
$$

Remark 15. In [11] a quantum weakest precondition calculus is developed using effects on a finite-dimensional Hilbert space as predicates and density matrices as states. The underlying duality can be made explicit in the current setting. Programs act on states and are thus modeled as "state transformer" maps $\mathcal{DM}(H) \to \mathcal{DM}(K)$. Here we ignore complete positivity aspects and simply consider these state transformers as affine maps, i.e. as maps in the category ***Conv***. Corresponding to such programs there are "predicate transformers" $\mathcal{E}f(K) \to \mathcal{E}f(H)$ going in the opposite direction. Naturally we consider them to be maps of effect modules. The (dual) correspondence between state transformers and predicate transformers can then be derived using the adjunction $\boldsymbol{Conv} \leftrightarrows \boldsymbol{EMod}^{\mathrm{op}}$ from Proposition 13 and the isomorphisms (15) from Theorem 14:

$$
\begin{array}{cl}
\mathcal{DM}(H) \longrightarrow \mathcal{DM}(K) & \text{in } \boldsymbol{Conv} \\
\hline\hline
\mathcal{DM}(H) \longrightarrow \boldsymbol{EMod}(\mathcal{E}f(K), [0,1]) & (15) \\
\hline\hline
\mathcal{E}f(K) \longrightarrow \boldsymbol{Conv}(\mathcal{DM}(H), [0,1]) & (\textit{Prop.}\ 13) \\
\hline\hline
\mathcal{E}f(K) \longrightarrow \mathcal{E}f(H) & (15) \quad \text{in } \boldsymbol{EMod}
\end{array}
$$

Such correspondences form the basis of Dijkstra's seminal work on program correctness, see e.g. [12]. For a state transformer $f\colon \mathcal{DM}(H) \to \mathcal{DM}(K)$ the corresponding predicate transformer $\mathrm{wp}(f,-)\colon \mathcal{E}f(K) \to \mathcal{E}f(H)$ is the "weakest precondition operation". It is given by:

$$
\begin{aligned}
\mathrm{wp}(f, A) &= \mathfrak{hs}_{\mathcal{E}f}{}^{-1}\big(\lambda B \in \mathcal{DM}(H).\, \mathfrak{hs}_{\mathcal{DM}}(f(B))(A)\big) \\
&= \mathfrak{hs}_{\mathcal{E}f}{}^{-1}\big(\lambda B \in \mathcal{DM}(H).\, \mathrm{tr}(f(B)A)\big),
\end{aligned}
$$

where we use the isomorphisms $\mathfrak{hs}_{\mathcal{DM}}\colon \mathcal{DM}(K) \xrightarrow{\cong} \boldsymbol{EMod}(\mathcal{E}f(K), [0,1])$ and $\mathfrak{hs}_{\mathcal{E}f}{}^{-1}\colon \boldsymbol{Conv}(\mathcal{DM}(H), [0,1]) \xrightarrow{\cong} \mathcal{E}f(H)$ from (15). By elaborating the formulas for the matrix entries $\mathrm{wp}(f, A)_{jk}$, the weakest precondition can be computed explicitly (for instance, by a computer algebra tool).

The dual adjunction $\boldsymbol{Conv} \leftrightarrows \boldsymbol{EMod}^{\mathrm{op}}$ from Proposition 13 can be restricted to a (dual) equivalence of categories, giving a probabilistic version of Gelfand duality, see [29]. One obtains an equivalence $\boldsymbol{CCH}_{\mathrm{obs}} \simeq \boldsymbol{BEMod}^{\mathrm{op}}$ between 'observable' convex compact Hausdorff spaces and Banach effect modules. The latter are suitably complete with respect to a definable norm. The map of adjunctions from Theorem 14 then restricts to a map of equivalences:

$$\begin{array}{ccc}
\boldsymbol{FdHilb}_{\mathrm{Un}} & \underset{(-)^{\dagger}}{\overset{(-)^{\dagger}}{\rightleftarrows}}\ \simeq & \boldsymbol{FdHilb}_{\mathrm{Un}}{}^{\mathrm{op}} \\
\downarrow{\scriptstyle \mathcal{DM}} & & \downarrow{\scriptstyle \mathcal{E}f} \\
\boldsymbol{CCH}_{\mathrm{obs}} & \underset{Hom(-,[0,1])}{\overset{Hom(-,[0,1])}{\rightleftarrows}}\ \simeq & \boldsymbol{BEMod}^{\mathrm{op}}
\end{array}$$

We refer to [29] for further details. This equivalence leads to a reformulation of Gleason's Theorem [19]. In original form it says that projections on a Hilbert space H (of dimension at least 3) correspond to measures:

$$\mathcal{DM}(H) \cong \qquad \boldsymbol{EA}\left(\mathcal{P}r(H),\ [0,1]\right).$$

In [29] it is shown that Gleason's theorem is equivalent to:

$$\mathcal{E}f(H) \cong \qquad [0,1] \otimes \mathcal{P}r(H).$$

This says that effects form the free effect module on projections. We can now summarise how the whole edifice of operators on a Hilbert space H can be obtained from its projections $\mathcal{P}r(H)$, see Table 1.

TABLE 1. Various operators on a Hilbert space H, constructed from the projections $\mathcal{P}r(H)$.

Operators	**Formula**	**Description**
effects	$\mathcal{E}f(H) \cong [0,1] \otimes \mathcal{P}r(H)$	Gleason's Theorem
density matrices	$\mathcal{DM}(H) \cong \boldsymbol{EMod}(\mathcal{E}f(H), [0,1])$	Theorem 14
positive operators	$\mathcal{P}os(H) \cong \mathcal{S}(\mathcal{DM}(H))$	Theorem 10
self-adjoint operators	$\mathcal{SA}(H) \cong \mathcal{R}(\mathcal{P}os(H))$	Theorem 6
bounded operators	$\mathcal{B}(H) \cong \mathcal{C}(\mathcal{SA}(H))$	Theorem 6

This concludes our overview of the categorical structure of the various operators on a (finite dimensional) Hilbert space.

Acknowledgements. The first steps of the research underlying this work was carried out during a sabbatical visit of the first author (BJ) to the Quantum Group at Oxford University in April and May 2010. Special thanks, for discussion and/or feedback, go to Bob Coecke, Rick Dejonghe, Chris Heunen, Klaas Landsman, Bas Spitters, and Dusko Pavlović.

REFERENCES

[1] S. Abramsky, *Domain theory in logical form*, ***Annals of Pure and Applied Logic***, vol. 51 (1991), pp. 1–77.

[2] S. Abramsky and B. Coecke, *A categorical semantics of quantum protocols*, ***Handbook of Quantum Logic and Quantum Structures: Quantum Logic*** (K. Engesser, D. M. Gabbay, and D. Lehmann, editors), North Holland, Elsevier, Computer Science Press, 2009, pp. 261–323.

[3] M. A. Arbib and E. G. Manes, ***Algebraic Approaches to Program Semantics***, Texts and Monographs in Computer Science, Springer, Berlin, 1986.

[4] S. Awodey, ***Category Theory***, Oxford Logic Guides, Oxford University Press, 2006.

[5] M. Barr and Ch. Wells, ***Toposes, Triples and Theories***, Springer, Berlin, 1985, revised and corrected version available from URL: www.cwru.edu/artsci/math/wells/pub/ttt.html.

[6] E. J. Beggs and S. Majid, *Bar categories and star operations*, ***Algebras and Representation Theory***, vol. 12 (2009), pp. 103–152.

[7] F. Borceux, ***Handbook of Categorical Algebra***, Encyclopedia of Mathematics, vol. 50, 51 and 52, Cambridge University Press, 1994.

[8] P. Busch, *Quantum states and generalized observables: a simple proof of Gleason's Theorem*, ***Physical Review Letters***, vol. 91 (2003), no. 12, p. 120403.

[9] B. Coecke (editor), ***New Structures for Physics***, Lecture Notes in Physics, vol. 813, Springer, Berlin, 2011.

[10] D. Coumans and B. Jacobs, *Scalars, monads and categories*, ***Compositional Methods in Physics and Linguistics*** (C. Heunen and M. Sadrzadeh, editors), Oxford University Press, 2012, See arxiv.org/abs/1003.0585.

[11] E. D'Hondt and P. Panangaden, *Quantum weakest preconditions*, ***Mathematical Structures in Computer Science***, vol. 16 (2006), pp. 429–451.

[12] E. W. Dijkstra and C. Scholten, ***Predicate Calculus and Program Semantics***, Springer, Berlin, 1990.

[13] E.-E. Doberkat, *Eilenberg-Moore algebras for stochastic relations*, ***Information and Computation***, vol. 204 (2006), pp. 1756–1781, Erratum and addendum in: 206(12):1476–1484, 2008.

[14] A. Dvurečenskij and S. Pulmannová, ***New Trends in Quantum Structures***, Kluwer, Dordrecht, 2000.

[15] A. Edalat, *An extension of Gleason's theorem for quantum computation*, ***International Journal of Theoretical Physics***, vol. 43 (2004), pp. 1827–1840.

[16] J. M. Egger, *On involutive monoidal categories*, ***Theory and Applications of Categories***, vol. 25 (2011), pp. 368–393.

[17] D. J. Foulis, *Observables, states and symmetries in the context of CB-effect algebras*, ***Reports on Mathematical Physics***, vol. 60 (2007), pp. 329–346.

[18] D. J. Foulis and M. K. Bennett, *Effect algebras and unsharp quantum logics*, ***Foundations of Physics***, vol. 24 (1994), pp. 1331–1352.

[19] A. Gleason, *Measures on the closed subspaces of a Hilbert space*, ***Journal of Mathematics and Mechanics***, vol. 6 (1957), pp. 885–893.

[20] J. S. Golan, ***Semirings and their Applications***, Kluwer, 1999.

[21] S. Gudder, *Examples, problems and results in effect algebras*, ***International Journal of Theoretical Physics***, vol. 35 (1996), pp. 2365–2376.

[22] T. Heinosaari and M. Ziman, ***The Mathematical Language of Quantum Theory. From Uncertainty to Entanglement***, Cambridge University Press, 2012.

[23] B. Jacobs, *Semantics of weakening and contraction*, ***Annals of Pure and Applied Logic***, vol. 69 (1994), pp. 73–106.

[24] ———, ***Categorical Logic and Type Theory***, North Holland, Amsterdam, 1999.

[25] ———, *Convexity, duality and effects*, ***IFIP Theoretical Computer Science 2010*** (Boston) (C. S. Calude and V. Sassone, editors), IFIP Advances in Information and Communication Technology, vol. 82, Springer, 2010, pp. 1–19.

[26] ———, *Probabilities, distribution monads and convex categories*, ***Theoretical Computer Science***, vol. 412 (2011), pp. 3323–3336.

[27] ———, *Involutive categories and monoids, with a GNS-correspondence*, ***Foundations of Physics***, vol. 42 (2012), pp. 874–895.

[28] B. Jacobs and J. Mandemaker, *Coreflections in algebraic quantum logic*, ***Foundations of Physics***, (10 May 2012), pp. 932–958, http://dx.doi.org/doi:10.1007/s10701-012-9654-8.

[29] ———, *The expectation monad in quantum foundations*, ***Quantum Physics and Logic (QPL) 2011*** (B. Jacobs, P. Selinger, and B. Spitters, editors), 2012, EPTCS, to appear; see arxiv.org/abs/1112.3805.

[30] B. Jacobs and A. Sokolova, *Exemplaric expressivity of modal logics*, ***Journal of Logic and Computation***, vol. 20 (2010), pp. 1041–1068.

[31] P. T. Johnstone, ***Stone Spaces***, Cambridge Studies in Advanced Mathematics, vol. 3, Cambridge University Press, 1982.

[32] K. Keimel, *The monad of probability measures over compact ordered spaces and its Eilenberg-Moore algebras*, ***Topology and its Applications***, vol. 156 (2008), pp. 227–239.

[33] A. Kock, *Closed categories generated by commutative monads*, ***Journal of the Australian Mathematical Society***, vol. XII (1971), pp. 405–424.

[34] S. Mac Lane, ***Categories for the Working Mathematician***, Springer, Berlin, 1971.

[35] E. G. Manes, ***Algebraic Theories***, Springer, Berlin, 1974.

[36] S. Pulmannová and S. Gudder, *Representation theorem for convex effect algebras*, ***Commentationes Mathematicae Universitatis Carolinae***, vol. 39 (1998), pp. 645–659, available from http://dml.cz/dmlcz/119041.

[37] P. Selinger, *Towards a quantum programming language*, ***Mathematical Structures in Computer Science***, vol. 14 (2004), pp. 527–586.

[38] M. H. Stone, *Postulates for the barycentric calculus*, ***Annali di Matematica Pura ed Applicata***, vol. 29 (1949), pp. 25–30.

[39] T. Swirszcz, *Monadic functors and convexity*, ***Bulletin de l'Acad. Polonaise des Sciences. Sér. des sciences math., astr. et phys.***, vol. 22 (1974), pp. 39–42.

INSTITUTE FOR COMPUTING AND INFORMATION SCIENCES (ICIS)
RADBOUD UNIVERSITY NIJMEGEN, THE NETHERLANDS
E-mail: bart@cs.ru.nl

TOPOS-BASED LOGIC FOR QUANTUM SYSTEMS AND BI-HEYTING ALGEBRAS

ANDREAS DÖRING

Abstract. To each quantum system, described by a von Neumann algebra of physical quantities, we associate a complete bi-Heyting algebra. The elements of this algebra represent contextualised propositions about the values of the physical quantities of the quantum system.

§1. **Introduction.** Quantum logic started with Birkhoff and von Neumann's seminal article [4]. Since then, non-distributive lattices with an orthocomplement (and generalisations thereof) have been used as representatives of the algebra of propositions about the quantum system at hand. There are a number of well-known conceptual and interpretational problems with this kind of 'logic'. For review of standard quantum logic(s), see the article [6].

In the last few years, a different form of logic for quantum systems based on generalised spaces in the form of presheaves and topos theory has been developed by Chris Isham and this author [12, 13, 14, 15, 9, 16, 11, 10]. This new form of logic for quantum systems is based on a certain Heyting algebra $\mathrm{Sub}_{\mathrm{cl}}\underline{\Sigma}$ of clopen, i.e., closed and open subobjects of the spectral presheaf $\underline{\Sigma}$. This generalised space takes the rôle of a state space for the quantum system. (All technical notions are defined in the main text.) In this way, one obtains a well-behaved intuitionistic form of logic for quantum systems which moreover has a topological underpinning.

In this article, we will continue the development of the topos-based form of logic for quantum systems. The main new observation is that the complete Heyting algebra $\mathrm{Sub}_{\mathrm{cl}}\underline{\Sigma}$ of clopen subobjects representing propositions is also a complete co-Heyting algebra. Hence, we relate quantum systems to complete bi-Heyting algebras in a systematic way. This includes two notions of implication and two kinds of negation, as discussed in the following sections.

The plan of the paper is as follows: in section 2, we briefly give some background on standard quantum logic and the main ideas behind the new topos-based form of logic for quantum systems. Section 3 recalls the definitions and main properties of Heyting, co-Heyting and bi-Heyting algebras,

Logic and Algebraic Structures in Quantum Computing
Edited by J. Chubb, A. Eskandarian and V. Harizanov
Lecture Notes in Logic, 45
© 2016, Association for Symbolic Logic

section 4 introduces the spectral presheaf $\underline{\Sigma}$ and the algebra $\mathrm{Sub_{cl}}\,\underline{\Sigma}$ of its clopen subobjects. In section 5, the link between standard quantum logic and the topos-based form of quantum logic is established and it is shown that $\mathrm{Sub_{cl}}\,\underline{\Sigma}$ is a complete bi-Heyting algebra. In section 6, the two kinds of negations associated with the Heyting resp. co-Heyting structure are considered. Heyting-regular and co-Heyting regular elements are characterised and a tentative physical interpretation of the two kinds of negation is given. Section 7 concludes.

Throughout, we assume some familiarity with the most basic aspects of the theory of von Neumann algebras and with basics of category and topos theory. The text is interspersed with some physical interpretations of the mathematical constructions.

§2. Background.

Von Neumann algebras. In this article, we will discuss structures associated with von Neumann algebras, see e.g. [28]. This class of algebras is general enough to describe a large variety of quantum mechanical systems, including systems with symmetries and/or superselection rules. The fact that each von Neumann algebra has 'sufficiently many' projections makes it attractive for quantum logic. More specifically, each von Neumann algebra is generated by its projections, and the spectral theorem holds in a von Neumann algebra, providing the link between self-adjoint operators (representing physical quantities) and projections (representing propositions).

The reader not familiar with von Neumann algebras can always take the algebra $\mathcal{B}(\mathcal{H})$ of all bounded operators on a separable, complex Hilbert space $\mathcal{H}$ as an example of a von Neumann algebra. If the Hilbert space $\mathcal{H}$ is finite-dimensional, $\dim \mathcal{H} = n$, then $\mathcal{B}(\mathcal{H})$ is nothing but the algebra of complex $n \times n$-matrices.

Standard quantum logic. From the perspective of quantum logic, the key thing is that the projection operators in a von Neumann algebra $\mathcal{N}$ form a complete orthomodular lattice $\mathcal{P}(\mathcal{N})$. Starting from Birkhoff and von Neumann [4], such lattices (and various kinds of generalisations, which we don't consider here) have been considered as quantum logics, or more precisely as algebras representing propositions about quantum systems.

The kind of propositions that we are concerned with (at least in standard quantum logic) are of the form "the physical quantity A has a value in the Borel set Δ of real numbers", which is written shortly as "$A \varepsilon \Delta$". These propositions are pre-mathematical entities that refer to the 'world out there'. In standard quantum logic, propositions of the form "$A \varepsilon \Delta$" are represented by projection operators via the spectral theorem. If, as we always assume, the physical quantity A is described by a self-adjoint operator $\hat{A}$ in a given von Neumann algebra $\mathcal{N}$, or is affiliated with $\mathcal{N}$ in the case that $\hat{A}$ is unbounded, then the projection corresponding to "$A \varepsilon \Delta$" lies in $\mathcal{P}(\mathcal{N})$. (For details on the spectral theorem see any book on functional analysis, e.g. [28].)

Following Birkhoff and von Neumann, one then interprets the lattice operations $\wedge, \vee$ in the projection lattice $\mathcal{P}(\mathcal{N})$ as logical connectives between the propositions represented by the projections. In this way, the meet $\wedge$ becomes a conjunction and the join $\vee$ a disjunction. Moreover, the orthogonal complement of a projection, $\hat{P}' := \hat{1} - \hat{P}$, is interpreted as negation. Crucially, meets and joins do not distribute over each other. In fact, $\mathcal{P}(\mathcal{N})$ is a distributive lattice if and only if $\mathcal{N}$ is abelian if and only if all physical quantities considered are mutually compatible, i.e., co-measurable.

Quantum systems always have some incompatible physical quantities, so $\mathcal{N}$ is never abelian and $\mathcal{P}(\mathcal{N})$ is non-distributive. This makes the interpretation of $\mathcal{P}(\mathcal{N})$ as an algebra of propositions somewhat dubious. There are many other conceptual difficulties with quantum logics based on orthomodular lattices, see e.g. [6].

Contexts and coarse-graining. The topos-based form of quantum logic that was established in [13] and developed further in [9, 16, 11, 10] is fundamentally different from standard quantum logic. For some conceptual discussion, see in particular [11]. Two key ideas are *contextuality* and *coarse-graining* of propositions. Contextuality has of course been considered widely in foundations of quantum theory, in particular since Kochen and Specker's seminal paper [29]. Yet, the systematic implementation of contextuality in the language of presheaves is comparatively new. It first showed up in work by Chris Isham and Jeremy Butterfield [21, 23, 24, 27, 25, 26, 22] and was substantially developed by this author and Isham. For recent, related work see also [18, 5, 20, 19] and [1, 2].

Physically, a context is nothing but a set of compatible, i.e., co-measurable physical quantities $(A_i)_{i \in I}$. Such a set determines and is determined by an abelian von Neumann subalgebra V of the non-abelian von Neumann algebra $\mathcal{N}$ of (all) physical quantities. Each physical quantity A_i in the set is represented by some self-adjoint operator $\hat{A}$ in V.[1] In fact, V is generated by the operators $(\hat{A}_i)_{i \in I}$ and the identity $\hat{1}$, in the sense that $V = \{\hat{1}, \hat{A}_i \mid i \in I\}''$, where $\{S\}''$ denotes the double commutant of a set S of operators (see e.g. [28]).[2] Each abelian von Neumann subalgebra V of $\mathcal{N}$ will be called a context, thus identifying the mathematical notion and its physical interpretation. The set of all contexts will be denoted $\mathcal{V}(\mathcal{N})$. Each context provides one of many 'classical perspectives' on a quantum system. We partially order the set of

[1] From here on, we assume that all the physical quantities A_i correspond to *bounded* self-adjoint operators that lie in $\mathcal{N}$. Unbounded self-adjoint operators affiliated with $\mathcal{N}$ can be treated in a straightforward manner.

[2] We will often use the notation V' for a subalgebra of V, which does *not* mean the commutant of V. We trust that this will not lead to confusion.

contexts $\mathcal{V}(\mathcal{N})$ by inclusion. A smaller context $V' \subset V$ represents a 'poorer', more limited classical perspective containing fewer physical quantities than V.

Each context $V \in \mathcal{V}(\mathcal{N})$ has a complete Boolean algebra $\mathcal{P}(V)$ of projections, and $\mathcal{P}(V)$ clearly is a sublattice of $\mathcal{P}(\mathcal{N})$. Propositions "$A \varepsilon \Delta$" about the values of physical quantities A in a (physical) context correspond to projections in the (mathematical) context V. Since $\mathcal{P}(V)$ is a Boolean algebra, there are Boolean algebra homomorphisms $\lambda : \mathcal{P}(V) \to \{0, 1\} \simeq \{\text{false}, \text{true}\}$, which can be seen as truth-value assignments as usual. Hence, there are consistent truth-value assignments for all propositions "$A \varepsilon \Delta$" for propositions about physical quantities *within* a context.

The key result by Kochen and Specker [29] shows that for $\mathcal{N} = \mathcal{B}(\mathcal{H})$, $\dim \mathcal{H} \geq 3$, there are no truth-value assignments for *all* contexts simultaneously in the following sense: there is no family of Boolean algebra homomorphisms $(\lambda_V : \mathcal{P}(V) \to \{0, 1\})_{V \in \mathcal{V}(\mathcal{N})}$ such that if $V' = V \cap \tilde{V}$ is a subcontext of both V and $\tilde{V}$, then $\lambda_{V'} = \lambda_V|_{V'} = \lambda_{\tilde{V}}|_{V'}$, where $\lambda_V|_{V'}$ is the restriction of λ_V to the subcontext V', and analogously $\lambda_{\tilde{V}}|_{V'}$. As Isham and Butterfield realised [23, 27], this means that a certain presheaf has no global elements. In [7], it is shown that this result generalises to all von Neumann algebras without a type I_2-summand.

In the topos approach to quantum theory, propositions are represented not by projections, but by suitable subobjects of a quantum state space. An obstacle arises since the Kochen-Specker theorem seems to show that such a quantum state space cannot exist. Yet, if one considers presheaves instead of sets, this problem can be overcome. The presheaves we consider are 'varying sets' $(\underline{S}_V)_{V \in \mathcal{V}(\mathcal{N})}$, indexed by contexts. Whenever $V' \subset V$, there is a function defined from $\underline{S}_V$, the set associated with the context V, to $\underline{S}_{V'}$, the set associated with the smaller context V'. This makes $\underline{S} = (\underline{S}_V)_{V \in \mathcal{V}(\mathcal{N})}$ into a contravariant, ***Set***-valued functor.

Since by contravariance we go from $\underline{S}_V$ to $\underline{S}_{V'}$, there is a built-in idea of *coarse-graining*. V is the bigger context, containing more self-adjoint operators and more projections than the smaller context V', so we can describe more physics from the perspective of V than from V'. Typically, the presheaves defined over contexts will mirror this fact: the component $\underline{S}_V$ at V contains more information (in a suitable sense, to be made precise in the examples in section 4) than $\underline{S}_{V'}$, the component at V'. Hence, the presheaf map $\underline{S}(i_{V'V}) : \underline{S}_V \to \underline{S}_{V'}$ will implement a form of coarse-graining of the information available at V to that available at V'.

The subobjects of the quantum state space, which will be called the spectral presheaf $\underline{\Sigma}$, form a (complete) Heyting algebra. This is typical, since the subobjects of any object in a topos form a Heyting algebra. Heyting algebras are the algebraic representatives of (propositional) intuitionistic logics. In fact, we will not consider *all* subobjects of the spectral presheaf, but rather the so-called clopen subobjects. The latter also form a complete Heyting algebra, as was first shown in [13] and is proven here in a different way, using Galois connections, in section 5. The difference between the set of all subobjects of the spectral presheaf and the set of clopen subobjects is analogous to the difference between all subsets of a classical state space and (equivalence classes modulo null subsets of) measurable subsets.

Together with the representation of states (which we will not discuss here, but see [13, 8, 11, 17]), these constructions provide an intuitionistic form of logic for quantum systems. Moreover, there is a clear topological underpinning, since the quantum state space $\underline{\Sigma}$ is a generalised space associated with the nonabelian algebra $\mathcal{N}$.

The construction of the presheaf $\underline{\Sigma}$ and its algebra of subobjects incorporates the concepts of contextuality and coarse-graining in a direct way, see sections 4 and 5.

§3. Bi-Heyting algebras. The use of bi-Heyting algebras in superintuitionistic logic was developed by Rauszer [36, 37]. Lawvere emphasised the importance of co-Heyting and bi-Heyting algebras in category and topos theory, in particular in connection with continuum physics [30, 31]. Reyes, with Makkai [35] and Zolfaghari [38], connected bi-Heyting algebras with modal logic. In a recent paper, Bezhanishvili et al. [3] prove (among other things) new duality theorems for bi-Heyting algebras based on bitopological spaces. Majid has suggested to use Heyting and co-Heyting algebras within a tentative representation-theoretic approach to the formulation of quantum gravity [33, 34].

As far as we are aware, nobody has connected quantum systems and their logic with bi-Heyting algebras before.

The following definitions are standard and can be found in various places in the literature; see e.g. [38].

A *Heyting algebra* H is a lattice with bottom element 0 and top element 1 which is a cartesian closed category. In other words, H is a lattice such that for any two elements $A, B \in H$, there exists an exponential $A \Rightarrow B$, called the *Heyting implication (from A to B)*, which is characterised by the adjunction

$$C \leq (A \Rightarrow B) \quad \text{if and only if} \quad C \wedge A \leq B. \tag{3.1}$$

This means that the product (meet) functor $A \wedge \_ : H \to H$ has a right adjoint $A \Rightarrow \_ : H \to H$ for all $A \in H$.

It is straightforward to show that the underlying lattice of a Heyting algebra is distributive. If the underlying lattice is complete, then the adjoint functor theorem for posets shows that for all $A \in H$ and all families $(A_i)_{i \in I} \subseteq H$, the following infinite distributivity law holds:

$$A \wedge \bigvee_{i \in I} A_i = \bigvee_{i \in I} (A \wedge A_i). \tag{3.2}$$

The *Heyting negation* is defined as

$$\begin{aligned} \neg : H &\longrightarrow H^{\mathrm{op}} \\ A &\longmapsto (A \Rightarrow 0). \end{aligned} \tag{3.3}$$

The defining adjunction shows that $\neg A = \bigvee\{B \in H \mid A \wedge B = 0\}$, i.e., $\neg A$ is the largest element in H such that $A \wedge \neg A = 0$. Some standard properties of the Heyting negation are:

$$A \leq B \text{ implies } \neg A \geq \neg B, \tag{3.4}$$

$$\neg\neg A \geq A, \tag{3.5}$$

$$\neg\neg\neg A = \neg A \tag{3.6}$$

$$\neg A \vee A \leq 1. \tag{3.7}$$

Interpreted in logical terms, the last property on this list means that in a Heyting algebra the law of excluded middle need not hold: in general, the disjunction between a proposition represented by $A \in H$ and its Heyting negation (also called Heyting complement, or pseudo-complement) $\neg A$ can be smaller than 1, which represents the trivially true proposition. Heyting algebras are algebraic representatives of (propositional) intuitionistic logics.

A canonical example of a Heyting algebra is the topology $\mathcal{T}$ of a topological space $(X, \mathcal{T})$, with unions of open sets as joins and intersections as meets.

A *co-Heyting algebra* (also called *Brouwer algebra*) J is a lattice with bottom element 0 and top element 1 such that the coproduct (join) functor $A \vee \_ : J \to J$ has a left adjoint $A \Leftarrow \_ : J \to J$, called the *co-Heyting implication (from A)*. It is characterised by the adjunction

$$(A \Leftarrow B) \leq C \text{ iff } A \leq B \vee C. \tag{3.8}$$

It is straightforward to show that the underlying lattice of a co-Heyting algebra is distributive. If the underlying lattice is complete, then the adjoint functor theorem for posets shows that for all $A \in J$ and all families $(A_i)_{i \in I} \subseteq J$, the following infinite distributivity law holds:

$$A \vee \bigwedge_{i \in I} A_i = \bigwedge_{i \in I} (A \vee A_i). \tag{3.9}$$

The *co-Heyting negation* is defined as

$$\sim : J \longrightarrow J^{\mathrm{op}} \tag{3.10}$$

$$A \longmapsto (1 \Leftarrow A). \tag{3.11}$$

The defining adjunction shows that $\sim A = \bigwedge\{B \in J \mid A \vee B = 1\}$, i.e., $\sim A$ is the smallest element in J such that $A \vee \sim A = 1$. Some properties of the co-Heyting negation are:

$$A \leq B \text{ implies } {\sim}A \geq {\sim}B, \tag{3.12}$$

$${\sim}{\sim}A \leq A, \tag{3.13}$$

$${\sim}{\sim}{\sim}A = {\sim}A \tag{3.14}$$

$${\sim}A \wedge A \geq 0. \tag{3.15}$$

Interpreted in logical terms, the last property on this list means that in a co-Heyting algebra the law of noncontradiction does not hold: in general, the conjunction between a proposition represented by $A \in J$ and its co-Heyting negation ${\sim}A$ can be larger than 0, which represents the trivially false proposition. Co-Heyting algebras are algebraic representatives of (propositional) paraconsistent logics.

We will not discuss paraconsistent logic in general, but in the final section 6, we will give and interpretation of the co-Heyting negation showing up in the form of quantum logic to be presented in this article.

A canonical example of a co-Heyting algebra is given by the closed sets $\mathcal{C}$ of a topological space, with unions of closed sets as joins and intersections as meets.

Of course, Heyting algebras and co-Heyting algebras are dual notions. The opposite H^{op} of a Heyting algebra is a co-Heyting algebra and vice versa.

A *bi-Heyting algebra* K is a lattice which is a Heyting algebra and a co-Heyting algebra. For each $A \in K$, the functor $A \wedge \_ : K \to K$ has a right adjoint $A \Rightarrow \_ : K \to K$, and the functor $A \vee \_ : K \to K$ has a left adjoint $K \Leftarrow \_ : K \to K$. A bi-Heyting algebra K is called *complete* if it is complete as a Heyting algebra and complete as a co-Heyting algebra.

A canonical example of a bi-Heyting algebra is a Boolean algebra $\mathcal{B}$. (Note that by Stone's representation theorem, each Boolean algebra is isomorphic to the algebra of clopen, i.e., closed and open, subsets of its Stone space. This gives the connection with the topological examples.) In a Boolean algebra, we have for the Heyting negation that, for all $A \in \mathcal{B}$,

$$A \vee \neg A = 1, \tag{3.16}$$

which is the characterising property of the co-Heyting negation. In fact, in a Boolean algebra, $\neg = {\sim}$.

§4. The spectral presheaf of a von Neumann algebra and clopen subobjects. With each von Neumann algebra $\mathcal{N}$, we associate a particular presheaf, the so-called spectral presheaf. A distinguished family of subobjects, the so-called clopen subobjects, are defined and their interpretation is given: clopen subobjects can be seen as families of local propositions, compatible with respect to coarse-graining. The constructions presented here summarise those discussed in [12, 13, 16, 17].

Let $\mathcal{N}$ be a von Neumann algebra, and let $\mathcal{V}(\mathcal{N})$ be the set of its abelian von Neumann subalgebras, partially ordered under inclusion. We only consider subalgebras $V \subset \mathcal{N}$ which have the same unit element as $\mathcal{N}$, given by the identity operator $\hat{1}$ on the Hilbert space on which $\mathcal{N}$ is represented. By convention, we exclude the trivial subalgebra $V_0 = \mathbb{C}\hat{1}$ from $\mathcal{V}(\mathcal{N})$. (This will play an important rôle in the discussion of the Heyting negation in section 6.) The poset $\mathcal{V}(\mathcal{N})$ is called the *context category of the von Neumann algebra* $\mathcal{N}$.

For $V', V \in \mathcal{V}(\mathcal{N})$ such that $V' \subset V$, the inclusion $i_{V'V} : V' \hookrightarrow V$ restricts to a morphism $i_{V'V}|_{\mathcal{P}(V')} : \mathcal{P}(V') \to \mathcal{P}(V)$ of complete Boolean algebras. In particular, $i_{V'V}$ preserves all meets, hence it has a left adjoint

$$\begin{aligned} \delta^o_{V,V'} : \mathcal{P}(V) &\longrightarrow \mathcal{P}(V') \\ \hat{P} &\longmapsto \delta^o_{V,V'}(\hat{P}) = \bigwedge\{\hat{Q} \in \mathcal{P}(V') \mid \hat{Q} \geq \hat{P}\} \end{aligned} \tag{4.1}$$

that preserves all joins, i.e., for all families $(\hat{P}_i)_{i\in I} \subseteq \mathcal{P}(V)$ it holds that

$$\delta^o_{V,V'}\left(\bigvee_{i\in I} \hat{P}_i\right) = \bigvee_{i\in I} \delta^o_{V,V'}(\hat{P}_i), \tag{4.2}$$

where the join on the left hand side is taken in $\mathcal{P}(V)$ and the join on the right hand side is in $\mathcal{P}(V')$. If $W \subset V' \subset V$, then $\delta^o_{V,W} = \delta^o_{V',W} \circ \delta^o_{V,V'}$, obviously.

We note that distributivity of the lattices $\mathcal{P}(V)$ and $\mathcal{P}(V')$ plays no rôle here. If $\mathcal{N}$ is a von Neumann algebra and $\mathcal{M}$ is any von Neumann subalgebra such that their unit elements coincide, $\hat{1}_{\mathcal{M}} = \hat{1}_{\mathcal{N}}$, then there is a join-preserving map

$$\begin{aligned} \delta^o_{\mathcal{N},\mathcal{M}'} : \mathcal{P}(\mathcal{N}) &\longrightarrow \mathcal{P}(\mathcal{M}) \\ \hat{P} &\longmapsto \delta^o_{\mathcal{N},\mathcal{M}}(\hat{P}) = \bigwedge\{\hat{Q} \in \mathcal{P}(\mathcal{M}) \mid \hat{Q} \geq \hat{P}\}. \end{aligned} \tag{4.3}$$

Recall that the Gel'fand spectrum $\Sigma(A)$ of an abelian C^*-algebra A is the set of algebra homomorphisms $\lambda : A \to \mathbb{C}$. Equivalently, the elements of the Gel'fand spectrum $\Sigma(A)$ are the pure states of A. The set $\Sigma(A)$ is given the relative weak*-topology (as a subset of the dual space of A), which makes it into a compact Hausdorff space. By Gel'fand-Naimark duality, $A \simeq C(\Sigma(A))$, that is, A is isometrically $*$-isomorphic to the abelian C^*-algebra $C(\Sigma(A))$ of continuous, complex-valued functions on $\Sigma(A)$, equipped with the supremum norm. If A is an abelian von Neumann algebra, then $\Sigma(A)$ is extremely disconnected.

We now define the main object of interest:

DEFINITION 1. Let $\mathcal{N}$ be a von Neumann algebra. The *spectral presheaf* $\underline{\Sigma}$ *of* $\mathcal{N}$ is the presheaf over $\mathcal{V}(\mathcal{N})$ given

(a) on objects: for all $V \in \mathcal{V}(\mathcal{N})$, $\underline{\Sigma}_V := \Sigma(V)$, the Gel'fand spectrum of V,

(b) on arrows: for all inclusions $i_{V'V} : V' \hookrightarrow V$,

$$\begin{aligned} \underline{\Sigma}(i_{V'V}) : \underline{\Sigma}_V &\longrightarrow \underline{\Sigma}_{V'} \\ \lambda &\longmapsto \lambda|_{V'}. \end{aligned} \tag{4.4}$$

The restriction maps $\underline{\Sigma}(i_{V'V})$ are well-known to be continuous, surjective maps with respect to the Gel'fand topologies on $\underline{\Sigma}_V$ and $\underline{\Sigma}_{V'}$, respectively. They are also open and closed, see e.g. [13].

We equip the spectral presheaf with a distinguished family of subobjects (which are subpresheaves):

DEFINITION 2. A subobject $\underline{S}$ of $\underline{\Sigma}$ is called *clopen* if for each $V \in \mathcal{V}(\mathcal{N})$, the set $\underline{S}_V$ is a clopen subset of the Gel'fand spectrum $\underline{\Sigma}_V$. The set of all clopen subobjects of $\underline{\Sigma}$ is denoted as $\mathrm{Sub}_{\mathrm{cl}}\,\underline{\Sigma}$.

The set $\mathrm{Sub}_{\mathrm{cl}}\,\underline{\Sigma}$, together with the lattice operations and bi-Heyting algebra structure defined below, is the algebraic implementation of the new topos-based form of quantum logic. The elements $\underline{S} \in \mathrm{Sub}_{\mathrm{cl}}\,\underline{\Sigma}$ represent propositions about the values of the physical quantities of the quantum system. The most direct connection with propositions of the form "$A \varepsilon \Delta$" is given by the map called daseinisation, see Def. 3 below.

We note that the concept of contextuality (cf. section 2) is implemented by this construction, since $\underline{\Sigma}$ is a presheaf over the context category $\mathcal{V}(\mathcal{N})$. Moreover, coarse-graining is mathematically realised by the fact that we use subobjects of presheaves. In the case of $\underline{\Sigma}$ and its clopen subobjects, this means the following: for each context $V \in \mathcal{V}(\mathcal{N})$, the component $\underline{S}_V \subseteq \underline{\Sigma}_V$ represents a *local proposition* about the value of some physical quantity. If $V' \subset V$, then $\underline{S}_{V'} \supseteq \underline{\Sigma}(i_{V'V})(\underline{S}_V)$ (since $\underline{S}$ is a subobject), so $\underline{S}_{V'}$ represents a local proposition at the smaller context $V' \subset V$ that is *coarser* than (i.e., a consequence of) the local proposition represented by $\underline{S}_V$.

A clopen subobject $\underline{S} \in \mathrm{Sub}_{\mathrm{cl}}\,\underline{\Sigma}$ can hence be interpreted as a collection of *local propositions*, one for each context, such that smaller contexts are assigned coarser propositions.

Clearly, the definition of clopen subobjects makes use of the Gel'fand topologies on the components $\underline{\Sigma}_V$, $V \in \mathcal{V}(\mathcal{N})$. We note that for each abelian von Neumann algebra V (and hence for each context $V \in \mathcal{V}(\mathcal{N})$), there is an isomorphism of complete Boolean algebras

$$\begin{aligned} \alpha_V : \mathcal{P}(V) &\longrightarrow \mathcal{C}l(\underline{\Sigma}_V) \\ \hat{P} &\longmapsto \{\lambda \in \underline{\Sigma}_V \mid \lambda(\hat{P}) = 1\}. \end{aligned} \tag{4.5}$$

Here, $\mathcal{C}l(\underline{\Sigma}_V)$ denotes the clopen subsets of $\underline{\Sigma}_V$.

There is a purely order-theoretic description of $\mathrm{Sub}_{\mathrm{cl}}\,\underline{\Sigma}$: let

$$\mathcal{P} := \prod_{V \in \mathcal{V}(\mathcal{N})} \mathcal{P}(V) \tag{4.6}$$

be the set of choice functions $f : \mathcal{V}(\mathcal{N}) \to \coprod_{V \in \mathcal{V}(\mathcal{N})} \mathcal{P}(V)$, where $f(V) \in \mathcal{P}(V)$ for all $V \in \mathcal{V}(\mathcal{N})$. Equipped with pointwise operations, $\mathcal{P}$ is a complete Boolean algebra, since each $\mathcal{P}(V)$ is a complete Boolean algebra. Consider the subset $\mathcal{S}$ of $\mathcal{P}$ consisting of those functions for which $V' \subset V$ implies $f(V') \geq f(V)$ (this comparison is taken in $\mathcal{P}(V)$, into which $\mathcal{P}(V')$ can be included). The subset $\mathcal{S}$ is closed under all meets and joins (in $\mathcal{P}$), and clearly, $\mathcal{S} \simeq \mathrm{Sub_{cl}}\,\underline{\Sigma}$.

We define a partial order on $\mathrm{Sub_{cl}}\,\underline{\Sigma}$ in the obvious way:

$$\forall \underline{S}, \underline{T} \in \mathrm{Sub_{cl}}\,\underline{\Sigma} : \quad \underline{S} \leq \underline{T} :\Longleftrightarrow (\forall V \in \mathcal{V}(\mathcal{N}) : \underline{S}_V \subseteq \underline{T}_V). \tag{4.7}$$

We define the corresponding (complete) lattice operations in a stagewise manner, i.e., at each context $V \in \mathcal{V}(\mathcal{N})$ separately: for any family $(\underline{S}_i)_{i \in I}$,

$$\forall V \in \mathcal{V}(\mathcal{N}) : \left(\bigwedge_{i \in I} \underline{S}_i \right)_V := \mathrm{int}\left(\bigcap_{i \in I} \underline{S}_{i;V} \right), \tag{4.8}$$

where $\underline{S}_{i;V} \subseteq \underline{\Sigma}_V$ is the component at V of the clopen subobject $\underline{S}_i$. Note that the lattice operation is not just componentwise set-theoretic intersection, but rather the interior (with respect to the Gel'fand topology) of the intersection. This guarantees that one obtains clopen subsets at each stage V, not just closed ones. Analogously,

$$\forall V \in \mathcal{V}(\mathcal{N}) : \left(\bigvee_{i \in I} \underline{S}_i \right)_V := \mathrm{cl}\left(\bigcup_{i \in I} \underline{S}_{i;V} \right), \tag{4.9}$$

where the closure of the union is necessary in order to obtain clopen sets, not just open ones. The fact that meets and joins are not given by set-theoretic intersections and unions also means that $\mathrm{Sub_{cl}}\,\underline{\Sigma}$ is not a sub-Heyting algebra of the Heyting algebra $\mathrm{Sub}\,\underline{\Sigma}$ of all subobjects of the spectral presheaf. The difference between $\mathrm{Sub}\,\underline{\Sigma}$ and $\mathrm{Sub_{cl}}\,\underline{\Sigma}$ is analogous to the difference between the power set PX of a set X and the complete Boolean algebra BX of measurable subsets (with respect to some measure) modulo null sets. For results on measures and quantum states from the perspective of the topos approach, see [8, 17].

In section 5, we will show that $\mathrm{Sub_{cl}}\,\underline{\Sigma}$ is a complete bi-Heyting algebra.

Example 1. For illustration, we consider a simple example: let $\mathcal{N}$ be an *abelian* von Neumann of diagonal matrices in 3 dimensions. This is given by

$$\mathcal{N} := \mathbb{C}\hat{P}_1 + \mathbb{C}\hat{P}_2 + \mathbb{C}\hat{P}_3, \tag{4.10}$$

where $\hat{P}_1, \hat{P}_2, \hat{P}_3$ are pairwise orthogonal rank-1 projections on a Hilbert space of dimension 3. The projection lattice $\mathcal{P}(\mathcal{N})$ of $\mathcal{N}$ has 8 elements,

$$\mathcal{P}(\mathcal{N}) = \{\hat{0}, \hat{P}_1, \hat{P}_2, \hat{P}_3, \hat{P}_1 + \hat{P}_2, \hat{P}_1 + \hat{P}_3, \hat{P}_2 + \hat{P}_3, \hat{1}\}. \tag{4.11}$$

Of course, $\mathcal{P}(\mathcal{N})$ is a Boolean algebra.

The algebra $\mathcal{N}$ has three non-trivial abelian (von Neumann) subalgebras,

$$V_i := \mathbb{C}\hat{P}_i + \mathbb{C}(\hat{1} - \hat{P}_i), \quad i = 1, 2, 3. \tag{4.12}$$

Hence, the context category $\mathcal{V}(\mathcal{N})$ is the 4-element poset with $\mathcal{N}$ as top element and $V_i \subset \mathcal{N}$ for $i = 1, 2, 3$.

The Gel'fand spectrum $\underline{\Sigma}_{\mathcal{N}}$ of $\mathcal{N}$ has three elements $\lambda_1, \lambda_2, \lambda_3$, where

$$\lambda_i(\hat{P}_j) = \delta_{ij}. \tag{4.13}$$

The Gel'fand spectrum $\underline{\Sigma}_{V_1}$ of V_1 has two elements $\lambda'_1, \lambda'_{2+3}$ such that

$$\lambda'_1(\hat{P}_1) = 1, \quad \lambda'_1(\hat{1} - \hat{P}_1) = 0, \quad \lambda'_{2+3}(\hat{P}_1) = 0, \quad \lambda'_{2+3}(\hat{1} - \hat{P}_1) = 1. \tag{4.14}$$

(Note that $\hat{1} - \hat{P}_1 = \hat{P}_2 + \hat{P}_3$.) Analogously, the spectrum $\underline{\Sigma}_{V_2}$ has two elements $\lambda'_{1+3}, \lambda'_2$, and the spectrum $\underline{\Sigma}_{V_3}$ has two elements $\lambda'_{1+2}, \lambda'_3$.

Consider the restriction map of the spectral presheaf from $\underline{\Sigma}_{\mathcal{N}}$ to $\underline{\Sigma}_1$:

$$\underline{\Sigma}(i_{V_1,\mathcal{N}})(\lambda_1) = \lambda'_1, \quad \underline{\Sigma}(i_{V_1,\mathcal{N}})(\lambda_2) = \underline{\Sigma}(i_{V_1,\mathcal{N}})(\lambda_3) = \lambda'_{2+3}. \tag{4.15}$$

The restriction maps from $\underline{\Sigma}_{\mathcal{N}}$ to $\underline{\Sigma}_{V_2}$ resp. $\underline{\Sigma}_{V_3}$ are defined analogously. This completes the description of the spectral presheaf $\underline{\Sigma}$ of the algebra $\mathcal{N}$.

We will now determine all clopen subobjects of $\underline{\Sigma}$. First, note that the Gel'fand spectra all are discrete sets, so topological questions are trivial here. We simply have to determine all subobjects of $\underline{\Sigma}$. We distinguish a number of cases:

(a) Let $\underline{S} \in \mathrm{Sub_{cl}}\,\underline{\Sigma}$ be a subobject such that $\underline{S}_{\mathcal{N}} = \underline{\Sigma}_{\mathcal{N}} = \{\lambda_1, \lambda_2, \lambda_3\}$. Then the restriction maps of $\underline{\Sigma}$ dictate that for each V_i, $i = 1, 2, 3$, we have $\underline{S}_{V_i} \supset \underline{\Sigma}(i_{V_i,\mathcal{N}})(\underline{S}_{\mathcal{N}}) = \underline{\Sigma}_{V_i}$, so $\underline{S}$ must be $\underline{\Sigma}$ itself.

(b) Let $\underline{S}$ be a subobject such that $\underline{S}_{\mathcal{N}}$ contains two elements, e.g. $\underline{S}_{\mathcal{N}} = \{\lambda_1, \lambda_2\}$. Then $\underline{S}_{V_1} = \underline{\Sigma}_{V_1}$ and $\underline{S}_{V_2} = \underline{\Sigma}_{V_2}$, but $\underline{S}_{V_3}$ can either be $\{\lambda'_{1+2}\}$ or $\{\lambda'_{1+2}, \lambda'_3\}$ (since $\underline{\Sigma}(i_{V_3,\mathcal{N}})(\underline{S}_{\mathcal{N}}) = \{\lambda'_{1+2}\}$ must be contained in, but not necessarily equal to $\underline{S}_{V_3}$ by the definition of subobjects), so there are two options. Moreover, there are three ways of picking two elements from the three-element set $\underline{\Sigma}_{\mathcal{N}}$, so we have $3 \cdot 2 = 6$ subobjects $\underline{S}$ with two elements in $\underline{S}_{\mathcal{N}}$.

(c) Let $\underline{S}$ be such that $\underline{S}_{\mathcal{N}}$ contains one element, e.g. $\underline{S}_{\mathcal{N}} = \{\lambda_1\}$. Then $\underline{S}_{V_1}$ can either be $\{\lambda'_1\}$ or $\{\lambda'_1, \lambda'_{2+3}\}$; $\underline{S}_{V_2}$ can either be $\{\lambda'_{1+3}\}$ or $\{\lambda'_{1+3}, \lambda'_2\}$; and $\underline{S}_{V_3}$ can either be $\{\lambda'_{1+2}\}$ or $\{\lambda'_{1+2}, \lambda'_3\}$. Hence, there are 2^3 options. Moreover, there are three ways of picking one element from $\underline{\Sigma}_{\mathcal{N}}$, so there are $3 \cdot 2^3 = 24$ subobjects $\underline{S}$ with one element in $\underline{S}_{\mathcal{N}}$.

(d) Finally, consider a subobject $\underline{S}$ such that $\underline{S}_{\mathcal{N}} = \emptyset$. Since the V_i are not contained in one another, there are no conditions arising from restriction maps of the spectral presheaf $\underline{\Sigma}$. Hence, we can pick an arbitrary subset of $\underline{\Sigma}_{V_i}$ for $i = 1, 2, 3$. Each $\underline{\Sigma}_{V_i}$ has 2 elements, hence there are 4 subsets of $\underline{\Sigma}_{V_i}$ for $i = 1, 2, 3$, so we have $4^3 = 64$ subobjects $\underline{S}$ with $\underline{S}_{\mathcal{N}} = \emptyset$.

In all, $\mathrm{Sub_{cl}}\,\underline{\Sigma}$ has $64+24+6+1 = 95$ elements. This can be compared with the 8-element Boolean algebra $\mathcal{P}(\mathcal{N})$ that we started from. The increased number of elements is not due to non-distributivity, but results from considering all subcontexts (i.e., Boolean subalgebras).

Apart from the spectral presheaf, there are a number of other presheaves that play a role in the new mathematical description of quantum system provided by the topos approach to quantum theory. We do not discuss these other presheaves here, see [16] for details. The topos in which the spectral presheaf and the other presheaves lie is the topos $\boldsymbol{Set}^{\mathcal{V}(\mathcal{N})^{\mathrm{op}}}$ of presheaves over the context category $\mathcal{V}(\mathcal{N})$.

§5. Representation of propositions and bi-Heyting algebra structure.

DEFINITION 3. Let $\mathcal{N}$ be a von Neumann algebra, and let $\mathcal{P}(\mathcal{N})$ be its lattice of projections. The map

$$\begin{aligned} \underline{\delta^o} : \mathcal{P}(\mathcal{N}) &\longrightarrow \mathrm{Sub_{cl}}\,\underline{\Sigma} \\ \hat{P} &\longmapsto \underline{\delta^o}(\hat{P}) := (\alpha_V(\delta^o_{\mathcal{N},V}(\hat{P})))_{V\in\mathcal{V}(\mathcal{N})} \end{aligned} \tag{5.1}$$

is called *outer daseinisation of projections.*

This map was introduced in [13] and discussed in detail in [11, 10]. It can be seen as a 'translation' map from standard quantum logic, encoded by the complete orthomodular lattice $\mathcal{P}(\mathcal{N})$ of projections, to a form of (super)intuitionistic logic for quantum systems, based on the clopen subobjects of the spectral presheaf $\underline{\Sigma}$, which conceptually plays the rôle of a quantum state space.

In standard quantum logic, the projections $\hat{P} \in \mathcal{P}(\mathcal{N})$ represent propositions of the form "$A \varepsilon \Delta$", that is, "the physical quantity A has a value in the Borel set Δ of real numbers". The connection between propositions and projections is given by the spectral theorem. Outer daseinisation can hence be seen as a map from propositions of the form "$A \varepsilon \Delta$" into the bi-Heyting algebra $\mathrm{Sub_{cl}}\,\underline{\Sigma}$ of clopen subobjects of the spectral presheaf. A projection $\hat{P}$, representing a proposition "$A \varepsilon \Delta$", is mapped to a collection $(\delta^o_{\mathcal{N},V}(\hat{P}))_{V\in\mathcal{V}(\mathcal{N})}$, consisting of one projection $\delta^o_{\mathcal{N},V}(\hat{P})$ for each context $V \in \mathcal{V}(\mathcal{N})$. (Each isomorphism α_V, $V \in \mathcal{V}(\mathcal{N})$, just maps the projection $\delta^o_{\mathcal{N},V}(\hat{P})$ to the corresponding clopen subset of $\underline{\Sigma}_V$, which does not affect the interpretation.)

Since we have $\delta^o_{\mathcal{N},V}(\hat{P}) \geq \hat{P}$ for all V, the projection $\delta^o_{\mathcal{N},V}(\hat{P})$ represents a coarser (local) proposition than "$A \varepsilon \Delta$" in general. For example, if $\hat{P}$ represents "$A \varepsilon \Delta$", then $\delta^o_{\mathcal{N},V}(\hat{P})$ may represent "$A \varepsilon \Gamma$" where $\Gamma \supset \Delta$.

The map $\underline{\delta^o}$ preserves all joins, as shown in section 2.D of [13] and in [11]. Here is a direct argument: being left adjoint to the inclusion of $\mathcal{P}(V)$ into $\mathcal{P}(\mathcal{N})$, the map $\delta^o_{\mathcal{N},V}$ preserves all colimits, which are joins. Moreover, α_V

is an isomorphism of complete Boolean algebras, so $\alpha_V \circ \delta^o_{\mathcal{N},V}$ preserves all joins. This holds for all $V \in \mathcal{V}(\mathcal{N})$, and joins in $\mathrm{Sub_{cl}}\,\underline{\Sigma}$ are defined stagewise, so $\underline{\delta^o}$ preserves all joins.

Moreover, $\underline{\delta^o}$ is order-preserving and injective, but not surjective. Clearly, $\underline{\delta^o}(\hat{0}) = \underline{0}$, the empty subobject, and $\underline{\delta^o}(\hat{1}) = \underline{\Sigma}$. For meets, we have

$$\forall \hat{P}, \hat{Q} \in \mathcal{P}(\mathcal{N}) : \underline{\delta^o}(\hat{P} \wedge \hat{Q}) \leq \underline{\delta^o}(\hat{P}) \wedge \underline{\delta^o}(\hat{Q}). \tag{5.2}$$

In general, $\underline{\delta^o}(\hat{P}) \wedge \underline{\delta^o}(\hat{Q})$ is not of the form $\underline{\delta^o}(\hat{R})$ for any projection $\hat{R} \in \mathcal{P}(\mathcal{N})$. See [13, 11] for proof of these statements.

Let $(\underline{S}_i)_{i \in I} \subseteq \mathrm{Sub_{cl}}\,\underline{\Sigma}$ be a family of clopen subobjects of $\underline{\Sigma}$, and let $\underline{S} \in \mathrm{Sub_{cl}}\,\underline{\Sigma}$. Then

$$\forall V \in \mathcal{V}(\mathcal{N}) : \left(\underline{S} \wedge \bigvee_{i \in I} \underline{S}_i \right)_V = \bigvee_{i \in I} (\underline{S}_V \wedge \underline{S}_{i;V}), \tag{5.3}$$

since $Cl(\underline{\Sigma}_V)$ is a distributive lattice (in fact, a complete Boolean algebra) in which finite meets distribute over arbitrary joins. Hence, for each $\underline{S} \in \mathrm{Sub_{cl}}\,\underline{\Sigma}$, the functor

$$\underline{S} \wedge \_ : \mathrm{Sub_{cl}}\,\underline{\Sigma} \longrightarrow \mathrm{Sub_{cl}}\,\underline{\Sigma} \tag{5.4}$$

preserves all joins, so by the adjoint functor theorem for posets, it has a right adjoint

$$\underline{S} \Rightarrow \_ : \mathrm{Sub_{cl}}\,\underline{\Sigma} \longrightarrow \mathrm{Sub_{cl}}\,\underline{\Sigma}. \tag{5.5}$$

This map, the *Heyting implication from* $\underline{S}$, makes $\mathrm{Sub_{cl}}\,\underline{\Sigma}$ into a complete Heyting algebra. This was shown before in [13]. The Heyting implication is given by the adjunction

$$\underline{R} \wedge \underline{S} \leq \underline{T} \quad \text{if and only if} \quad \underline{R} \leq (\underline{S} \Rightarrow \underline{T}). \tag{5.6}$$

(Note that $\underline{S} \wedge \_ = \_ \wedge \underline{S}$.) This implies that

$$(\underline{S} \Rightarrow \underline{T}) = \bigvee \{ \underline{R} \in \mathrm{Sub_{cl}}\,\underline{\Sigma} \mid \underline{R} \wedge \underline{S} \leq \underline{T} \}. \tag{5.7}$$

The stagewise definition is: for all $V \in \mathcal{V}(\mathcal{N})$,

$$(\underline{S} \Rightarrow \underline{T})_V = \{ \lambda \in \underline{\Sigma}_V \mid \forall V' \subseteq V : \text{ if } \lambda|_{V'} \in \underline{S}_{V'}, \text{ then } \lambda|_{V'} \in \underline{T}_{V'} \}. \tag{5.8}$$

As usual, the *Heyting negation* $\neg$ is defined for all $\underline{S} \in \mathrm{Sub_{cl}}\,\underline{\Sigma}$ by

$$\neg \underline{S} := (\underline{S} \Rightarrow \underline{0}). \tag{5.9}$$

That is, $\neg \underline{S}$ is the largest element of $\mathrm{Sub_{cl}}\,\underline{\Sigma}$ such that

$$\underline{S} \wedge \neg \underline{S} = \underline{0}. \tag{5.10}$$

The stagewise expression for $\neg \underline{S}$ is

$$(\neg \underline{S})_V = \{ \lambda \in \underline{\Sigma}_V \mid \forall V' \subseteq V : \lambda|_{V'} \notin \underline{S}_{V'} \}. \tag{5.11}$$

In $\mathrm{Sub}_{\mathrm{cl}}\underline{\Sigma}$, we also have, for all families $(\underline{S}_i)_{i\in I} \subseteq \mathrm{Sub}_{\mathrm{cl}}\underline{\Sigma}$ and all $\underline{S} \in \mathrm{Sub}_{\mathrm{cl}}\underline{\Sigma}$,

$$\forall V \in \mathcal{V}(\mathcal{N}) : \Big(\underline{S} \vee \bigwedge_{i\in I} \underline{S}_i\Big)_V = \bigwedge_{i\in I}(\underline{S}_V \vee \underline{S}_{i;V}), \tag{5.12}$$

since finite joins distribute over arbitrary meets in $\mathcal{C}l(\underline{\Sigma}_V)$. Hence, for each $\underline{S}$ the functor

$$\underline{S} \vee \_ : \mathrm{Sub}_{\mathrm{cl}}\underline{\Sigma} \longrightarrow \mathrm{Sub}_{\mathrm{cl}}\underline{\Sigma} \tag{5.13}$$

preserves all meets, so it has a left adjoint

$$\underline{S} \Leftarrow \_ : \mathrm{Sub}_{\mathrm{cl}}\underline{\Sigma} \longrightarrow \mathrm{Sub}_{\mathrm{cl}}\underline{\Sigma} \tag{5.14}$$

which we call *co-Heyting implication.* This map makes $\mathrm{Sub}_{\mathrm{cl}}\underline{\Sigma}$ into a complete co-Heyting algebra. It is characterised by the adjunction

$$(\underline{S} \Leftarrow \underline{T}) \leq \underline{R} \quad \text{iff} \quad \underline{S} \leq \underline{T} \vee \underline{R}, \tag{5.15}$$

so

$$(\underline{S} \Leftarrow \underline{T}) = \bigwedge\{\underline{R} \in \mathrm{Sub}_{\mathrm{cl}}\underline{\Sigma} \mid \underline{S} \leq \underline{T} \vee \underline{R}\}. \tag{5.16}$$

One can think of $\underline{S} \Leftarrow \_$ as a kind of 'subtraction' (see e.g. [38]): $\underline{S} \Leftarrow \underline{T}$ is the smallest clopen subobject $\underline{R}$ for which $\underline{T} \vee \underline{R}$ is bigger then $\underline{S}$, so it encodes how much is 'missing' from $\underline{T}$ to cover $\underline{S}$.

We define a *co-Heyting negation* for each $\underline{S} \in \mathrm{Sub}_{\mathrm{cl}}\underline{\Sigma}$ by

$$\sim\underline{S} := (\underline{\Sigma} \Leftarrow \underline{S}). \tag{5.17}$$

(Note that $\underline{\Sigma}$ is the top element in $\mathrm{Sub}_{\mathrm{cl}}\underline{\Sigma}$.) Hence, $\sim\underline{S}$ is the smallest clopen subobject such that

$$\sim\underline{S} \vee \underline{S} = \underline{\Sigma} \tag{5.18}$$

holds. We have shown:

PROPOSITION 1. $(\mathrm{Sub}_{\mathrm{cl}}\underline{\Sigma}, \wedge, \vee, \underline{0}, \underline{\Sigma}, \Rightarrow, \neg, \Leftarrow, \sim)$ *is a complete bi-Heyting algebra.*

We give direct arguments for the following two facts (which also follow from the general theory of bi-Heyting algebras):

LEMMA 1. *For all $\underline{S} \in \mathrm{Sub}_{\mathrm{cl}}\underline{\Sigma}$, we have $\neg\underline{S} \leq\ \sim\underline{S}$.*

PROOF. For all $V \in \mathcal{V}(\mathcal{N})$, it holds that $(\neg\underline{S})_V \subseteq \underline{\Sigma}_V \backslash \underline{S}_V$, since $(\neg\underline{S} \wedge \underline{S})_V = (\neg\underline{S})_V \cap \underline{S}_V = \emptyset$, while $(\sim\underline{S})_V \supseteq \underline{\Sigma}_V \backslash \underline{S}_V$ since $(\sim\underline{S} \vee \underline{S})_V = (\sim\underline{S})_V \cup \underline{S}_V = \underline{\Sigma}_V$. ⊣

The above lemma and the fact that $\neg\underline{S}$ is the largest subobject such that $\neg\underline{S} \wedge \underline{S} = \underline{0}$ imply

COROLLARY 1. *In general, $\sim\underline{S} \wedge \underline{S} \geq \underline{0}$.*

This means that the co-Heyting negation does not give a system in which a central axiom of most logical systems, viz. freedom from contradiction, holds. We have a glimpse of *paraconsistent logic*.

In fact, a somewhat stronger result holds: for any von Neumann algebra except for $\mathbb{C}\hat{1} = M_1(\mathbb{C})$ and $M_2(\mathbb{C})$, we have $\sim\underline{S} > \neg\underline{S}$ and $\sim\underline{S} \wedge \underline{S} > \underline{0}$ for all clopen subobjects except $\underline{0}$ and $\underline{\Sigma}$. This follows easily from the representation of clopen subobjects as families of projections, see beginning of next section.

§6. Negations and regular elements. In this section, we will examine the Heyting negation $\neg$ and the co-Heyting negation $\sim$ more closely. We will determine regular elements with respect to the Heyting and the co-Heyting algebra structure.

Throughout, we will make use of the isomorphism $\alpha_V : \mathcal{P}(V) \to \mathcal{C}l(\underline{\Sigma}_V)$ (defined in (4.5)) between the complete Boolean algebras of projections in an abelian von Neumann algebra V and the clopen subsets of its spectrum $\underline{\Sigma}_V$. Given a projection $\hat{P} \in \mathcal{P}(V)$, we will use the notation $S_{\hat{P}} := \alpha_V(\hat{P})$. Conversely, for $S \in \mathcal{C}l(\underline{\Sigma}_V)$, we write $\hat{P}_S := \alpha_V^{-1}(S)$.

Given a clopen subobject $\underline{S} \in \mathrm{Sub_{cl}}\,\underline{\Sigma}$, it is useful to think of it as a collection of projections; consider

$$(\hat{P}_{\underline{S}_V})_{V\in\mathcal{V}(\mathcal{N})} = (\alpha_V(\underline{S}_V))_{V\in\mathcal{V}(\mathcal{N})}, \tag{6.1}$$

which consists of one projection for each context V. The fact that $\underline{S}$ is a subobject then translates to the fact that if $V' \subset V$, then $\hat{P}_{\underline{S}_{V'}} \geq \hat{P}_{\underline{S}_V}$. (This is another instance of coarse-graining.)

If $\lambda \in \underline{\Sigma}_V$ and $\hat{P} \in \mathcal{P}(V)$, then

$$\lambda(\hat{P}) = \lambda(\hat{P}^2) = \lambda(\hat{P})^2 \in \{0, 1\}, \tag{6.2}$$

where we used that $\hat{P}$ is idempotent and that λ is multiplicative.

Heyting negation and Heyting-regular elements. We consider the stagewise expression (see eq. (5.11)) for the Heyting negation:

$$\begin{aligned}(\neg\underline{S})_V &= \{\lambda \in \underline{\Sigma}_V \mid \forall V' \subseteq V : \lambda|_{V'} \notin \underline{S}_{V'}\} &(6.3)\\ &= \{\lambda \in \underline{\Sigma}_V \mid \forall V' \subseteq V : \lambda|_{V'}(\hat{P}_{\underline{S}_{V'}}) = 0\} &(6.4)\\ &= \{\lambda \in \underline{\Sigma}_V \mid \forall V' \subseteq V : \lambda(\hat{P}_{\underline{S}_{V'}}) = 0\} &(6.5)\\ &= \left\{\lambda \in \underline{\Sigma}_V \,\middle|\, \lambda\Big(\bigvee_{V'\subseteq V} \hat{P}_{\underline{S}_{V'}}\Big) = 0\right\} &(6.6)\end{aligned}$$

As we saw above, the smaller the context V', the larger the associated projection $\hat{P}_{\underline{S}_{V'}}$. Hence, for the join in the above expression, only the *minimal* contexts V' contained in V are relevant. A minimal context is generated by a single projection $\hat{Q}$ and the identity,

$$V_{\hat{Q}} := \{\hat{Q}, \hat{1}\}'' = \mathbb{C}\hat{Q} + \mathbb{C}\hat{1}. \tag{6.7}$$

Here, it becomes important that we excluded the trivial context $V_0 = \{\hat{1}\}'' = \mathbb{C}\hat{1}$. Let

$$m_V := \{V' \subseteq V \mid V' \text{ minimal}\} = \{V_{\hat{Q}} \mid \hat{Q} \in \mathcal{P}(V)\}. \tag{6.8}$$

We obtain

$$(\neg\underline{S})_V = \left\{\lambda \in \underline{\Sigma}_V \;\middle|\; \lambda\Big(\bigvee_{V' \in m_V} \hat{P}_{\underline{S}_{V'}}\Big) = 0\right\} \tag{6.9}$$

$$= \left\{\lambda \in \underline{\Sigma}_V \;\middle|\; \lambda\Big(\hat{1} - \bigvee_{V' \in m_V} \hat{P}_{\underline{S}_{V'}}\Big) = 1\right\} \tag{6.10}$$

$$= S_{\hat{1} - \bigvee_{V' \in m_V} \hat{P}_{\underline{S}_{V'}}}. \tag{6.11}$$

This shows:

PROPOSITION 2. *Let $\underline{S} \in \mathrm{Sub_{cl}}\,\underline{\Sigma}$, and let $V \in \mathcal{V}(\mathcal{N})$. Then*

$$\hat{P}_{(\neg\underline{S})_V} = \hat{1} - \bigvee_{V' \in m_V} \hat{P}_{\underline{S}_{V'}}, \tag{6.12}$$

where $m_V = \{V' \subseteq V \mid V'$ minimal$\}$.

We can now consider double negation $(\neg\neg\underline{S})_V = S_{\hat{1} - \bigvee_{V' \in m_V} \hat{P}_{(\neg\underline{S})_{V'}}}$, so

$$\hat{P}_{(\neg\neg\underline{S})_V} = \hat{1} - \bigvee_{V' \in m_V} \hat{P}_{(\neg\underline{S})_{V'}}. \tag{6.13}$$

For a $V' \in m_V$, we have $\hat{P}_{(\neg\underline{S})_{V'}} = \hat{1} - \bigvee_{W \in m_{V'}} \hat{P}_{\underline{S}_W}$, but $m_{V'} = \{V'\}$, since V' is minimal, so $\hat{P}_{(\neg\underline{S})_{V'}} = \hat{1} - \hat{P}_{\underline{S}_{V'}}$. Thus,

$$\hat{P}_{(\neg\neg\underline{S})_V} = \hat{1} - \bigvee_{V' \in m_V} (\hat{1} - \hat{P}_{\underline{S}_{V'}}) = \bigwedge_{V' \in m_V} \hat{P}_{\underline{S}_{V'}}. \tag{6.14}$$

Since $\hat{P}_{\underline{S}_{V'}} \geq \hat{P}_{\underline{S}_V}$ for all $V' \in m_V$ (because $\underline{S}$ is a subobject), we have

$$\hat{P}_{(\neg\neg\underline{S})_V} = \bigwedge_{V' \in m_V} \hat{P}_{\underline{S}_{V'}} \geq \hat{P}_{\underline{S}_V} \tag{6.15}$$

for all $V \in \mathcal{V}(\mathcal{N})$, so $\neg\neg\underline{S} \geq \underline{S}$ as expected. We have shown:

PROPOSITION 3. *An element $\underline{S}$ of $\mathrm{Sub_{cl}}\,\underline{\Sigma}$ is Heyting-regular, i.e., $\neg\neg\underline{S} = \underline{S}$, if and only if for all $V \in \mathcal{V}(\mathcal{N})$, it holds that*

$$\hat{P}_{\underline{S}_V} = \bigwedge_{V' \in m_V} \hat{P}_{\underline{S}_{V'}}, \tag{6.16}$$

where $m_V = \{V' \subseteq V \mid V'$ minimal$\}$.

DEFINITION 4. A clopen subobject $\underline{S} \in \mathrm{Sub_{cl}}\,\underline{\Sigma}$ is called *tight* if

$$\underline{\Sigma}(i_{V'V})(\underline{S}_V) = \underline{S}_{V'} \tag{6.17}$$

for all $V', V \in \mathcal{V}(\mathcal{N})$ such that $V' \subseteq V$.

For arbitrary subobjects, we only have $\underline{\Sigma}(i_{V'V})(\underline{S}_V) \subseteq \underline{S}_{V'}$. Let $\underline{S} \in \mathrm{Sub_{cl}}\,\underline{\Sigma}$ be an arbitrary clopen subobject, and let $V, V' \in \mathcal{V}(\mathcal{N})$ such that $V' \subset V$. Then $\underline{\Sigma}(i_{V'V})(\underline{S}_V) \subseteq \underline{S}_{V'} \subseteq \underline{\Sigma}_{V'}$, so $\hat{P}_{\underline{\Sigma}(i_{V'V})(\underline{S}_V)} \in \mathcal{P}(V')$. Thm. 3.1 in [13] shows that

$$\hat{P}_{\underline{\Sigma}(i_{V'V})(\underline{S}_V)} = \delta^o_{V,V'}(\hat{P}_{\underline{S}_V}). \tag{6.18}$$

This key formula relates the restriction maps $\underline{\Sigma}(i_{V'V}) : \underline{\Sigma}_V \to \underline{\Sigma}_{V'}$ of the spectral presheaf to the maps $\delta^o_{V,V'} : \mathcal{P}(V) \to \mathcal{P}(V')$. Using this, we see that

PROPOSITION 4. *A clopen subobject $\underline{S} \in \mathrm{Sub_{cl}}\,\underline{\Sigma}$ is tight if and only if $\hat{P}_{\underline{S}_{V'}} = \delta^o_{V,V'}(\hat{P}_{\underline{S}_V})$ for all $V', V \in \mathcal{V}(\mathcal{N})$ such that $V' \subseteq V$.*

It is clear that all clopen subobjects of the form $\underline{\delta}^o(\hat{P})$, $\hat{P} \in \mathcal{P}(\mathcal{N})$, are tight (see Def. 3).

PROPOSITION 5. *For a tight subobject $\underline{S} \in \mathrm{Sub_{cl}}\,\underline{\Sigma}$, it holds that $\neg\neg\underline{S} = \underline{S}$, i.e., tight subobjects are Heyting-regular.*

PROOF. We saw in equation (6.14) that $\hat{P}_{(\neg\neg\underline{S})_V} = \bigwedge_{V' \in m_V} \hat{P}_{\underline{S}_{V'}}$ for all $V \in \mathcal{V}(\mathcal{N})$. Moreover, $\hat{P}_{(\neg\neg\underline{S})_V} \geq \hat{P}_{\underline{S}_V}$ from equation (6.15). Consider the minimal subalgebra $V_{\hat{P}_{\underline{S}_V}} = \{\hat{P}_{\underline{S}_V}, \hat{1}\}''$ of V. Then, since $\underline{S}$ is tight, we have

$$\delta^o_{V,V_{\hat{P}_{\underline{S}_V}}}(\hat{P}_{\underline{S}_V}) = \bigwedge\{\hat{Q} \in \mathcal{P}(V_{\hat{P}_{\underline{S}_V}}) \mid \hat{Q} \geq \hat{P}_{\underline{S}_V}\} = \hat{P}_{\underline{S}_V}, \tag{6.19}$$

so, for all $V \in \mathcal{V}(\mathcal{N})$,

$$\hat{P}_{(\neg\neg\underline{S})_V} = \bigwedge_{V' \in m_V} \hat{P}_{\underline{S}_{V'}} = \hat{P}_{\underline{S}_V}. \tag{6.20}$$

⊣

COROLLARY 2. *Outer daseinisation $\underline{\delta}^o : \mathcal{P}(\mathcal{N}) \to \mathrm{Sub_{cl}}\,\underline{\Sigma}$ maps projections into the Heyting-regular elements of $\mathrm{Sub_{cl}}\,\underline{\Sigma}$.*

We remark that in order to be Heyting-regular, an element $\underline{S} \in \mathrm{Sub_{cl}}\,\underline{\Sigma}$ need not be tight.

Co-Heyting negation and co-Heyting regular elements. For any $\underline{S} \in \mathrm{Sub_{cl}}\,\underline{\Sigma}$, by its defining property $\sim\underline{S}$ is the smallest element of $\mathrm{Sub_{cl}}\,\underline{\Sigma}$ such that $\underline{S} \vee \sim\underline{S} = \underline{\Sigma}$.

Let V be a maximal context, i.e., a maximal abelian subalgebra (masa) of the non-abelian von Neumann algebra $\mathcal{N}$. Then clearly

$$(\sim\underline{S})_V = \underline{\Sigma}_V \backslash \underline{S}_V. \tag{6.21}$$

Let $V \in \mathcal{V}(\mathcal{N})$, not necessarily maximal. We define

$$M_V := \{\tilde{V} \supseteq V \mid \tilde{V} \text{ maximal}\}. \tag{6.22}$$

PROPOSITION 6. *Let* $\underline{S} \in \mathrm{Sub_{cl}}\,\underline{\Sigma}$, *and let* $V \in \mathcal{V}(\mathcal{N})$. *Then*

$$\hat{P}_{(\sim\underline{S})_V} = \bigvee_{\tilde{V} \in M_V} (\delta^o_{\tilde{V},V}(\hat{1} - \hat{P}_{\underline{S}_{\tilde{V}}})), \tag{6.23}$$

where $M_V = \{\tilde{V} \supseteq V \mid \tilde{V}$ *maximal*$\}$.

PROOF. $\sim\underline{S}$ is a (clopen) subobject, so we must have

$$\hat{P}_{(\sim\underline{S})_V} \geq \bigvee_{\tilde{V} \in M_V} (\delta^o_{\tilde{V},V}(\hat{1} - \hat{P}_{\underline{S}_{\tilde{V}}})), \tag{6.24}$$

since $(\sim\underline{S})_V$, the component at V, must contain all the restrictions of the components $(\sim\underline{S})_{\tilde{V}}$ for $\tilde{V} \in M_V$ (and the above inequality expresses this using the corresponding projections).

On the other hand, $\sim\underline{S}$ is the *smallest* clopen subobject such that $\underline{S} \vee \sim\underline{S} = \underline{\Sigma}$. So it suffices to show that for $\hat{P}_{(\sim\underline{S})_V} = \bigvee_{\tilde{V} \in M_V} (\delta^o_{\tilde{V},V}(\hat{1} - \hat{P}_{\underline{S}_{\tilde{V}}}))$, we have $\hat{P}_{(\sim\underline{S})_V} \vee \hat{P}_{\underline{S}_V} = \hat{1}$ for all $V \in \mathcal{V}(\mathcal{N})$, and hence $\sim\underline{S} \vee \underline{S} = \underline{\Sigma}$.

If V is maximal, then $\hat{P}_{(\sim\underline{S})_V} = \delta^o_{V,V}(\hat{1} - \hat{P}_{\underline{S}_V}) = \hat{1} - \hat{P}_{\underline{S}_V}$ and hence $\hat{P}_{(\sim\underline{S})_V} \vee \hat{P}_{\underline{S}_V} = \hat{1}$. If V is non-maximal and $\tilde{V}$ is any maximal context containing V, then $\hat{P}_{(\sim\underline{S})_V} \geq \hat{P}_{(\sim\underline{S})_{\tilde{V}}}$ and $\hat{P}_{\underline{S}_V} \geq \hat{P}_{\underline{S}_{\tilde{V}}}$, so $\hat{P}_{(\sim\underline{S})_V} \vee \hat{P}_{\underline{S}_V} \geq \hat{P}_{(\sim\underline{S})_{\tilde{V}}} \vee \hat{P}_{\underline{S}_{\tilde{V}}} = \hat{1}$. ⊣

For the double co-Heyting negation, we obtain

$$\hat{P}_{(\sim\sim\underline{S})_V} = \bigvee_{\tilde{V} \in M_V} \delta^o_{\tilde{V},V}(\hat{1} - \hat{P}_{(\sim\underline{S})_{\tilde{V}}}) \tag{6.25}$$

$$= \bigvee_{\tilde{V} \in M_V} \delta^o_{\tilde{V},V}\left(\hat{1} - \bigvee_{W \in M_{\tilde{V}}} \delta^o_{W,\tilde{V}}(\hat{1} - \hat{P}_{\underline{S}_W})\right). \tag{6.26}$$

Since $\tilde{V}$ is maximal, we have $M_{\tilde{V}} = \{\tilde{V}\}$, and the above expression simplifies to

$$\hat{P}_{(\sim\sim\underline{S})_V} = \bigvee_{\tilde{V} \in M_V} \delta^o_{\tilde{V},V}(\hat{1} - (\hat{1} - \hat{P}_{\underline{S}_{\tilde{V}}})) \tag{6.27}$$

$$= \bigvee_{\tilde{V} \in M_V} \delta^o_{\tilde{V},V}(\hat{P}_{\underline{S}_{\tilde{V}}}). \tag{6.28}$$

Note that the fact that $\underline{S}$ is a subobject implies that

$$\hat{P}_{(\sim\sim\underline{S})_V} \leq \hat{P}_{\underline{S}_V} \tag{6.29}$$

for all $V \in \mathcal{V}(\mathcal{N})$, so $\sim\sim\underline{S} \leq \underline{S}$ as expected. We have shown:

PROPOSITION 7. *An element* $\underline{S}$ *of* $\mathrm{Sub_{cl}}\,\underline{\Sigma}$ *is co-Heyting-regular, i.e.,* $\sim\sim\underline{S} = \underline{S}$, *if and only if for all* $V \in \mathcal{V}(\mathcal{N})$ *it holds that*

$$\hat{P}_{\underline{S}_V} = \bigvee_{\tilde{V} \in M_V} \delta^o_{\tilde{V},V}(\hat{P}_{\underline{S}_{\tilde{V}}}), \tag{6.30}$$

where $M_V = \{\tilde{V} \supseteq V \mid \tilde{V}$ *maximal*$\}$.

PROPOSITION 8. *If* $\underline{S} \in \mathrm{Sub_{cl}}\,\underline{\Sigma}$ *is tight, then* $\sim\sim\underline{S} = \underline{S}$, *i.e., tight subobjects are co-Heyting regular.*

PROOF. If $\underline{S}$ is tight, then for all $V \in \mathcal{V}(\mathcal{N})$ and for all $\tilde{V} \in M_V$, we have $\hat{P}_{\underline{S}_V} = \delta^o_{\tilde{V},V}(\hat{P}_{\underline{S}_{\tilde{V}}})$, so $\bigvee_{\tilde{V} \in M_V} \delta^o_{\tilde{V},V}(\hat{P}_{\underline{S}_{\tilde{V}}}) = \hat{P}_{\underline{S}_V}$. By Prop. 7, the result follows. ⊣

COROLLARY 3. *Outer daseinisation* $\underline{\delta^o} : \mathcal{P}(\mathcal{N}) \to \mathrm{Sub_{cl}}\,\underline{\Sigma}$ *maps projections into the co-Heyting-regular elements of* $\mathrm{Sub_{cl}}\,\underline{\Sigma}$.

Physical interpretation. We conclude this section by giving a tentative physical interpretation of the two kinds of negation. For this interpretation, it is important to think of an element $\underline{S} \in \mathrm{Sub_{cl}}\,\underline{\Sigma}$ as a collection of local propositions $\underline{S}_V$ (resp. $\hat{P}_{\underline{S}_V}$), one for each context V. Moreover, if $V' \subset V$, then the local proposition represented by $\underline{S}_{V'}$ is coarser than the local proposition represented by $\underline{S}_V$.

Let $\underline{S} \in \mathrm{Sub_{cl}}\,\underline{\Sigma}$ be a clopen subobject, and let $\neg\underline{S}$ be its Heyting complement. As shown in Prop. 2, the local expression for components of $\neg\underline{S}$ is given by

$$\hat{P}_{(\neg\underline{S})_V} = \hat{1} - \bigvee_{V' \in m_V} \hat{P}_{\underline{S}_{V'}}, \tag{6.31}$$

where m_V is the set of all minimal contexts contained in V. The projection $\hat{P}_{(\neg\underline{S})_V}$ is always smaller than or equal to $\hat{1} - \hat{P}_{\underline{S}_V}$, since $\hat{P}_{\underline{S}_{V'}} \geq \hat{P}_{\underline{S}_V}$ for all $V' \in m_V$. For the Heyting negation of the local proposition in the context V, represented by $\underline{S}_V$ or equivalently by the projection $\hat{P}_{\underline{S}_V}$, one has to consider all the coarse-grainings of this proposition to minimal contexts (which are the 'maximal' coarse-grainings). The Heyting complement $\neg\underline{S}$ is determined at each stage V as the complement of the join of all the coarse-grainings $\hat{P}_{\underline{S}_{V'}}$ of $\hat{P}_{\underline{S}_V}$. In other words, the component of the Heyting complement $\neg\underline{S}$ at V is not simply the complement of $\underline{S}_V$, but the complement of the disjunction of all the coarse-grainings of this local proposition to all smaller contexts. The coarse-grainings of $\underline{S}_V$ are specified by the clopen subobject $\underline{S}$ itself.

The component of the co-Heyting complement $\sim\underline{S}$ at a context V is given by

$$\hat{P}_{(\sim\underline{S})_V} = \bigvee_{\tilde{V} \in M_V} (\delta^o_{\tilde{V},V}(\hat{1} - \hat{P}_{\underline{S}_{\tilde{V}}})), \tag{6.32}$$

where M_V is the set of maximal contexts containing V. The projection $\hat{P}_{(\sim\underline{S})_V}$ is always larger than or equal to $\hat{1} - \hat{P}_{\underline{S}_V}$, as was argued in the proof of Prop. 6.

This means that the co-Heyting complement $\sim\underline{S}$ has a component $(\sim\underline{S})_V$ at V that may overlap with the component $\underline{S}_V$, hence the corresponding local propositions are not mutually exclusive in general. Instead, $\hat{P}_{(\sim\underline{S})_V}$ is the disjunction of all the coarse-grainings of complements of (finer, i.e., stronger) local propositions at contexts $\tilde{V} \supset V$.

The co-Heyting negation hence gives local propositions that for each context V take into account all those contexts $\tilde{V}$ from which one can coarse-grain to V. The component $(\sim\underline{S})_V$ is defined in such a way that all the stronger local propositions at maximal contexts $\tilde{V} \supset V$ are complemented in the usual sense, i.e., $\hat{P}_{(\sim\underline{S})_{\tilde{V}}} = \hat{1} - \hat{P}_{\underline{S}_{\tilde{V}}}$ for all maximal contexts $\tilde{V}$. At smaller contexts V, we have some coarse-grained local proposition, represented by $\hat{P}_{(\sim\underline{S})_V}$, that will in general not be disjoint from (i.e., mutually exclusive with) the local proposition represented by $\hat{P}_{\underline{S}_V}$.

§7. Conclusion and outlook. Summing up, we have shown that to each quantum system described by a von Neumann algebra $\mathcal{N}$ of physical quantities one can associate a (generalised) quantum state space, the spectral presheaf $\underline{\Sigma}$, together with a complete bi-Heyting algebra $\mathrm{Sub_{cl}}\,\underline{\Sigma}$ of clopen subobjects. Elements $\underline{S}$ can be interpreted as families of local propositions, where 'local' refers to contextuality; each component $\underline{S}_V$ of a clopen subobject represents a proposition about the value of a physical quantity in the context (i.e., abelian von Neumann subalgebra) V of $\mathcal{N}$. Since $\underline{S}$ is a subobject, there is a built-in form of coarse-graining which guarantees that if $V' \subset V$ is a smaller context, then the local proposition represented by $\underline{S}_{V'}$ is coarser than the proposition represented by $\underline{S}_V$.

The map called outer daseinisation of projections (see Def. 3) is a convenient bridge between the usual Hilbert space formalism and the new topos-based form of quantum logic. Daseinisation maps a propositions of the form "$A \varepsilon \Delta$", represented by a projection $\hat{P}$ in the complete orthomodular lattice $\mathcal{P}(\mathcal{N})$ of projections in the von Neumann algebra $\mathcal{N}$, to an element $\underline{\delta^o}(\hat{P})$ of the bi-Heyting algebra $\mathrm{Sub_{cl}}\,\underline{\Sigma}$.

We characterised the two forms of negation arising from the Heyting and the co-Heyting structure on $\mathrm{Sub_{cl}}\,\underline{\Sigma}$ by giving concrete stagewise expressions (see Props. 2 and 6), considered double negation and characterised Heyting regular elements of $\mathrm{Sub_{cl}}\,\underline{\Sigma}$ (Prop. 3) as well as co-Heyting regular elements (Prop. 7). It turns out that daseinisation maps projections into Heyting regular and co-Heyting regular elements of the bi-Heyting algebra of clopen subobjects.

The main thrust of this article is to replace the standard algebraic representation of quantum logic in projection lattices of von Neumann algebras by a better behaved form based on bi-Heyting algebras. Instead of having a non-distributive orthomodular lattice of projections, which comes with a host of well-known conceptual and interpretational problems, one can consider a complete bi-Heyting algebra of propositions. In particular, this provides a distributive form of quantum logic. Roughly speaking, a non-distributive lattice with an orthocomplement has been traded for a distributive one with two different negations.

We conclude by giving some open problems for further study:

(a) It will be interesting to see how far the constructions presented in this article can be generalised beyond the case of von Neumann algebras. A generalisation to complete orthomodular lattices is immediate, but more general structures used in the study of quantum logic(s) remain to be considered.

(b) Bi-Heyting algebras are related to bitopological spaces, see [3] and references therein. But the spectral presheaf $\underline{\Sigma}$ is not a topological (or bitopological) space in the usual sense. Rather, it is a presheaf which has no global elements. Hence, there is no direct notion of points available, which makes it impossible to define a set underlying the topology (or topologies). Generalised notions of topology such as frames will be useful to study the connections with bitopological spaces.

(c) All the arguments given in this article are topos-external. There is an internal analogue of the bi-Heyting algebra $\mathrm{Sub}_{\mathrm{cl}}\,\underline{\Sigma}$ in the form of the power object $P\underline{O}$ of the so-called outer presheaf, see [17], so one can study many aspects internally in the topos $\boldsymbol{Set}^{\mathcal{V}(\mathcal{N})^{\mathrm{op}}}$ associated with the quantum system. This also provides the means to go beyond propositional logic to predicate logic, since each topos possesses an internal higher-order intuitionistic logic.

Acknowledgements. I am very grateful to the ASL, and to Reed Solomon, Valentina Harizanov and Jennifer Chubb personally, for giving me the opportunity to organise a Special Session on "Logic and Foundations of Physics" for the 2010 North American Meeting of the ASL, Washington D.C., March 17–20, 2010. I would like to thank Chris Isham and Rui Soares Barbosa for discussions and support. Many thanks to Dan Marsden, who read the manuscript at an early stage and made some valuable comments and suggestions. The anonymous referee also provided some very useful suggestions, which I incorporated. Finally, Dominique Lambert's recent talk at *Categories and Physics 2011* at Paris 7 served as an eye-opener on paraconsistent logic (and made me lose my fear of contradictions ;-)).

REFERENCES

[1] S. Abramsky and A. Brandenburger, *The sheaf-theoretic structure of non-locality and contextuality*, ***New Journal of Physics***, vol. 13 (2011), p. 113036.

[2] S. Abramsky, S. Mansfield, and R. Soares Barbosa, *The cohomology of non-locality and contextuality*, ***Proceedings 8th International Workshop on Quantum Physics and Logic, Nijmegen,***

Netherlands, October 27-29, 2011 (B. Jacobs, P. Selinger, and B. Spitters, editors), Electronic Proceedings in Theoretical Computer Science, vol. 95, Open Publishing Association, 2012, eprint available at arXiv:1111.3620 [quant-ph], pp. 1–14.

[3] G. Bezhanishvili et al., *Bitopological duality for distributive lattices and heyting algebras*, ***Mathematical Structures in Computer Science***, vol. 20 (2010), pp. 359–393.

[4] G. Birkhoff and J. von Neumann, *The logic of quantum mechanics*, ***Annali di Matematica Pura ed Applicata***, vol. 37 (1936), pp. 823–843.

[5] M. Caspers, C. Heunen, N. P. Landsman, and B. Spitters, *Intuitionistic quantum logic of an n-level system*, ***Foundations of Physics***, vol. 39 (2009), pp. 731–759.

[6] M. L. Dalla Chiara and R. Giuntini, *Quantum logics*, ***Handbook of Philosophical Logic*** (G. Gabbay and F. Guenthner, editors), vol. VI, Kluwer, Dordrecht, 2002, pp. 129–228.

[7] A. Döring, *Kochen-specker theorem for von Neumann algebras*, ***International Journal of Theoretical Physics***, vol. 44 (2005), pp. 139–160.

[8] ———, *Quantum states and measures on the spectral presheaf*, ***Adv. Sci. Lett.***, vol. 2 (2009), pp. 291–301, special issue on "Quantum Gravity, Cosmology and Black Holes", ed. M. Bojowald.

[9] ———, *Topos theory and 'neo-realist' quantum theory*, ***Quantum Field Theory, Competitive Models*** (B. Fauser, J. Tolksdorf, and E. Zeidler, editors), Birkhäuser, Basel, Boston, Berlin, 2009.

[10] ———, *The physical interpretation of daseinisation*, ***Deep Beauty*** (H. Halvorson, editor), Cambridge University Press, 2011, pp. 207–238.

[11] ———, *Topos quantum logic and mixed states*, ***Proceedings of the 6th International Workshop on Quantum Physics and Logic (QPL 2009), Oxford***, vol. 270, Electronic Notes in Theoretical Computer Science, no. 2, 2011.

[12] A. Döring and C. J. Isham, *A topos foundation for theories of physics: I. Formal languages for physics*, ***Journal of Mathematical Physics***, vol. 49 (2008), p. 053515.

[13] ———, *A topos foundation for theories of physics: II. Daseinisation and the liberation of quantum theory*, ***Journal of Mathematical Physics***, vol. 49 (2008), p. 053516.

[14] ———, *A topos foundation for theories of physics: III. Quantum theory and the representation of physical quantities with arrows* $\breve{\delta}(\hat{A}) : \underline{\Sigma} \rightarrow \underline{\mathbb{R}^{\leftrightarrow}}$, ***Journal of Mathematical Physics***, vol. 49 (2008), p. 053517.

[15] ———, *A topos foundation for theories of physics: IV. Categories of systems*, ***Journal of Mathematical Physics***, vol. 49 (2008), p. 053518.

[16] ———, *'What is a Thing?': Topos Theory in the Foundations of Physics*, ***New Structures for Physics*** (B. Coecke, editor), Lecture Notes in Physics, vol. 813, Springer, Heidelberg, Dordrecht, London, New York, 2011, pp. 753–937.

[17] ———, *Classical and quantum probabilities as truth values*, ***Journal of Mathematical Physics***, vol. 53 (2012), p. 032101.

[18] C. Heunen, N. P. Landsman, and B. Spitters, *A topos for algebraic quantum theory*, ***Communications in Mathematical Physics***, vol. 291 (2009), pp. 63–110.

[19] ———, *Bohrification*, ***Deep Beauty*** (H. Halvorson, editor), Cambridge University Press, 2011, pp. 271–313.

[20] ———, *Bohrification of von Neumann algebras and quantum logic*, ***Synthese***, (2011), pp. 719–752, Online first, DOI: 10.1007/s11229-011-9918-4.

[21] C. J. Isham, *Topos theory and consistent histories: The internal logic of the set of all consistent sets*, ***International Journal of Theoretical Physics***, vol. 36 (1997), pp. 785–814.

[22] ———, *Is it true; or is it false; or somewhere in between? the logic of quantum theory*, ***Contemporary Physics***, vol. 46 (2005), pp. 207–219.

[23] C. J. Isham and J. Butterfield, *A topos perspective on the Kochen-Specker theorem: I. Quantum states as generalised valuations*, ***International Journal of Theoretical Physics***, vol. 37 (1998), pp. 2669–2733.

[24] ———, *A topos perspective on the Kochen-Specker theorem: II. Conceptual aspects and classical analogues*, ***International Journal of Theoretical Physics***, vol. 38 (1999), pp. 827–859.

[25] ———, *Some possible roles for topos theory in quantum theory and quantum gravity*, ***Foundations of Physics***, vol. 30 (2000), pp. 1707–1735.

[26] ———, *A topos perspective on the Kochen-Specker theorem: IV. Interval valuations*, ***International Journal of Theoretical Physics***, vol. 41 (2002), pp. 613–639.

[27] C. J. Isham, J. Hamilton, and J. Butterfield, *A topos perspective on the Kochen-Specker theorem: III. Von Neumann algebras as the base category*, ***International Journal of Theoretical Physics***, vol. 39 (2000), pp. 1413–1436.

[28] R. V. Kadison and J. R. Ringrose, ***Fundamentals of the Theory of Operator Algebras***, vol. I, II, Academic Press, New York, 1983.

[29] S. Kochen and E. P. Specker, *The problem of hidden variables in quantum mechanics*, ***Journal of Mathematics and Mechanics***, vol. 17 (1967), pp. 59–87.

[30] F. W. Lawvere, *Introduction*, ***Categories in Continuum Physics (Buffalo 1982)***, Lecture Notes in Mathematics, vol. 1174, Springer, Berlin, Heidelberg, New York, Tokyo, 1986, pp. 1–16.

[31] ———, *Intrinsic co-heyting boundaries and the leibniz rule in certain toposes*, ***Category Theory, Proceedings, Como 1990*** (A. Carboni, M. C. Pedicchio, and G. Rosolini, editors), Lecture Notes in Mathematics, vol. 1488, Springer, Berlin, Heidelberg, New York, 1991, pp. 279–281.

[32] S. Mac Lane and I. Moerdijk, ***Sheaves in Geometry and Logic: A First Introduction to Topos Theory***, Springer, New York, Berlin, Heidelberg, 1992.

[33] S. Majid, ***Foundations of Quantum Group Theory***, Cambridge University Press, 1995.

[34] ———, *Quantum spacetime and physical reality*, ***On Space and Time*** (S. Majid, editor), Cambridge University Press, 2008, pp. 56–140.

[35] M. Makkai and G. E. Reyes, *Completeness results for intuitionistic and modal logic in a categorical setting*, ***Annals of Pure and Applied Logic***, vol. 72 (1995), pp. 25–101.

[36] C. Rauszer, *Semi-boolean algebras and their applications to intuitionistic logic with dual operations*, ***Fundamenta Mathematicae***, vol. 83 (1973), pp. 219–249.

[37] ———, *Model theory for an extension of intuitionistic logic*, ***Studia Logica***, vol. 36 (1977), pp. 73–87.

[38] G. E. Reyes and H. Zolfaghari, *Bi-heyting algebras, toposes and modalities*, ***Journal of Philosophical Logic***, vol. 25 (1996), pp. 25–43.

E-mail: andreas.doering@posteo.de

THE LOGIC OF QUANTUM MECHANICS – TAKE II

BOB COECKE

Abstract. We put forward a new take on the logic of quantum mechanics, following Schrödinger's point of view that it is composition which makes quantum theory what it is, rather than its particular propositional structure due to the existence of superpositions, as proposed by Birkhoff and von Neumann. This gives rise to an intrinsically quantitative kind of logic, which truly deserves the name 'logic' in that it also models meaning in natural language, the latter being the origin of logic, that it supports automation, the most prominent practical use of logic, and that it supports probabilistic inference.

§1. The physics and the logic of quantum-ish logic. In 1932 John von Neumann formalized Quantum Mechanics in his book "Mathematische Grundlagen der Quantenmechanik". This was effectively the official birth of the quantum mechanical formalism which until now, some 75 years later, has remained the same. Quantum theory underpins so many things in our daily lives including chemical industry, energy production and information technology, which arguably makes it the most technologically successful theory of physics ever.

However, in 1935, merely three years after the birth of his brainchild, von Neumann wrote in a letter to American mathematician Garrett Birkhoff: "I would like to make a confession which may seem immoral: I do not believe absolutely in Hilbert space no more." (sic)—for more details see [73].

Soon thereafter they published a paper entitled "The *Logic* of Quantum Mechanics" [13]. Their 'quantum logic' was cast in order-theoretic terms, very much in the spirit of the then reigning algebraic view of logic, with the distributive law being replaced with a weaker (ortho)modular law.

The work presented here is supported by the British Engineering and Physical Research Council (EPSRC), the US Office of Naval Research (ONR) and the Foundational Questions Institute (FQXi). The content of this paper reflects a series of seminars in 2010–2012 with as titles: "Monoidal categories as an axiomatic foundation", "In the beginning God created tensor ... then matter ... then speech", "How computer science helps bringing quantum mechanics to the masses", "Selling categories to the masses", or the actual title of this paper itself. We thank David Corfield and Pascal Vaudrevange for feedback on a previous version.

Logic and Algebraic Structures in Quantum Computing
Edited by J. Chubb, A. Eskandarian and V. Harizanov
Lecture Notes in Logic, 45
© 2016, Association for Symbolic Logic

This resulted in a research community of *quantum logicians* [68, 71, 47, 30]. However, despite von Neumann's reputation, and the large body of research that has been produced in the area, one does not find a trace of this activity neither in the mainstream physics, mathematics, nor logic literature. Hence, 75 years later one may want to conclude that this activity was a failure.

What went wrong?

1.1. The mathematics of it. Let us consider the raison d'être for the Hilbert space formalism. So why would one need all this 'Hilbert space stuff', i.e. the continuum structure, the field structure of complex numbers, a vector space over it, inner-product structure, etc. Why? According to von Neumann, he simply used it because it happened to be 'available'. The use of linear algebra and complex numbers in so many different scientific areas, as well as results in model theory, clearly show that quite a bit of modeling can be done using Hilbert spaces. On the other hand, we can also model any movie by means of the data stream that runs through your cables when watching it. But does this mean that these data streams make up the stuff that makes a movie? Clearly not, we should rather turn our attention to the stuff that is being taught at drama schools and directing schools. Similarly, von Neumann turned his attention to the actual physical concepts behind quantum theory, more specifically, the notion of a physical property and the structure imposed on these by the peculiar nature of quantum observation. His quantum logic gave the resulting 'algebra of physical properties' a privileged role. All of this leads us to ...

1.2. ... the physics of it. Birkhoff and von Neumann crafted quantum logic in order to emphasize the notion of quantum *superposition* In terms of states of a physical system and properties of that system, superposition means that the strongest property which is true for two distinct states is also true for states other than the two given ones. In order-theoretic terms this means, representing states by the atoms of a lattice[1] of properties [69], that the join $p \vee q$ of two atoms p and q is also above other atoms. From this it easily follows that the distributive law[2] breaks down: given atom[3] $r \neq p, q$ with $r < p \vee q$ we have $r \wedge (p \vee q) = r$ while $(r \wedge p) \vee (r \wedge q) = 0 \vee 0 = 0$. Birkhoff and von Neumann as well as many others believed that understanding the deep structure of superposition is the key to obtaining a better understanding of quantum theory as a whole. But as already mentioned, 75 years later quantum logic did not break through.

---

[1]I.e. a partially ordered set with a minimal element 0 and maximal element 1, and in which each pair of elements has a supremum and an infimum. In fact, there are physical resons for assuming that this lattice is *complete* [71, 69], i.e. arbitrary suprema and infima exist.

[2]*Distributivity* means that for any elements a, b, c of the lattice we have that $a \wedge (b \vee c) = (a \wedge b) \vee (a \wedge c)$ and that $a \vee (b \wedge c) = (a \vee b) \wedge (a \vee c)$.

[3]An *atom*is an element $p \neq 0$ which is such that whenever $a < p$ then $a = 0$.

The Achilles' heel of quantum logic is the fact that it fails to elegantly capture 'composition of quantum systems', that is, how do we describe multiple quantum systems given that we know how to describe the individual quantum systems. On the other hand, also in 1935, Schrödinger pushed forward the idea that the stuff which truly characterizes quantum behavior is precisely the manner in which quantum systems compose [74]. Over the past 30 years or so we have seen ample evidence for this claim. So-called 'quantum non-locality' was experimentally confirmed, and the focus on quantum information processing has revealed a wide range of quantum phenomena which all crucially depend on the manner in which quantum systems compose, most notably exponential quantum computational speed-up which led to the quantum computing paradigm [79].

Now reversing the roles, rather than explaining all of quantum theory in terms of superposition, can we maybe explain all of quantum theory in term of the manner in which quantum systems compose, including superposition?

1.3. The game plan. Here is the list of tasks we've set ourselves:

- **Task 0.** First we want to solve:

$$\frac{\text{tensor product structure}}{\text{the other Hilbert space stuff}} = ???$$

  that is, we want to know what remains of the Hilbert space formalism if we 'remove all of its structure except for the manner in which systems compose'. In other words, we want to axiomatize composition of systems, which we denote by $\otimes$, without any reference to underlying spaces.
- **Task 1.** Next we investigate which additional assumptions on $\otimes$ are needed in order to deduce experimentally observed phenomena? That is, given that the structure deduced in Task 0 applies to a wide range of theories (as we shall see below in Section 2) what extra structure do we need to add such that the resulting framework allows us to derive typical quantum behaviors.
- **Task 2.** Once this 'typically quantum' structure has been identified, we take on the challenge to find this same structure elsewhere in what we usually conceive as 'our classical reality'. This may involve looking at this classical reality through a 'novel pair of glasses'.

And, ... here are the resulting outcomes:

- **Outcome 0:** That was an easy one. The solution to this has been around for quite a while. It is called *symmetric monoidal category*[11]. In fact, as discused in [23, 32], physical processes themselves form a *strict* symmetric monoidal category, while set theory based models such as the Hilbert space model are typically non-strict, which invokes so-called 'coherence conditions' [67] between 'natural transformations' [45]. But one can show that an arbitrary symmetric monoidal category is always 'categorically equivalent' to a strict symmetric monoidal category, which means that, up to isomorphisms, whatever one can do with a non-strict one, one can do with a strict one too. Hence, here we will only spell out strict symmetric monoidal categories, in terms of their *graphical language* [70, 59], that is, a language which is such that an equational statement holds in it if and only if it follows from the axioms of a strict symmetric monoidal category.

- **Outcome 1a:** Quoting Princeton philosopher Hans Halvorson in his editorial to the volume *Deep Beauty: Understanding the Quantum World through Mathematical Innovation* which marked 75 years since the publications of von Neumann's quantum formalism [52]: "What is perhaps most striking about Coecke's approach is the sheer ratio of results to assumptions." As we shall see below, with very little additional structure one can already derive a wide range of quantum phenomena, and the required computations are utterly trivial. This is in sharp contrast with Birkhoff-von Neumann quantum logic where one couldn't derive much; and in the case that one could derive something physically relevant one had to work really hard.
- **Outcome 1b:** Moreover, exposing this structure has already helped to solve standing open problems in quantum information, e.g. [44, 57, 14], and provided novel insights in the nature of quantum non-locality [27, 26].
- **Outcome 1c:** The diagrammatic framework underpinning strict symmetric monoidal categories has meanwhile been adopted by several leading researchers in quantum foundations e.g. [49, 17, 54]; quoting Lucien Hardy in [54]: "[. . .] we join the *quantum picturalism* revolution [23]".
- **Outcome 2a:** Observe the following similar looking pictures:

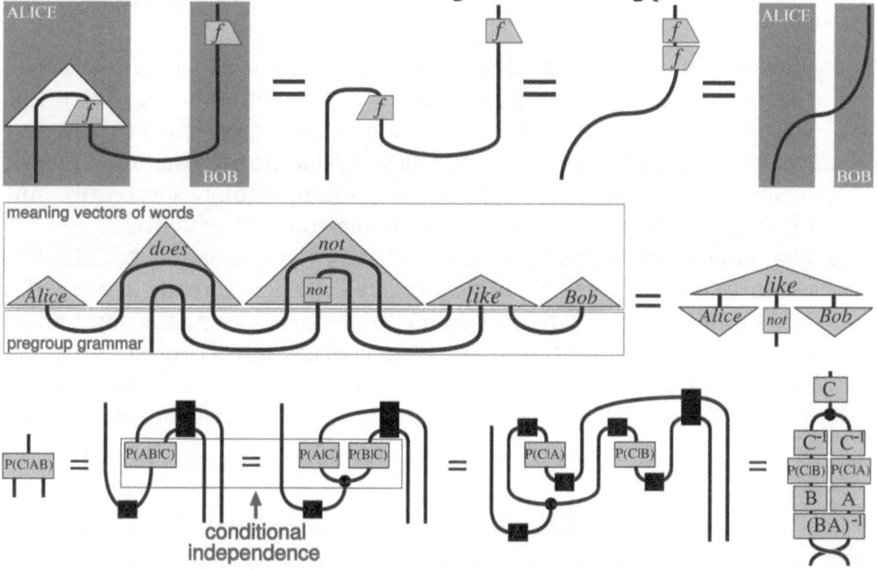

> These are respectively taken from a physics paper on the flow of information in quantum protocols [1, 20, 23], a linguistics paper on how to compute the meaning of a given the meaning of its words [19, 37], and a probability theory paper that axiomatizes Bayesian inference [38]. The graphical calculi are in each case very similar, which points at a common reasoning system in each of these very distinct areas. Note in particular that in each case the data of interest is of a fundamentally quantitative nature. Could this be pointing at the existence of some sort of *quantitative logic*, which is not typical to these areas but of a more universal nature?

So let us now consider . . .

1.4. . . . the logic of it. What is logic? The previous century has known a huge proliferation of logics of various kinds, and there probably are as many opinions of what logic actually is. Rather than making a case for one or another logical paradigm we will take a pragmatic stance and conceive logic in terms of its origin and its most prominent practical use:

- **Origin: structure in natural language.** The origin of logic, tracing back to Aristotle, is that it is about 'arguments in natural language'. Consider for example the sentence:"Alice and Bob either ate everything or nothing, then got sick." By using connectives, quantifiers, variable f referring to food, constants a(lice) and b(ob), and predicates $Sick(\texttt{person})$ and $Eat(\texttt{person, some kind of food})$ we can formalize this as follows:

$$(\forall f : Eat(a, f) \wedge Eat(b, f)) \vee \neg(\exists f : Eat(a, f) \vee Eat(b, f)) \Rightarrow Sick(a), Sick(b)$$

  However, statements like this are still tightly related to a truth-concept, that is, we classify statements in terms of these either being true or not. Clearly there is a lot more to the *meaning* of a sentence than it either being true or false. This leads us to the following questions: What do we mean by meaning? What is the logic governing meaning, more specifically, how do meanings of words interact to form meanings of sentences?
- **Use: automated reasoning.** Logic now forms the foundation for fields like automated proof checking and automated theorem proving in computer science, which are key to modern methods for verifying the correctness of new software and hardware. Logic also controls robot behaviors in artificial intelligence. Even more adventurous is *automated theory exploration*, where one does not only try to automatically prove theorems, but also generate them, which is a much harder task (cf. P vs. NP)—see also Figure 1.

Our diagrammatic framework appeals to both of these senses of logic, and in doing so produced important new applications in each of these areas:

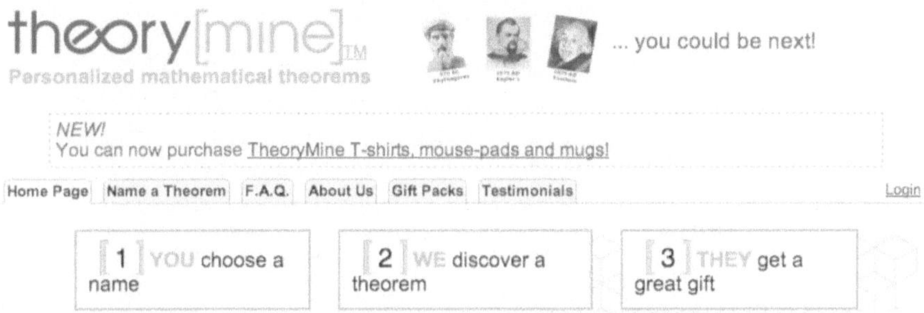

FIGURE 1. The theory[mine] website which allows one to buy an automatically generated theorem and name it after someone. It is a novelty gift spin-off from the automated theory exploration expertise at Edinburgh University—see [15] for the science.

- The above depicted framework for modeling how meanings of words interact to form meanings of sentences, introduced by Clark, Sadrzadeh and the author in [19, 37], is the first to do so based on a clear conceptual underpinning. It was a cover heading feature in New Scientist [5] and meanwhile greatly improved performance of several natural language processing tasks [51]. We explain this framework in Section 4, as well as its structural relationship the graphical quantum formalism.
- The diagrammatic formalism underpins the automated reasoning software `quantomatic` developed at Oxford and Google by Dixon, Kissinger, Merry, Duncan, Soloviev and Frot—see also Figure 2. More recently, work on automated theory exploration of graphical theories also started at Oxford [61], building further on the work done at Edinburgh [58]. We won't discuss this here; details are in [40, 42, 41] and on the `quantomatic` website.

§2. Minimal process logic. By a process logic we simply mean any strict symmetric monoidal category, and by minimal we mean that at this stage we consider no structure (yet) other than the strict symmetric monoidal structure. We explain this structure in terms of its graphical language.

We could as well have given a symbolic presentation. We refer the reader to [23, 32] for such a symbolic presentation, exemplified for the specific case of cooking processes, and how they compose to make up recipes—[23] also discusses how a process logic explains why tigers have stripes while lions don't.

2.1. Graphical language. The data of a minimal process logic consists of processes, represented by boxes, each of which takes some type of systems as

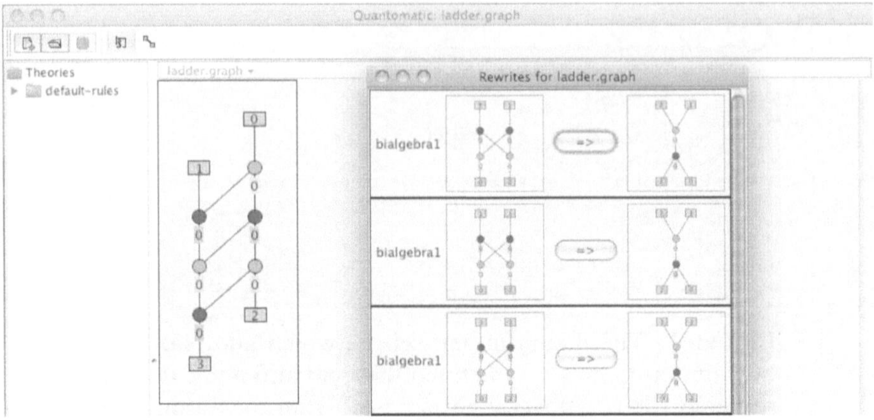

FIGURE 2. Screenshot of the quantomatic software developed in a collaboration between Oxford and Google, which can be downloaded from http://sites.google.com/site/quantomatic/.

its input, represented by (an) input wire(s), and some type of system as its output:

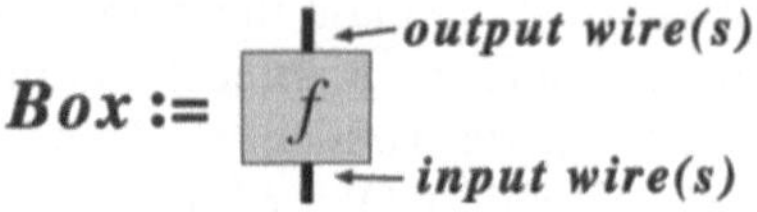

These types may be compound, or trivial, i.e. representing 'no system':

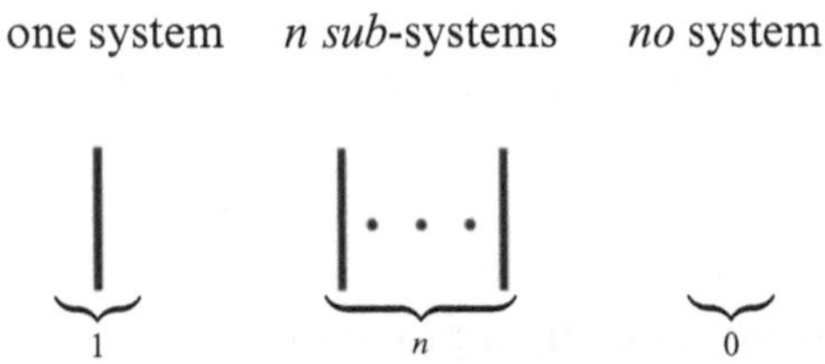

Examples of types could be a particular quantum system, classical data of a certain size, grammatical types, e.g. the type of a noun, verb, or a sentence, etc. A process with no input wire is called a *state*—one can think of these as 'preparation processes'. Those with neither an input type nor an output type are called *values*. A process without an output type is called a *valuation*.

The connectives of a minimal process logic constitute composition of processes. There are two modes: *sequential* or *causal* or *connected* composition,

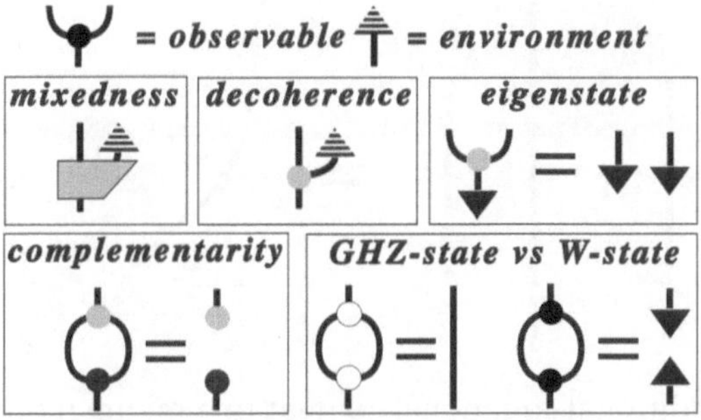

FIGURE 3. Examples of quantum mechanical concepts that can be expressed in purely topological terms, with the help of some new graphical elements. They are taken from [22, 36, 34, 25, 28].

and, *parallel* or *acausal* or *disconnected* composition , respectively depicted as:

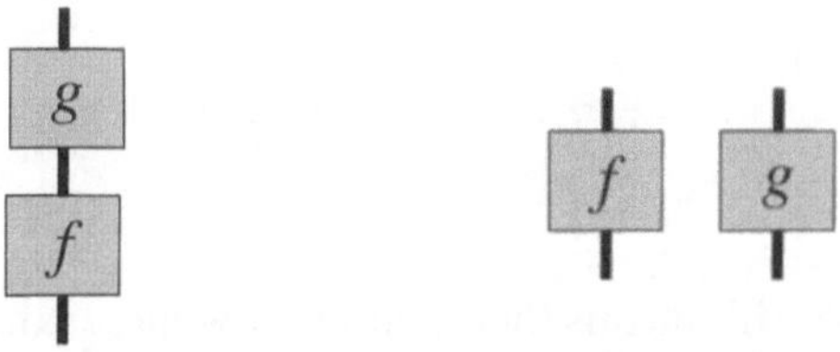

So by post-composing a state with a valuation one obtains a value. Note that sequential composition requires the output type of f to be equal to the input type of g while no such restriction exists for parallel composition .

The formal paradigm underpinning minimal process logic is a topological one:

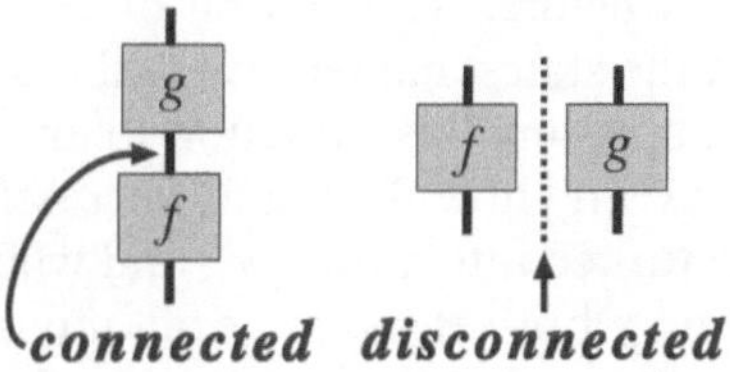

The topology captures 'what interacts with what', a wire standing for interaction while no wire stands for no interaction. It is surprising how many concepts can be expressed purely in these topological terms—e.g. see Figure 3 for some topologically characterized quantum mechanical concepts.

The computational content of minimal process logic boils down to the simple intuitive rule that topologically equivalent diagrams are equal. Hence computation proceeds by topological deformations:

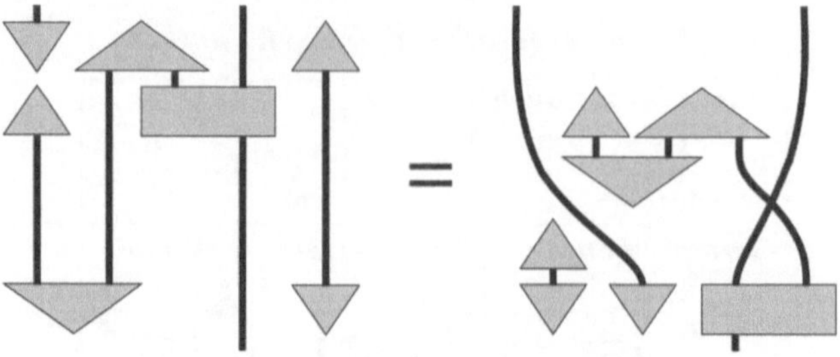

There is no additional equational content to a minimal process logic. This may sound surprising, since a strict symmetric monoidal category is subject to a number of axioms. The explanation is that in the graphical language all these equations become tautologies. For example, denoting sequential composition by $\circ$ and parallel composition by $\otimes$, the 'bifunctoriality equation' $(g \circ f) \otimes (k \circ h) = (g \otimes k) \circ (f \otimes h)$ of monoidal categories becomes:

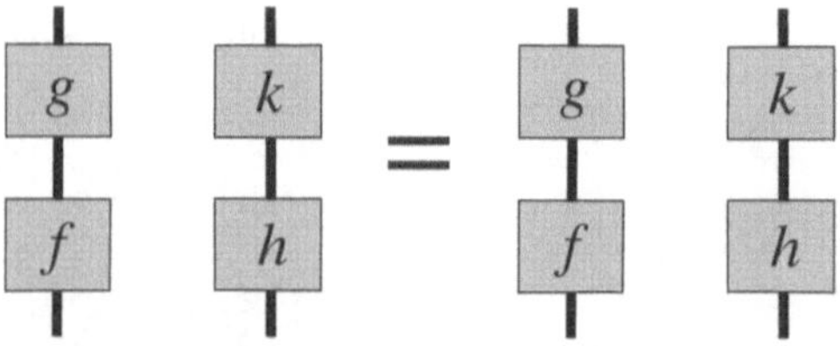

In terms of processes this means that 'g after f, while, k after h' is the same as 'g while k, after, f while h'.

§3. Quantum process logic – Take IIa. Our next goal is to derive some non-trivial quantum phenomena by endowing a minimal process logic with a tiny bit of extra structure, identified by Abramsky and the author in [1, 2].

3.1. Dagger compact structure. The first bit of extra structure will induce some kind of metric on the states, namely, we will ask that each state can be turned into a valuation; applying this valuation to any other state will yield a value. Note that this is exactly how the highly successful Dirac notation [39] works: a ket $|\psi\rangle$ can be turned into a bra $\langle\psi|$, and when composing $\langle\psi|$ with another ket $|\phi\rangle$ we obtain a bra-ket $\langle\psi|\phi\rangle$ i.e. an inner-product. Since states may themselve arise by composing processes other than states, we will allow for the inputs and the outputs of any process to be 'flipped':

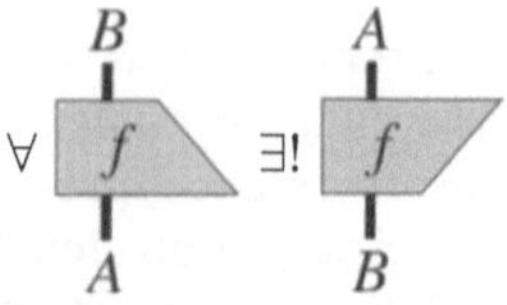

Note that flipping twice yields the original box, so flipping is involutive, and it is also clear that it preserves parallel composition , while it reverses sequential composition . We refer to flipping as the *adjoint* or *dagger*[4]

So far we haven't said anything specific about the parallel composition . Now we will truly follow Schrödinger's path and specify in which manner quantum systems compose differently than classical systems. In other words, we will assert that pure quantum states admit entanglement, diagrammatically:

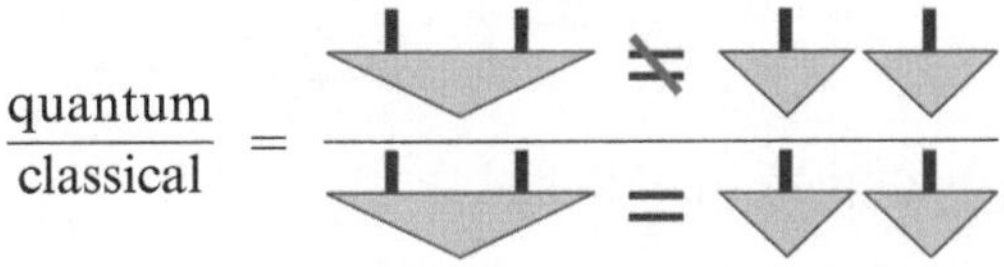

That is, a quantum state of two systems can in general not be described by describing the state of its parts. Note that this is also not the case for probabilistic classical data: a situation of two systems which comes with the promise that the states of the system are the same but unknown, can also not be described by independently describing the state of each system. However, in quantum theory this already occurs for states on which there is no uncertainty, that is, for which there exists a measurement that yields a particular outcome with certainty.

So how do we provide a constructive witness for the fact that the state of two systems does not 'disconnect' in two separate one-system states? Simply by explicitly introducing a special two system state which is obtained by (internally) connecting its two outputs with a cup-shaped wire:[5]

Sticking to our topological paradigm, such a cup-shaped state for example obeys:

(1)

The equivalent symbolic expression for this equation would be:

$$(\cap \otimes 1) \circ (1 \otimes \cup) = 1$$

[4]From the perspective of Birkhoff-von Neumann quantum logic, one could conceive this as the analog to an orthocomplementation on the lattice structure. That is, an order-reversing involution. Note in particularly that for non-Boolean lattices an orthocomplementation is a structure, not a property, as there can exist many different ones on the same lattice. In lattice theoretic terms the linear algebraic adjoint indeed arises as an expression involving Galois adjoints $(-)^*$and orthocomplementation $(-)'$, namely $f^\dagger(a) = (f^*(a'))'$ [46, 29].

[5]Note that a state of two systems doesn't have inputs, so this is more like "internal wiring" of its outputs e.g. the bad way to fix an old fashion fuse by means of a copper wire.

where 1 stands for a single straight wire and  is obtained simply by flipping  i.e. its adjoint. We obtain a strict *dagger-compact category* [1, 2].

3.2. Deriving physical phenomena. We assumed the existence of an adjoint for any box and represent it via flipping. Cup- and cap-shaped wires also enable us to 'define' the *transpose* which we depict by rotating a box 180°:

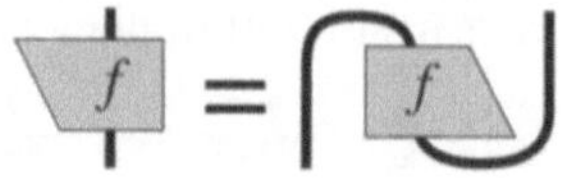

It then immediately follows that we have:

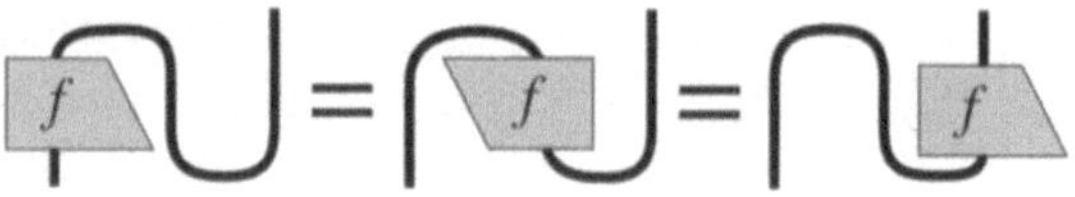

that is, we can slide boxes across cup- and cap-shaped wires. Going berserk,

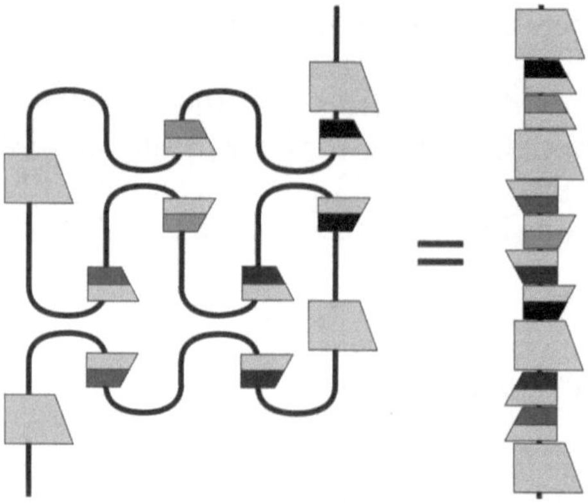

that is, we can treat the entire graphical calculus for dagger compact categories in terms of beads which slide on wires. Now for some physics. We have:

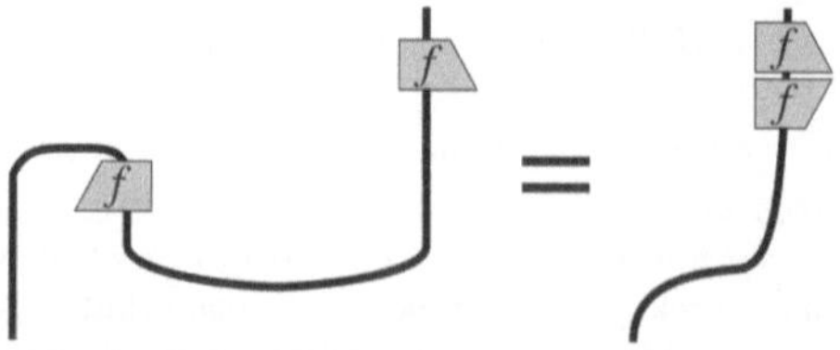

and we choose f such that its composite with its adjoint yields the identity, something to which we refer as *unitarity* Hence:

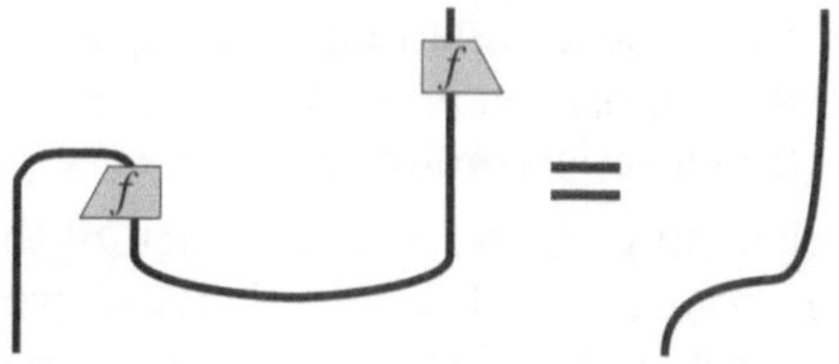

Introducing agents Alice and Bob yields *quantum teleportation*:

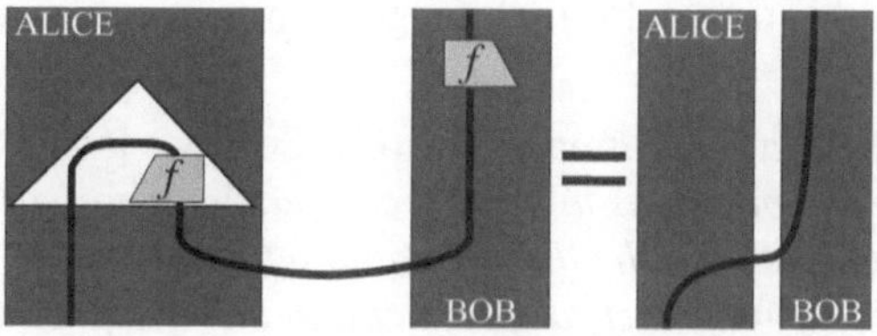

Note that, given that the quantum mechanical formalism was born in 1932, that this phenomenon took 60 years to be discovered [12]. The standard quantum mechanical formalism provides no indication whatsoever that something like this would be possible, so one had to rely on sheer luck to discover it.

A more detailed discussion of this graphical derivation and its physical interpretation is in [20, 23]. Similarly we derive another quantum mechanical feature, the *entanglement swapping* protocol [81]:

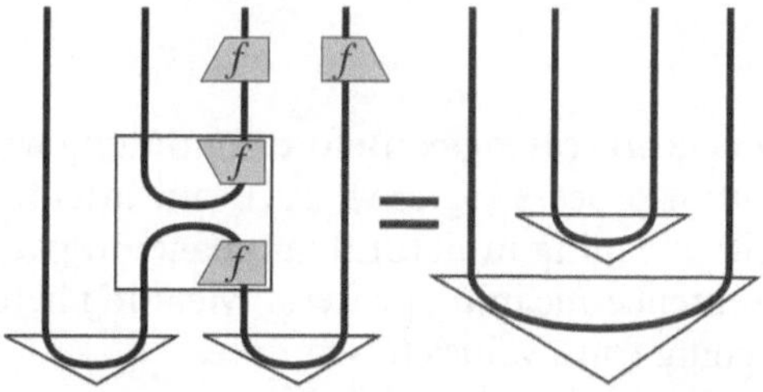

So how much quantum mechanics can we derive in this calculus?

3.3. Logical completeness wrt Hilbert spaces. The diagrammatic language presented above is directly related to the symbolic notion of a dagger compact category as follows:

THEOREM 1 (Kelly-Laplaza; Selinger [60, 77]). *An equational statement between expressions in the dagger compact categorical language holds if and only if it is derivable in the above described graphical calculus.*

Evidently there are many dagger compact categories, to mention two:

- Wires represent finite dimensional Hilbert spaces, boxes linear maps, the dagger is the linear algebraic adjoint, sequential composition is ordinary function composition , and the parallel composition of wires is the tensor product while parallel composition of boxes is the Kronecker product.

- Wires represent sets, boxes relations, the dagger is the relational converse, sequential composition is composition of relations, and parallel composition is the cartesian product.

The description of the compact structure for each of these as well as some more examples can be found in [32]. Evidently these two examples have very different spaces and one would evidently not associate sets and relations with quantum processes. Hence one could wonder how much one can actually derive in (the graphical calculus for) dagger compact categories. The answer is surprising.

THEOREM 2 (Hasegawa-Hofmann-Plotkin; Selinger [55, 78]). *An equational statement between expressions in dagger compact categorical language holds if and only if it is derivable in the dagger compact category of finite dimensional Hilbert spaces, linear maps, tensor product and linear algebraic adjoints.*

To put this in more quantum physics related terms, any equation involving:

- states, operations, effects, ...
- Bell-state, Bell-effect, transposition, conjugation, ...
- inner-product, linear-algebraic trace, Hilbert-Schmidt norm, ...
- adjoints (e.g. self-adjointness and unitarity), projections, positivity, ...
- complete positivity (cf. [77]), ...

holds in quantum theory if and only if it can be derived in the graphical language.

§4. Natural language process logic. Before continuing with the further development of quantum process logic, we turn our attention on something completely different: meaning in natural language, in particular, the from-word-meaning-to-sentence-meaning process. Meaning here manifestly goes beyond simply assigning truth values to sentences.

4.1. From word meaning to sentence meaning. Consider as given the meanings of words. This can mean many things, for example, one has a dictionary available. On the other hand, there are no dictionaries for entire sentences. So how do we know what a sentence means? There must be some kind of mechanism, used by all of us, for transforming the meaning of words into the meaning of a sentence, since surely, we all understand sentences that we may have never heard before in our lives, provided we understand all of its words.

There is a technological side to this. Search engines such as google and other natural language processing tools also have an understanding of meanings of words which they use to provide us with the most relevant outputs for our queries. The model of word meaning which these engines employ enables them to produce outputs that include words that are closely related to the words in our query, i.e. there doesn't have to be an exact match.

However, searching on Google for "I want something that allows me to go faster than when I only use my legs" returns among its top hits: "Difference Between Oxycontin and Oxycodone", "What are good ways for a girl to [XXX]", "How to Sprint Faster: 6 steps – wikiHow', "My Story – Onelegtim.com – Retired Police Officer & [...]" and "Golf Swing Power: What Your Legs Should Be Doing [...]". Neither of these point me in the direction of appropriate vehicles that would serve my purpose, so clearly there is no understanding of the meaning of my query. The reason is the lack of a theory that produces the meaning of a sentence from the meanings of its words, whatever the manner is in which we describe the meaning of words.

Now, representing grammatical types of words by wires and their meanings by state-boxes we can depict a string of words as:

But the overall type, i.e. the overall wire structure, depends on the grammatical structure of the sentence. However, sentences with different grammatical structure may have the same meaning, and more generally, we would like to have a fixed type for the meaning of all sentences. Hence there is some process, the from-word-meaning-to-sentence-meaning process, which transforms the meanings of the string of words in the meaning of the sentence made up from these:

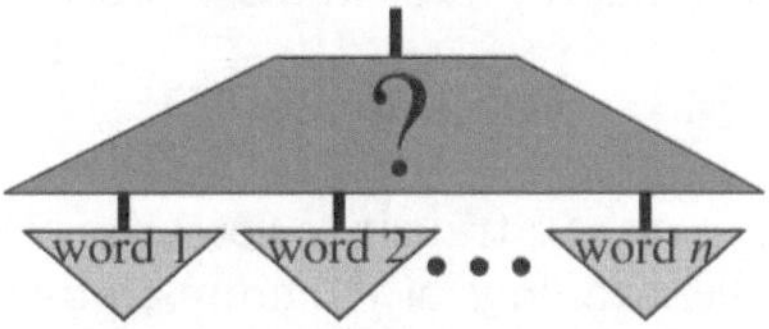

What drives this process? That is, given a string of words, what mediates their interaction? The answer is obvious:

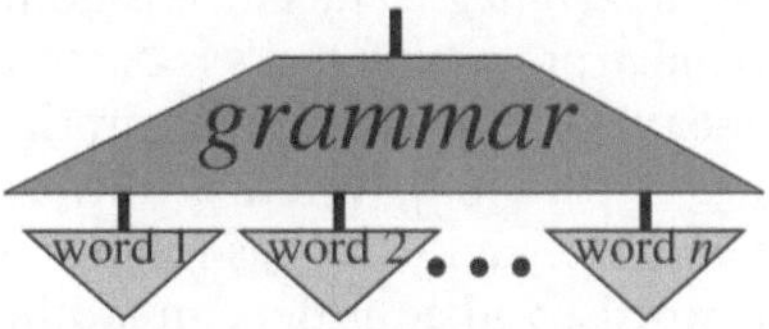

since grammatically incorrect sentences have no clear meaning anyway.

We can now describe the problem for from-word-meaning-to-sentence-meaning processes in more precise terms:

- Given a theory of word meaning, and given a theory of grammar, how can we combine these into an algorithm which produces the meaning of sentences from the meanings of its words?

As already mentioned, this problem was addressed by Clark, Sadrzadeh and the author in [19, 37]. Let's stay at an abstract level a bit longer, before we will describe concrete theories of word meaning and grammar. What is a verb? A transitive verb is something that requires an object and a subject in order to yield a grammatically correct sentence. So we can think of a transitive verb as a process with three wires, two respectively requiring an object and a subject, and one producing the sentence:

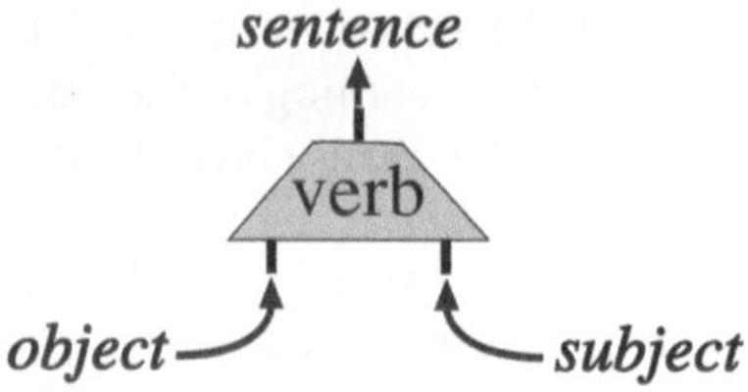

Since we rather represent a verb as a state we can use transposition, as defined above, to turn inputs into outputs and represent the verb as:

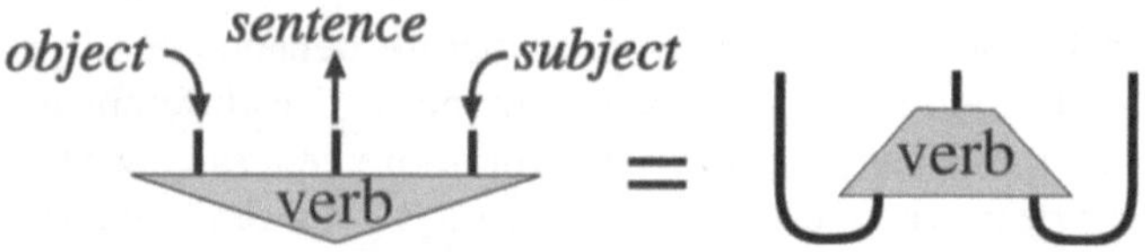

You may ask where these cups suddenly come from, but here we already anticipate the description of grammatical structure that we discuss below.

Note in particular also that for these kinds of word-states we again have:

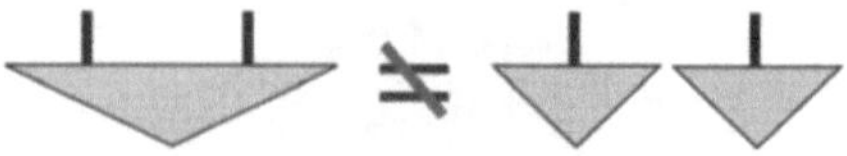

since otherwise, for the case of a transitive verb, the meaning of the sentence would not depend on the meanings of the nouns, which could have dramatic consequences. For example, considering the verb 'hate', it would be sufficient for one person to hate another person in order for everyone to hate everyone.

4.1.1. *A theory for word meaning.* The current dominant theory of word meaning for natural language processing tasks is the so-called *distributional* or *vector space model* of meaning [75]. It takes inspiration from Wittgensteins' philosophy of 'meaning is use' [80], whereby meanings of words can be determined from their context, and works as follows. One fixes a collection of n words, the *context* words, and considers an n dimensional vector space with chosen basis where each basis vector represents one of the context words. Then one selects a huge body of written text, the *corpus*. E.g. the internet, all editions of a certain newspaper, all novels, the British National Corpus[6] which is a 100 million word collection of samples of written and spoken language from a wide range of sources, etc. Next one decides on a *scope*, that is, a small

[6]This can be accesses at http://www.natcorp.ox.ac.uk/.

integer k, and for each context word x one counts how many times $N_x(a)$ a word a to which one wants assign a meaning occurs at a distance of at most k words from x. One obtains a vector $(N_1(a), \ldots, N_n(a))$, which one normalizes in order to obtain $(\phi_1(a), \ldots, \phi_n(a))$, the *meaning vector* of a. Now, in order to compare meanings of words, in particular, how closely their meanings are related, one can simply compute the inner-product of their meaning vectors.

4.1.2. *A theory of grammar.* Algebraic gadgets that govern grammatical types have been around for quite a bit longer [4, 10, 18, 64]. There are several variants available, each with their pro's and con's; here we will focus on Lambek's *pregroups* [65]. Philosophically, these algebraic gadgets trace back to Freges' principle that the meaning of a sentence is a function of the meaning of its parts [48]. However, this is only manifest in that these algebras all have a composition operation that allows to build larger strings of words from smaller strings of words. These algebras also have a relation $\leq$ where $a \cdot \cdots \cdot z \leq t$ means that the string of types $a \ldots z$ has as its overall type t. For example, $n \cdot tv \cdot n \leq s$ expresses the fact that a noun, a transitive verb and a noun make up a sentence s. Finally, there are additional operations subject to certain laws which make up the actual structure of the algebra, and these would allow one to derive correct statements such as $n \cdot tv \cdot n \leq s$.

For the specific case of pregroups, these additional operations are a *left inverse* ${}^{-1}(-)$ and a *right inverse* $(-)^{-1}$, subject to $x \cdot {}^{-1}(x) \leq 1$ and $(x)^{-1} \cdot x \leq 1$ where 1 is the unit for the composition operation, as well as to $1 \leq {}^{-1}(x) \cdot x$ and $1 \leq x \cdot (x)^{-1}$. Now we have to assign grammatical types to the elements of a pregroup. Some will be atomic, i.e. indecomposable, while others like transitive verbs will be assigned compound types. Concretely, $tv = {}^{-1}(n) \cdot s \cdot (n)^{-1}$, hence

$$n \cdot tv \cdot n = n \cdot \left({}^{-1}(n) \cdot s \cdot (n)^{-1}\right) \cdot n = \left(n \cdot {}^{-1}(n)\right) \cdot s \cdot \left((n)^{-1} \cdot n\right) \leq 1 \cdot s \cdot 1 = s,$$

so the string of types 'noun transitive verb noun' indeed makes up a grammatically correct sentence. We can depict this computation graphically as follows. We start with five systems of respective types n, ${}^{-1}(n)$, s, $(n)^{-1}$ and n:

Then, we use caps to indicate that n and ${}^{-1}(n)$, and, $(n)^{-1}$ and n, cancel out:

so that at the end the only remaining system is the sentence type. The caps here represent the equations $x \cdot {}^{-1}(x) \leq 1$ and $(x)^{-1} \cdot x \leq 1$. In fact, this is not just an analogy with the graphical language of compact categories. Pregroups *are* in fact compact categories! To see this, any partial order is a category, the composition provides the tensor, and while equations $x \cdot {}^{-1}(x) \leq 1$ and $(x)^{-1} \cdot x \leq 1$ provide caps, equations $1 \leq {}^{-1}(x) \cdot x$ and $1 \leq x \cdot (x)^{-1}$ provide cups. More details on this are in [37]. The reason that there are two kinds of caps and cups is the fact that we are not allowed to change the order of words in a sentence while two physical systems do not come with some ordering. In category-theoretic terms, here we are dealing with a *non-symmetric tensor*

4.2. Combining theories. The structural similarity between the pregroup theory of grammar, and the vector spaces for word meaning when organized as a dagger compact category, is exactly what we will exploit to explicitly construct the from-word-meaning-to-sentence-meaning process. We consider the graphical representation of the proof of grammatical correctness of a sentence, substitute the sentence types by meaning vectors of the particular words we are interested in, and substitute the caps by the vector space caps, so we obtain:

where the dotted line indicates the linear map that when applied to the vector $\overrightarrow{Alice} \otimes \overrightarrow{hates} \otimes \overrightarrow{Bob}$ produces the vector that we take to be the meaning of a sentence. By rewriting this using transposition, as in Section 4.1, the verb now acts as a fuction on the object and the subject:

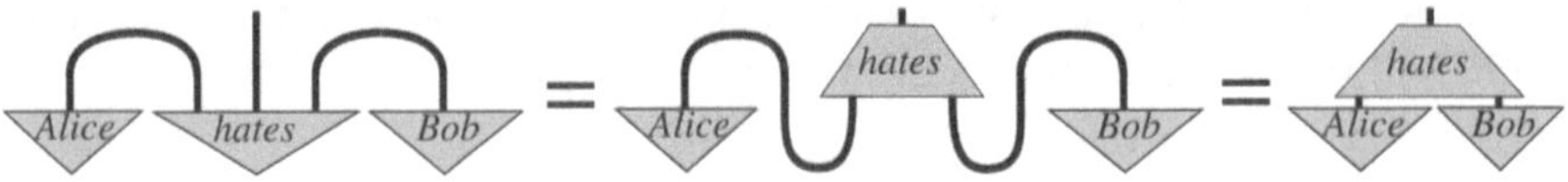

The meanings of all sentences live in the same vector space so we can again simply use the inner-product to measure their similarity. Grefenstette and Sadrzadeh have recently exploited this theory for standard natural processing tasks and their method outperforms all existing ones [51].

What about the cups? They can be used to model 'special words' like "does" and "not", which have a clear 'logical' meaning. Here is an example of this:

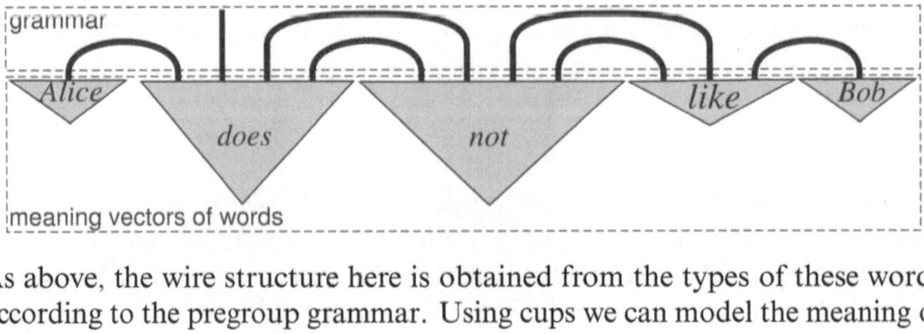

As above, the wire structure here is obtained from the types of these words according to the pregroup grammar. Using cups we can model the meaning of 'does', that is, 'does nothing really', and 'not', that is, 'negates meaning', for which we use an input-output *not*-box that does just that:

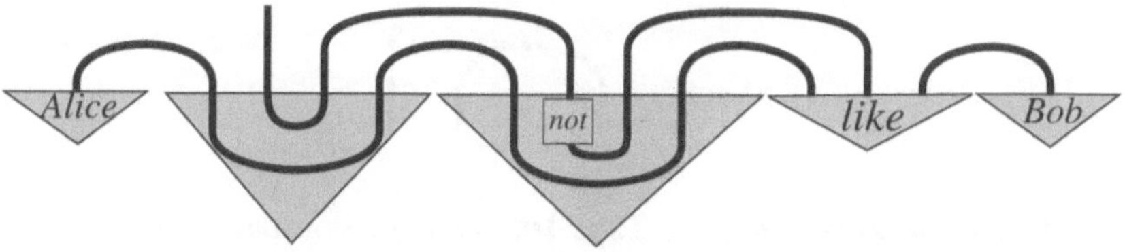

and then we can simply use homotopy to compute:

which is exactly what we would expect the meaning of 'Alice does not like Bob' to be: the negation of Alice liking Bob. This example also shows how the wires are mediating the 'flow' of word meaning in sentences. They allow for the words Alice and like, while far apart in the sentence, to interact.

Turning things upside-down, one can now ask the question: why are there algebraic gadgets that describe grammatical correctness, i.e. why do these even exist. Our theory of word meaning explains this: they witness the manner of how word meanings interact to form the meaning of a sentence.

4.3. An aside: quantizing grammar. An interesting analogy arises, which was first observed by Louis Crane, and which is discussed in detail in [72]. An important area of contemporary mathematics is the study of Topological Quantum Field Theory (TQFT) [6, 8, 7]. While it takes its inspiration from quantum field theory, it has become an area of research in its own right, mainly within topology. The object of study is a monoidal functor:

$$F : n\mathbf{Cob} \to \mathbf{FVect}_{\mathbb{K}} ::$$ 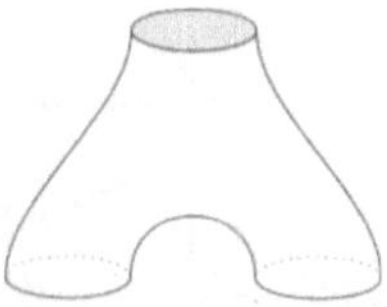 $$\mapsto V$$

from the compact category of closed $(n-1)$-dimensional manifolds with diffeomorphism classes of n-dimensional manifolds connecting the closed $(n-1)$-dimensional manifolds as morphisms, to the compact category of vector spaces over some field $\mathbb{K}$.[7] Now, rather than taking a category of topological structures as domain, we can take a pregroup as domain, i.e. a category of grammatical structures, and obtain a *grammatical quantum field*

[7] If the field has a non-trivial involution then this category has a dagger too ($\neq$ transposition).

theory:

$F : Pregroup \to \mathbf{FVect}_{\mathbb{R}_+} ::$  $\mapsto V$

§5. Quantum process logic – Take IIb. Dagger compact categories capture a substantial number of quantum mechanical concepts, and the dagger compact category **FdHilb** related to the von Neumann model described in Section 3.3, is complete with respect to them. But they are by no means universal with respect to quantum theory, by which we can mean two different things:

- that they do not capture all quantum mechanical concepts, and,
- that the language is not rich enough to describe all processes in **FdHilb**.

Examples of concepts that are not captured by dagger compact language are the classical data obtained in measurements, observables themselves, and relationships between these e.g. complementarity. Examples of **FdHilb**-processes not expressible in dagger compact language are basic quantum computational gates such as the CNOT-gate, phase-gates etc. We will now present an extended graphical language which does capture all of these. This was established in a series of papers by Pavlovic, Paquette, Duncan, and the author in [34, 33, 24, 25]. The calculus was also rich enough to address a number of concrete quantum computational and quantum foundational problems e.g. see [44, 27, 57, 14, 26].

Rather than only allowing for wires we allow for 'dots' at which wires branch into multiple wires, or none. We refer to these dots as ...

'spiders' $=$ 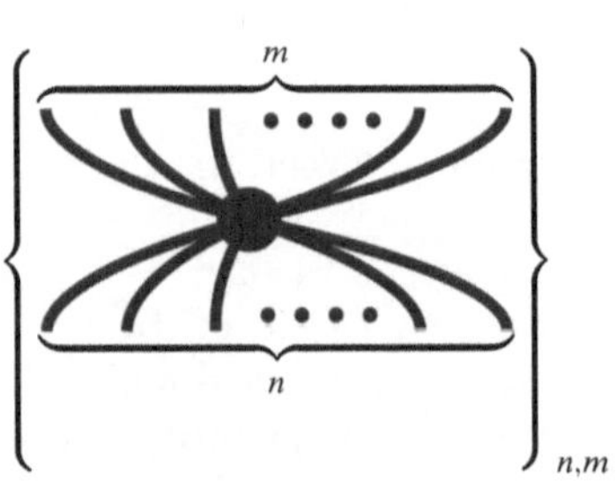

So what is the analogue of the topological calculus with cups and caps, and in particular, eq.(1)? Similarly to 'however one bends a wire, it still remains just a wire that acts as an identity', any web of spiders with the same overall number of inputs and outputs, independent of how the web is build up, is again the same. So for any $k > 0$:

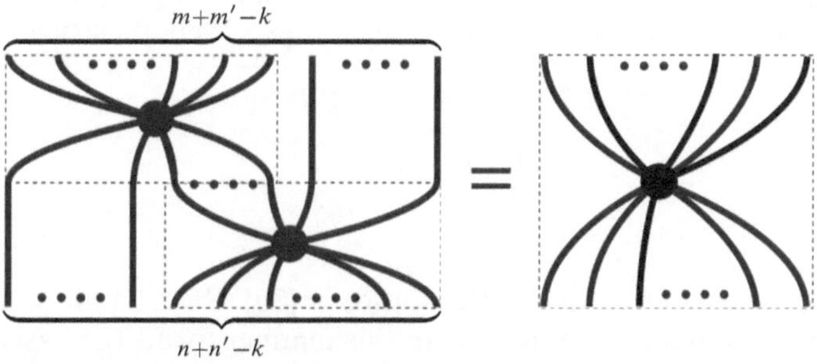

Hence, the rule governing spider calculus is that if two spiders 'shake legs', they fuse together. Again in other words, it only matters what is connected to what, but not the manner in which this connection is realized.

This in particular implies that for the specific spiders:

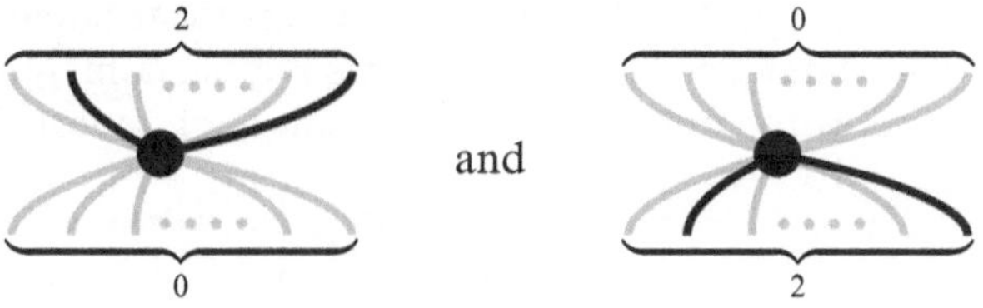

we obtain eq.(1):

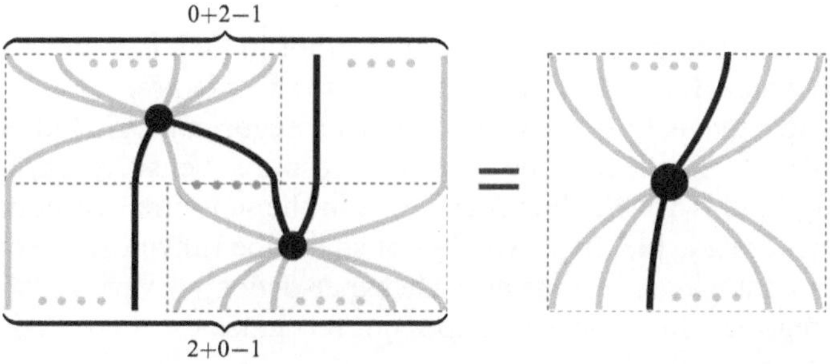

so *reasoning with spiders strictly generalizes reasoning with wires*

In **FdHilb** a family of spiders of the above kind on-the-nose captures an orthonormal basis, which is a non-trival result. Firstly, one can show that reasoning with those spiders is equivalent to working with a so-called dagger special commutative Frobenius algebra [62, 63, 31]. Next one shows that these dagger special commutative Frobenius algebras in **FdHilb** are the same thing as orthonormal bases [35]. Since bases allow to represent observables and classical data, we almost reached our goal, except for the fact that quantum theory only becomes interesting if we consider several 'incompatible' bases.

So now we consider two different families of spiders, represented by a different gray scale. What happens if a dark gray and a light gray spider which represent complementary observables 'shake legs'? Well, their 'legs fall off':

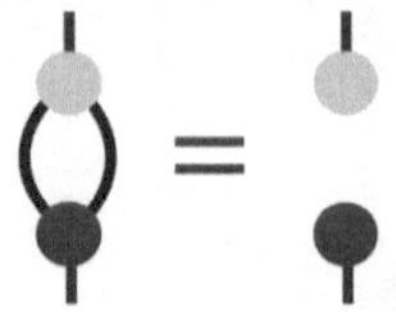

This was shown by Duncan and the author in [25]. Such a pair of differently colored spider families that interact in this manner forms the basis of a rich calculus with many more extra features than the ones described here. We refer the interested reader to [23, 25, 26] for more details and concrete applications.

§6. **The remaining challenge.** In this paper we pushed forward the idea that the diagrammatic languages describing quantum phenomena as well as meaning-related linguistic phenomena may constitute some new kind of quantitative logic. The same logic also governs Bayesian inference, Bayesian inversion boiling down to nothing but transposition for appropriately chosen cups and caps:

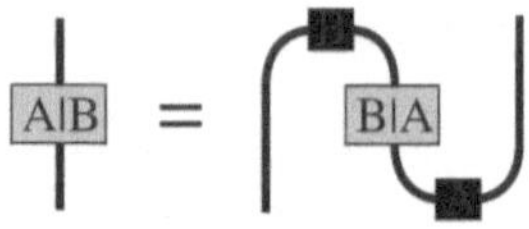

This was established by Spekkens and the author in [38], to which we refer for details. So where does traditional logic fit into this picture?

One perspective is to start with standard categorical logic [66, 3, 9]. The compact structure can then be seen as a resource sensitive variant (as in Linear Logic [50, 76]) which is degenerate in the sense that conjunction and disjunction coincide [43, 21].[8] We do not subscribe (anymore) to conceiving the diagrammatic logic as a 'degenerate hyper-deductive variant' of standard logic in categorical form since this does not recognize the quantitative nor the process content.

Rather, we would like to conceive the quantitative diagrammatic logic as 'the default thing' from which traditional qualitative logic arises via some kind of structural collapse. There are several results that could be taken as a starting point in this direction, for example, the generalization in [33] of Carboni and Walters' axiomatization of the category of relations [16]. But since this still belongs to the world of speculation, we leave this to future writings.

[8]There is also ongoing work on relating traditional quantum logic with dagger compact categories or related structures at a purely structural level e.g. [53, 56].

REFERENCES

[1] S. Abramsky and B. Coecke, *A categorical semantics of quantum protocols*, ***Proceedings of the 19th Annual IEEE Symposium on Logic in Computer Science (LICS)***, IEEE Computer Society, 2004, Extended version: arXiv:quant-ph/0402130, pp. 415–425.

[2] ———, *Abstract physical traces*, ***Theory and Applications of Categories***, vol. 14(6) (2005), pp. 111–124.

[3] S. Abramsky and N. Tzevelekos, *Introduction to categories and categorical logic*, ***New Structures for Physics*** (B. Coecke, editor), Lecture Notes in Physics, Springer, 2011, pp. 3–94.

[4] K. Ajdukiewicz, *Die syntaktische Konnexität*, ***Studia Philosophica***, vol. 1 (1937), pp. 1–27.

[5] J. Aron, *Quantum links let computers read*, ***New Scientist***, vol. 208(2790) (2010), pp. 10–11.

[6] M. Atiyah, *Topological quantum field theories*, ***Publications Mathématiques de l'IHÉS***, vol. 68(1) (1988), pp. 175–186.

[7] J. C. Baez, *Quantum quandaries: a category-theoretic perspective*, ***The Structural Foundations of Quantum Gravity*** (D. Rickles, S. French, and J. T. Saatsi, editors), Oxford University Press, 2006, arXiv:quant-ph/0404040, pp. 240–266.

[8] J. C. Baez and J. Dolan, *Higher-dimensional algebra and topological quantum field theory*, ***Journal of Mathematical Physics***, vol. 36 (1995), p. 6073, arXiv:q-alg/9503002.

[9] J. C. Baez and M. Stay, *Physics, topology, logic and computation: a Rosetta stone*, ***New Structures for Physics*** (B. Coecke, editor), Lecture Notes in Physics, Springer, 2011, pp. 95–172.

[10] Y. Bar-Hillel, *A quasiarithmetical notation for syntactic description*, ***Language***, vol. 29 (1953), pp. 47–58.

[11] J. Benabou, *Categories avec multiplication*, ***Comptes Rendus des Séances de l'Académie des Sciences. Paris***, vol. 256 (1963), pp. 1887–1890.

[12] C. H. Bennett, G. Brassard, C. Crepeau, R. Jozsa, A. Peres, and W. K. Wootters, *Teleporting an unknown quantum state via dual classical and einstein-podolsky-rosen channels*, ***Physical Review Letters***, vol. 70(13) (1993), pp. 1895–1899.

[13] G. Birkhoff and J. von Neumann, *The logic of quantum mechanics*, ***Annals of Mathematics***, vol. 37 (1936), pp. 823–843.

[14] S. Boixo and C. Heunen, *Entangled and sequential quantum protocols with dephasing*, ***Physical Review Letters***, vol. 108 (2012), p. 120402.

[15] A. Bundy, F. Cavallo, L. Dixon, M. Johansson, and R. McCasland, *The Theory behind TheoryMine*.

[16] A. Carboni and R. F. C. Walters, *Cartesian bicategories I*, ***Journal of Pure and Applied Algebra***, vol. 49 (1987), pp. 11–32.

[17] G. Chiribella, G. M. D'Ariano, and P. Perinotti, *Informational derivation of quantum theory*, ***Physical Review A***, vol. 84 (2011), no. 1, p. 012311.

[18] N. Chomsky, *Tree models for the description of language*, ***I.R.E. Transactions on Information Theory***, vol. IT-2 (1956), pp. 113–124.

[19] S. Clark, B. Coecke, and M. Sadrzadeh, *A compositional distributional model of meaning*, ***Proceedings of the Second Quantum Interaction Symposium (QI-2008)***, 2008, pp. 133–140.

[20] B. Coecke, *Kindergarten quantum mechanics — lecture notes*, ***Quantum Theory: Reconsiderations of the Foundations III*** (A. Khrennikov, editor), AIP Press, 2005, arXiv:quant-ph/0510032, pp. 81–98.

[21] ———, *Automated quantum reasoning: Non logic – semi-logic – hyper-logic*, ***AAAI Spring Symposium: Quantum Interaction***, AAAI, 2007, pp. 31–38.

[22] ———, *Axiomatic description of mixed states from Selinger's CPM-construction*, ***Electronic Notes in Theoretical Computer Science***, vol. 210 (2008), pp. 3–13.

[23] ———, *Quantum picturalism*, ***Contemporary Physics***, vol. 51 (2009), pp. 59–83, arXiv:0908.1787.

[24] B. Coecke and R. Duncan, *Interacting quantum observables*, ***Proceedings of the 37th International Colloquium on Automata, Languages and Programming (ICALP)***, Lecture Notes in Computer Science, 2008.

[25] ———, *Interacting quantum observables: categorical algebra and diagrammatics*, ***New Journal of Physics***, vol. 13 (2011), p. 043016, arXiv:quant-ph/09064725.

[26] B. Coecke, R. Duncan, A. Kissinger, and Q. Wang, *Strong complementarity and non-locality in categorical quantum mechanics*, ***Proceedings of the 27th Annual IEEE Symposium on Logic in Computer Science (LICS)***, IEEE Computer Society, 2012, arXiv:1203.4988.

[27] B. Coecke, B. Edwards, and R. W. Spekkens, *Phase groups and the origin of non-locality for qubits*, ***Electronic Notes in Theoretical Computer Science***, vol. 270(2) (2011), arXiv:1003.5005.

[28] B. Coecke and A. Kissinger, *The compositional structure of multipartite quantum entanglement*, ***Automata, Languages and Programming***, Lecture Notes in Computer Science, Springer, 2010, Extended version: arXiv:1002.2540, pp. 297–308.

[29] B. Coecke and D. J. Moore, *Operational Galois adjunctions*, ***Current Research in Operational Quantum Logic: Algebras, Categories and Languages*** (D. J. Moore B. Coecke and A. Wilce, editors), Fundamental Theories of Physics, vol. 111, Springer-Verlag, 2000, pp. 195–218.

[30] B. Coecke, D. J. Moore, and A. Wilce, *Operational quantum logic: An overview*, ***Current Research in Operational Quantum Logic: Algebras, Categories and Languages*** (B. Coecke, D. J. Moore, and A. Wilce, editors), Fundamental Theories of Physics, vol. 111, Springer-Verlag, 2000, arXiv:quant-ph/0008019, pp. 1–36.

[31] B. Coecke and E. O. Paquette, *POVMs and Naimark's theorem without sums*, ***Electronic Notes in Theoretical Computer Science***, vol. 210 (2008), pp. 15–31, arXiv:quant-ph/0608072.

[32] ———, *Categories for the practicing physicist*, ***New Structures for Physics*** (B. Coecke, editor), Lecture Notes in Physics, Springer, 2011, arXiv:0905.3010, pp. 167–271.

[33] B. Coecke, E. O. Paquette, and D. Pavlovic, *Classical and quantum structuralism*, ***Semantic Techniques in Quantum Computation*** (S. Gay and I. Mackie, editors), Cambridge University Press, 2010, arXiv:0904.1997, pp. 29–69.

[34] B. Coecke and D. Pavlovic, *Quantum measurements without sums*, ***Mathematics of Quantum Computing and Technology*** (L. Kauffman G. Chen and S. Lamonaco, editors), Taylor and Francis, 2007, arXiv:quant-ph/0608035, pp. 567–604.

[35] B. Coecke, D. Pavlovic, and J. Vicary, *A new description of orthogonal bases*, ***Mathematical Structures in Computer Science***, 2011, to appear; arXiv:quant-ph/0810.1037.

[36] B. Coecke and S. Perdrix, *Environment and classical channels in categorical quantum mechanics*, ***Proceedings of the 19th EACSL Annual Conference on Computer Science Logic (CSL)***, Lecture Notes in Computer Science, vol. 6247, 2010, Extended version: arXiv:1004.1598, pp. 230–244.

[37] B. Coecke, M. Sadrzadeh, and S. Clark, *Mathematical foundations for a compositional distributional model of meaning*, ***Linguistic Analysis***, vol. 36 (2010), pp. 345–384.

[38] B. Coecke and R. W. Spekkens, *Picturing classical and quantum Bayesian inference*, ***Synthese***, (2011), pp. 1–46, arXiv:1102.2368.

[39] P. A. M. Dirac, ***The principles of quantum mechanics (third edition)***, Oxford University Press, 1947.

[40] L. Dixon and R. Duncan, *Graphical reasoning in compact closed categories for quantum computation*, ***Annals of Mathematics and Artificial Intelligence***, vol. 56(1) (2009), pp. 23–42.

[41] L. Dixon, R. Duncan, B. Frot, A. Merry, A. Kissinger, and M. Soloviev, quantomatic, 2011, http://dream.inf.ed.ac.uk/projects/quantomatic/.

[42] L. Dixon and A. Kissinger, *Open graphs and monoidal theories*, ***Mathematical Structures in Computer Science***, 2011, to appear; arXiv:1011.4114.

[43] R. Duncan, ***Types for Quantum Computation***, Ph.D. thesis, Oxford University, 2006.

[44] R. Duncan and S. Perdrix, *Rewriting measurement-based quantum computations with generalised flow*, ***Proceedings of ICALP***, Lecture Notes in Computer Science, Springer, 2010, pp. 285–296.

[45] S. Eilenberg and S. Mac Lane, *General theory of natural equivalences*, ***Transactions of the American Mathematical Society***, vol. 58(2) (1945), p. 231.

[46] Cl-.A. Faure, D. J. Moore, and C. Piron, *Deterministic evolutions and Schrödinger flows*, ***Helvetica Physica Acta***, vol. 68(2) (1995), pp. 150–157.

[47] D. J. Foulis and C. H. Randall, *Operational statistics. I. Basic concepts*, ***Journal of Mathematical Physics***, vol. 13(11) (1972), pp. 1667–1675.

[48] G. Frege, *Über Sinn und Bedeutung*, ***Zeitschrift für Philosophie und Philosophische Kritik***, vol. 1007 (1892), pp. 25–50.

[49] G. M. D'Ariano G. Chiribella and P. Perinotti, *Probabilistic theories with purification*, ***Physical Review A***, vol. 81 (2010), no. 6, p. 062348.

[50] J.-Y. Girard, *Linear logic*, ***Theoretical Computer Science***, vol. 50(1) (1987), pp. 1–101.

[51] E. Grefenstette and M. Sadrzadeh, *Experimental support for a categorical compositional distributional model of meaning*, ***EMNLP***, ACL, 2011, pp. 1394–1404.

[52] H. Halvorson, ***Deep Beauty: Understanding the Quantum World Through Mathematical Innovation***, Cambridge University Press, 2011.

[53] J. Harding, *A link between quantum logic and categorical quantum mechanics*, ***International Journal of Theoretical Physics***, vol. 48(3) (2009), pp. 769–802.

[54] L. Hardy, *A formalism-local framework for general probabilistic theories including quantum theory*, ***arXiv:1005.5164***, (2010).

[55] M. Hasegawa, M. Hofmann, and G. D. Plotkin, *Finite dimensional vector spaces are complete for traced symmetric monoidal categories*, ***Pillars of Computer Science*** (A. Avron, N. Dershowitz, and A. Rabinovich, editors), Lecture Notes in Computer Science, vol. 4800, Springer, 2008, pp. 367–385.

[56] C. Heunen and B. Jacobs, *Quantum logic in dagger kernel categories*, ***Order***, vol. 27(2) (2010), pp. 177–212.

[57] C. Horsman, *Quantum picturalism for topological cluster-state computing*, ***New Journal of Physics***, vol. 13 (2011), p. 095011, arXiv:1101.4722.

[58] M. Johansson, L. Dixon, and A. Bundy, *Conjecture synthesis for inductive theories*, ***Journal of Automated Reasoning***, vol. 47(3) (2011), pp. 251–289.

[59] A. Joyal and R. Street, *The geometry of tensor calculus I*, ***Advances in Mathematics***, vol. 88 (1991), pp. 55–112.

[60] G. M. Kelly and M. L. Laplaza, *Coherence for compact closed categories*, ***Journal of Pure and Applied Algebra***, vol. 19 (1980), pp. 193–213.

[61] A. Kissinger, *Synthesising graphical theories*, ***arXiv:1202.6079***, (2012).

[62] J. Kock, ***Frobenius Algebras and 2D Topological Quantum Field Theories***, vol. 59, Cambridge University Press, 2004.

[63] S. Lack, *Composing PROPs*, ***Theory and Applications of Categories***, vol. 13 (2004), pp. 147–163.

[64] J. Lambek, *The mathematics of sentence structure*, ***American Mathematics Monthly***, vol. 65 (1958), pp. 154–170.

[65] ———, *Type grammar revisited*, ***Logical Aspects of Computational Linguistics***, Lecture Notes in Computer Science, vol. 1582, 1999, pp. 1–27.

[66] J. Lambek and P. J. Scott, ***Introduction to Higher Order Categorical Logic***, Cambridge University Press, 1988.

[67] S. Mac Lane, *Natural associativity and commutativity*, ***The Rice University Studies***, vol. 49(4) (1963), pp. 28–46.

[68] G. M. Mackey, ***The Mathematical Foundations of Quantum Mechanics***, W. A. Benjamin, New York, 1963.

[69] D. J. MOORE, *On state spaces and property lattices*, ***Studies in History and Philosophy of Modern Physics***, vol. 30(1) (March 1999), pp. 61–83.

[70] R. PENROSE, *Applications of negative dimensional tensors*, ***Combinatorial Mathematics and its Applications***, Academic Press, 1971, pp. 221–244.

[71] C. PIRON, ***Foundations of Quantum Physics***, W. A. Benjamin, 1976.

[72] A. PRELLER and M. SADRZADEH, *Bell states and negative sentences in the distributed model of meaning*, ***Electronic Notes in Theoretical Computer Science***, vol. 270(2) (2011), pp. 141–153.

[73] M. REDEI, *Why John von Neumann did not like the Hilbert space formalism of quantum mechanics (and what he liked instead)*, ***Studies in History and Philosophy of Modern Physics***, vol. 27(4) (1996), pp. 493–510.

[74] E. SCHRÖDINGER, *Discussion of probability relations between separated systems*, ***Cambridge Philosophical Society***, vol. 31 (1935), pp. 555–563.

[75] H. SCHÜTZE, *Automatic word sense discrimination*, ***Computational Linguistics***, vol. 24(1) (1998), pp. 97–123.

[76] R. A. G. SEELY, *Linear logic, ∗-autonomous categories and cofree algebras*, ***Contemporary Mathematics***, vol. 92 (1989), pp. 371–382.

[77] P. SELINGER, *Dagger compact closed categories and completely positive maps*, ***Electronic Notes in Theoretical Computer Science***, vol. 170 (2007), pp. 139–163.

[78] ———, *Finite dimensional Hilbert spaces are complete for dagger compact closed categories (extended abstract)*, ***Electronic Notes in Theoretical Computer Science***, vol. 270(1) (2011), pp. 113–119.

[79] P. W. SHOR, *Polynomial-time algorithms for prime factorization and discrete logarithms on a quantum computer*, ***SIAM Journal on Computing***, vol. 26(5) (1997), pp. 1484–1509.

[80] L. WITTGENSTEIN, ***Philosophical Investigations***, Basil & Blackwell, 1972.

[81] M. ZUKOWSKI, A. ZEILINGER, M. A. HORNE, and A. K. EKERT, *Event-ready-detectors Bell experiment via entanglement swapping*, ***Physical Review Letters***, vol. 71 (1993), pp. 4287–4290.

UNIVERSITY OF OXFORD,
DEPARTMENT OF COMPUTER SCIENCE,
QUANTUM GROUP
E-mail: coecke@cs.ox.ac.uk

REASONING ABOUT MEANING IN NATURAL LANGUAGE WITH COMPACT CLOSED CATEGORIES AND FROBENIUS ALGEBRAS

DIMITRI KARTSAKLIS, MEHRNOOSH SADRZADEH, STEPHEN PULMAN, AND BOB COECKE

Abstract. Compact closed categories have found applications in modeling quantum information protocols by Abramsky-Coecke. They also provide semantics for Lambek's pregroup algebras, applied to formalizing the grammatical structure of natural language, and are implicit in a distributional model of word meaning based on vector spaces. Specifically, in previous work Coecke-Clark-Sadrzadeh used the product category of pregroups with vector spaces and provided a distributional model of meaning for sentences. We recast this theory in terms of strongly monoidal functors and advance it via Frobenius algebras over vector spaces. The former are used to formalize topological quantum field theories by Atiyah and Baez-Dolan, and the latter are used to model classical data in quantum protocols by Coecke-Pavlovic-Vicary. The Frobenius algebras enable us to work in a single space in which meanings of words, phrases, and sentences of any structure live. Hence we can compare meanings of different language constructs and enhance the applicability of the theory. We report on experimental results on a number of language tasks and verify the theoretical predictions.

§1. Introduction. Compact closed categories were first introduced by Kelly [19] in early 1970's. Some thirty years later they found applications in quantum mechanics [1], whereby the vector space foundations of quantum mechanics were recasted in a higher order language and quantum protocols such as *teleportation* found succinct conceptual proofs. Compact closed categories are complete with regard to a pictorial calculus [19, 35]; this calculus is used to depict and reason about information flows in entangled quantum states modeled in tensor spaces, the phenomena that were considered to be mysteries of quantum mechanics and the Achilles heel of quantum logic [4]. The pictorial calculus revealed the multi-linear algebraic level needed for proving quantum information protocols and simplified the reasoning thereof to a great extent, by hiding the underlying additive vector space structure.

Most quantum protocols rely on classical, as well as quantum, data flow. In the work of [1], this classical data flow was modeled using bi-products defined over a compact closed category. However, the pictorial calculus could not

Support by EPSRC grant EP/F042728/1 is acknowledged by the first two authors.

Logic and Algebraic Structures in Quantum Computing
Edited by J. Chubb, A. Eskandarian and V. Harizanov
Lecture Notes in Logic, 45
© 2016, Association for Symbolic Logic

extend well to bi-products, and their categorical axiomatization was not as clear as the built-in monoidal tensor of the category. Later, Frobenius algebras, originally used in group theory [14] and later widely applied to other fields of mathematics and physics such as topological quantum field theory (TQFT) [2, 21, 3], proved useful. It turned out that the operations of such algebras on vector spaces with orthonormal basis correspond to a uniform copying and deleting of the basis, a property that only holds for, hence can be used to axiomatize, classical states [8].

Compact closed categories have also found applications in two completely orthogonal areas of computational linguistics: formalizing grammar and reasoning about lexical meanings of words. The former application is through Lambek's pregroup grammars [23], which are compact closed categories [31] and have been applied to formalizing grammars of a wide range of natural languages, for instance see [24]. The other application domain, referred to as *distributional* models of meaning, formalizes meanings of words regardless of their grammatical roles and via the context of their occurrence [13]. These models consist of vector spaces whose basis are sets of context words and whose vectors represent meanings of target words. Distributional models have been widely studied and successfully applied to a variety of language tasks [34, 25, 26] and in particular to automatic word-synonymy detection [10].

Whereas the type-logical approaches to language do not provide a convincing model of word meaning, the distributional models do not scale to meanings of phrases and sentences. The long standing challenge of combining these two models was addressed in previous work [6, 9, 32]. The solution was based on a cartesian product of the pregroup category and the category of finite dimensional vector spaces. The theoretical predictions of the model were made concrete in [17], then implemented and verified in [15]. In this article, we first recast the theoretical setting of [9] using a succinct functorial passage from a free pregroup of basic types to the category of finite dimensional vector spaces. Then, we further advance the theory and show how Frobenius algebras over vector spaces provide solutions for the problem of the concrete construction of linear maps for predicative words with complex types. As a result, we are able to compare meanings of phrases and sentences with different structures, and moreover compare these with lexical vectors of words. This enhances the domain of application of our model: we show how the theoretical predictions of the model, and in particular the Frobenius algebraic constructions, can be empirically verified by performing three experiments: the disambiguation task of [15], comparing meanings of transitive and intransitive sentences, and a new term/definition classification task.

§2. Recalling some categorical definitions. We start by recalling some definitions. A *monoidal category* [19] is a category $\mathcal{C}$ with a monoidal tensor $\otimes$, which is associative. That is, for all objects $A, B, C \in \mathcal{C}$, we have that

$A \otimes (B \otimes C) \cong (A \otimes B) \otimes C$. Moreover there exists an object $I \in \mathcal{C}$, which serves as the unit of the tensor, that is, $A \otimes I \cong A \cong I \otimes A$. These isomorphisms need to satisfy the usual coherence conditions.

A monoidal category is called *symmetric* whenever we have $A \otimes B \cong B \otimes A$, again satisfying the standard conditions. Furthermore, a monoidal category is *compact closed* whenever any object $A \in \mathcal{C}$ has a left A^l and a right adjoint A^r, that is, the following morphisms exist:

$$\varepsilon^r_A \colon A \otimes A^r \to I \qquad \eta^r_A \colon I \to A^r \otimes A$$
$$\varepsilon^l_A \colon A^l \otimes A \to I \qquad \eta^l_A \colon I \to A \otimes A^l$$

and they satisfy the following *yanking* conditions:

$$(1_A \otimes \varepsilon^l_A) \circ (\eta^l_A \otimes 1_A) = 1_A \qquad (\varepsilon^r_A \otimes 1_A) \circ (1_A \otimes \eta^r_A) = 1_A$$
$$(\varepsilon^l_A \otimes 1_{A^l}) \circ (1_{A^l} \otimes \eta^l_A) = 1_{A^l} \qquad (1_{A^r} \otimes \varepsilon^r_A) \circ (\eta^r_A \otimes 1_{A^r}) = 1_{A^r}$$

In a *symmetric compact closed* category, the left and right adjoints collapse into one, that is we have $A^* := A^l = A^r$ and the above four equalities collapse to the following two:

$$(\varepsilon_A \otimes 1_A) \circ (1_A \otimes \eta_A) = 1_A \qquad (1_{A^*} \otimes \varepsilon_A) \circ (\eta_A \otimes 1_{A^*}) = 1_{A^*}$$

A functor $\mathcal{F}$ from a monoidal category $\mathcal{C}$ to a monoidal category $\mathcal{D}$ is a *monoidal functor* [20], whenever $\mathcal{F}$ is a functor and moreover there exists a morphism $I \to \mathcal{F}(I)$ and the following is a natural transformation:

$$\mathcal{F}(A) \otimes \mathcal{F}(B) \to \mathcal{F}(A \otimes B)$$

satisfying the corresponding coherence conditions. A monoidal functor is *strongly monoidal* [20], whenever the above morphism and natural transformation are invertible.

A strongly monoidal functor on two compact closed categories $\mathcal{C}$ and $\mathcal{D}$ preserves the compact structure, that is $\mathcal{F}(A^l) = \mathcal{F}(A)^l$ and $\mathcal{F}(A^r) = \mathcal{F}(A)^r$. To see this, consider the case of the left adjoint, for which we have the following two compositions of morphisms:

$$\mathcal{F}(A^l) \otimes \mathcal{F}(A) \to \mathcal{F}(A^l \otimes A) \to \mathcal{F}(I) \to I$$
$$I \to \mathcal{F}(I) \to \mathcal{F}(A \otimes A^l) \to \mathcal{F}(A) \otimes \mathcal{F}(A^l)$$

From these, and since adjoints are unique, it follows that $\mathcal{F}(A^l)$ must be left adjoint to $\mathcal{F}(A)$. The case for the right adjoint is similar.

An example of a compact closed category is a Lambek pregroup [23], denoted by $(P, \leq, 1, \cdot, (-)^l, (-)^r)$; we refer to this category by **Preg**. This is a partially ordered monoid where each element of the partial order has a left and a right adjoint, that is we have the following inequalities, which are the partial order versions of the yanking conditions of a compact closed category:

$$p \cdot p^r \leq 1 \leq p^r \cdot p \qquad p^l \cdot p \leq 1 \leq p \cdot p^l$$

An example of a pregroup is the set of all unbounded monotone functions on integers, with function composition as the monoidal tensor and the identity function as its unit. The left and right adjoints are defined using the standard definition of adjoints and in terms of the min and max operations on the integers as follows, for $f \in \mathbb{Z}^{\mathbb{Z}}$ and $m, n \in \mathbb{Z}$:

$$f^r(n) = \bigvee\{m \in \mathbb{Z} \mid f(m) \leq n\} \qquad f^l(n) = \bigwedge\{m \in \mathbb{Z} \mid n \leq f(m)\}$$

An example of a symmetric compact closed category is the category of finite dimensional vector spaces and linear maps over a field (which for our purposes we take to be the set of real numbers $\mathbb{R}$); we refer to this category by **FVect**. The monoidal tensor is the tensor product of vector spaces whose unit is the field. The adjoint of each vector space is its dual, which, by fixing a basis $\{r_i\}_i$, becomes isomorphic to the vector space itself, that is we have $A^* \cong A$ (note that this isomorphism is not natural). The ε and η maps, given by the inner product and maximally entangled states or *Bell pairs*, are defined as follows:

$$\varepsilon_A \colon A \otimes A \to \mathbb{R} \qquad \text{given by} \quad \sum_{ij} c_{ij}\, r_i \otimes r_j \;\mapsto\; \sum_{ij} c_{ij} \langle r_i \mid r_j \rangle$$

$$\eta_A \colon \mathbb{R} \to A \otimes A \qquad \text{given by} \quad 1 \;\mapsto\; \sum_i r_i \otimes r_i$$

An example of a monoidal functor is Atiyah's definition of a topological quantum field theory (TQFT). This is a representation of category of manifolds and cobordisms **Cob** (representing, respectively possible choices of space and spacetime) over the category of finite dimensional vector spaces **FVect**. This representation is formalized using a strongly monoidal functor from **Cob** to **FVect** by Baez and Dylon [3], and assigns a vector space of states to each manifold and a linear operation to each cobordism.

§3. Category theory in linguistics. We briefly review two orthogonal models of meaning in computational linguistics: pregroup grammars and *distributional* models of meaning, and we show how one can interpret the former in the latter using a strongly monoidal functor.

3.1. Type-logical pregroup grammars. Consider the simple grammar generated by the following set of rules:

$S \to Np\ Vp$	$itV \to$ smile
$Vp \to tV\ Np \mid N$	$tV \to$ build
$Np \to Adj\ Np \mid N$	$Adj \to$ strong
	$N \to$ man, woman, house

The above rules are referred to as *generative rules*. The rules on the left describe the formation of a grammatical sentence S in terms of other non-terminals. According to these rules, a sentence is a noun phrase Np followed by a verb phrase *Vp*, where a verb phrase itself is a transitive verb *tV* followed either by a *Np* or a noun N, and a noun phrase is an adjective *Adj* followed either by a *Np* or a noun N. The rules on the right instantiate all but one (S) of the non-terminals to terminals. According to these, 'smile' is an intransitive verb, 'build' is a transitive verb, 'strong' is an adjective, and 'man', 'woman', and 'house' are nouns. We treat these words as lemmas and take freedom in conjugating them in our example sentences.

In a predicative approach, the non-terminals of the above grammar (except for S) are interpreted as unary or binary predicates to produce meaning for phrases and sentences. There are various options when interpreting these non-terminals: for instance, according to the first rule, we can either interpret a verb phrase as a binary predicate that inputs a noun phrase and outputs a sentence, or we can interpret a noun phrase as a binary predicate that inputs a verb phrase and outputs a sentence. We adhere to the more popular (among computational linguistics) verb-centric view and follow the former option. The types of the resulting predicates, obtained by recursively unfolding the rules, form an algebra of types, referred to as a *type-logical grammar*.

A pregroup type-logical grammar, or a pregroup grammar for short, is the pregroup freely generated over a set of basic types which, for the purpose of this paper, we take to be $\{n, s\}$. We refer to this free pregroup grammar by $\mathbf{Preg}_F$. Here, n is the type representing a noun phrase and s is the type representing a sentence. The complex types of this pregroup represent the predicates. For instance, $n^r \cdot s$ is the type of an intransitive verb, interpreted as a unary predicate that inputs a noun phrase and outputs a sentence. Explicit in this type is also the fact that the intransitive verb has to be on the right hand side of the noun phrase. This fact is succinctly expressed by the adjoint r of the type n. Similarly, $n^r \cdot s \cdot n^l$ is the type of a transitive verb, which is a binary predicate that inputs two noun phrases, has to be to the right of one and to the left of the other, and outputs a sentence. Finally, $n \cdot n^l$ is the type of an adjective in attributive position, a unary predicate that inputs a noun phrase and outputs another noun phrase; furthermore, it has to be to the left of its input noun phrase. These types are then assigned to the vocabulary of a language, that is to the non-terminals of the generative rules, via a relation referred to as a *type dictionary*. Our example type dictionary is as follows:

man	woman	houses	strong	smiled	built
n	n	n	$n \cdot n^l$	$n^r \cdot s$	$n^r \cdot s \cdot n^l$

Every sequence of words $w_1 w_2 \cdots w_n$ from the vocabulary has an associated *type reduction*, to which we refer to by $\alpha_{w_1 w_1 \cdots w_n}$. This type reduction represents the grammatical structure of the sequence. In a pregroup grammar, a type reduction is the result of applying the partial order, monoid, and adjunction axioms to the multiplication of the types of the words of the sequence. For example, the type reduction $\alpha_{\text{strong house}}$ associated to the sequence 'strong house' is computed by multiplying the types of 'strong' and 'house', that is $n \cdot n^l \cdot n$, then applying to it the adjunction and monoid axioms, hence obtaining $n \cdot n^l \cdot n \leq n$. Similarly, the type reduction of the sentence 'strong man built houses' is as follows:

$$\alpha_{\text{strong man built houses}} \quad : \quad n \cdot n^l \cdot n \cdot n^r \cdot s \cdot n^l \cdot n \ \leq \ n \cdot n^r \cdot s \ \leq \ s$$

In categorical terms, the type reduction is a morphism of the category $\mathbf{Preg}_F$, denoted by tensors and compositions of the ε and identity maps. For instance, the morphisms corresponding to the above adjective-noun phrase and sentence are as follows:

strong man	strong man built houses
$1_n \otimes \varepsilon_n^l$	$(\varepsilon_n^r \otimes 1_s) \circ (1_n \otimes \varepsilon_n^l \otimes 1_{n^r \cdot s} \otimes \varepsilon_n^l)$

The generative rules formalize the grammar of a natural language and their consequent type-logical grammars provide a predicative interpretation for the words with complex types. However, all the words with the same type have the same interpretation, and even worse, words with basic types are only interpreted as atomic symbols. In the next section, we will see how *distributional* models of meaning address this problem.

3.2. Distributional models of word meaning. Meanings of some words can be determined by their denotations. For instance, meaning of the word 'house' can be the set of all houses or their images; and the answer to the question 'what is a house?' can be provided by pointing to a house. Matters get complicated when it comes to words with complex types such as adjectives and verbs. It is not so clear what is the denotation of the adjective 'strong' or the verb 'build'. The problem is resolved by adhering to a meaning-as-use model of meaning, whereby one can assign meaning to all words, regardless of their grammatical type, according to the context in which they often appear. This context-based approach to meaning is the main idea behind the *distributional* models of meaning.

First formalized by Firth in 1957 [13] and about half a century later implemented and applied to word sense disambiguation by Schütze [34], distributional models of meaning interpret words as vectors in a highly dimensional (but finite) vector space with a fixed orthonormal basis over real numbers. A basis for this vector space is a set of *target* words, which in principle can be the set of all lemmatized words of a corpus of documents or a dictionary. In practice, the basis vectors are often restricted to the few thousands most occurring words of the corpus, or a set of specialized words depending on the application domain, e.g. a music dictionary. Alternatively, they can be *topics* obtained from a dimensionality reduction algorithm such as *single value decomposition* (SVD). We refer to such a vector space with an orthonormal basis $\{w_i\}_i$, no matter how it is built, as our basic distributional vector space W; and to **FVect** restricted to tensor powers of W as $\mathbf{FVect}_W$.

In this model, to each word is associated a vector, which serves as the meaning of the word. The weights of this vector are obtained by counting how many times the word has appeared 'close' to a basis word, where 'close' is a window of n (usually equal to 5) words. This number is usually normalized, often in the form of TF-IDF values which show how important is a specific basis word by taking into account not only the number of times it has occurred to the document, but also the number of documents in which it appears in the corpus.

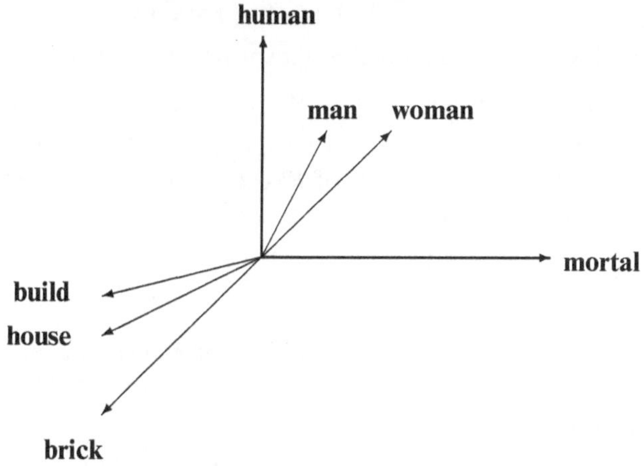

FIGURE 1. A toy distributional model of meaning.

As an example, consider the toy vector space of Figure 1. The set {human, mortal, brick} is the basis of this vector space and the words 'man', 'woman', 'house' and 'build' each have a vector assigned to them. The words that have often appeared in the same context have a smaller angle between their vectors. For instance, 'house' and 'build' often appear close to 'brick', whereas 'man' and 'woman' often appear close to 'mortal' and 'human'. The cosine of the angle between the word vectors has proved to be a good measure in predicting synonymy of words [10]. Despite these good predictions, the distributional models of meaning cannot serve as the definite models of natural language, as there is more to a language than the contexts of its words and these models on their own do not scale up to the interpretations of phrases and sentences. In the next section, we will see how a combination of type-logical and distributional models overcome both of their corresponding shortcomings.

3.3. Quantizing the grammar. We provide a mapping of the free pregroup grammar $\mathbf{Preg}_F$ to $\mathbf{FVect}_W$ via a strongly monoidal functor $\mathcal{F}$. This functor assigns the basic vector space W to both of the basic types, that is, we have:

$$\mathcal{F}(n) = \mathcal{F}(s) = W$$

By functoriality, the partial orders between the basic types (for example those presented in [23]) are mapped to linear maps from W to W. The adjoints of basic types are also mapped to W, since for $x \in \{n, s\}$ we have the following, motivated by the above mentioned fact that $W^* \cong W$:

$$\mathcal{F}(x^l) = \mathcal{F}(x^r) = \mathcal{F}(x)$$

Since $W^{**} \cong W^* \cong W$, the iterated adjoint types are also mapped to W:

$$\mathcal{F}(x^{ll}) = \mathcal{F}(x^{rr}) = \mathcal{F}(x)$$

The complex types are mapped to tensor products of vector spaces, that is:

$$\mathcal{F}(n \cdot n^l) = \mathcal{F}(n^r \cdot s) = W \otimes W \qquad \mathcal{F}(n^r \cdot s \cdot n^l) = W \otimes W \otimes W$$

Similarly, the type reductions are mapped to the compositions of tensor products of identity and ε maps of $\mathbf{FVect}_W$, for instance the type reduction of a transitive sentence is mapped as follows:

$$\begin{aligned}\mathcal{F}(\alpha_{\text{sbj verb obj}}) &= \mathcal{F}(\varepsilon_n^r \otimes 1_s \otimes \varepsilon_n^l)\\ &= \varepsilon_W \otimes 1_W \otimes \varepsilon_W \colon W \otimes (W \otimes W \otimes W) \otimes W \longrightarrow W\end{aligned}$$

Now we can use the definition of [9] to provide a meaning for phrases and sentences of our grammar. The meaning of a sequence of words $w_1 w_2 \cdots w_n$ with type reduction $\alpha_{w_1 w_2 \cdots w_n}$ is:

$$\textbf{Definition(*)} \quad \mathcal{F}(\alpha_{w_1 w_2 \cdots w_n})\,(\overrightarrow{w_1} \otimes \overrightarrow{w_2} \otimes \cdots \otimes \overrightarrow{w_n})$$

As an example, take:

$$\overrightarrow{men} = \sum_i c_i^{men} \overrightarrow{w_i} \quad \overrightarrow{houses} = \sum_k c_k^{houses} \overrightarrow{w_k} \quad \overline{built} = \sum_{ijk} c_{ijk}^{built} (\overrightarrow{w_i} \otimes \overrightarrow{w_j} \otimes \overrightarrow{w_k})$$

Substituting these in Definition(*), we obtain the following for the meaning of the sentence 'men built houses':

$$\begin{aligned}&\mathcal{F}\left(\varepsilon_n^r \otimes 1_s \otimes \varepsilon_n^l\right)\left(\overrightarrow{men} \otimes \overline{built} \otimes \overrightarrow{houses}\right)\\ &= (\varepsilon_W \otimes 1_W \otimes \varepsilon_W)\left(\overrightarrow{men} \otimes \overline{built} \otimes \overrightarrow{houses}\right)\\ &= \sum_{ijk} c_{ijk}^{built} \langle \overrightarrow{men} | \overrightarrow{w_i} \rangle \langle \overrightarrow{w_k} | \overrightarrow{houses} \rangle \overrightarrow{w_j}\end{aligned}$$

This definition ensures that the interpretations of noun phrases and sentences of any grammatical structure, for instance intransitive or transitive, will be a vector in W, hence we can measure the cosine of the angle between them and compute their synonymy. In order to determine that this measure of synonymy provides good predictions, we need to run some experiments. However, whereas we know very well how to build vectors in W for words with basic types such as 'man' and 'house', our method further requires interpretations of words with complex types to be in tensor spaces, and there is no known standard procedure to construct these. In the next section we show how the notion of a Frobenius algebra over a vector space can be of use in addressing this matter.

§4. **Frobenius algebras.** Frobenius algebras were originally introduced in 1903 by F. G. Frobenius in the context of proving representation theorems for group theory [14]. Since then, they have found applications in other fields of mathematics and physics, e.g. in topological quantum field theories [21] and in categorical quantum mechanics [8]. The general categorical definitions recalled below are due to Carboni and Walters [5]. Their concrete instantiations to algebras over vector spaces were developed in [8].

A Frobenius algebra over a symmetric monoidal category $(\mathcal{C}, \otimes, I)$ is a tuple $(F, \sigma, \iota, \mu, \zeta)$, where for an F object of $\mathcal{C}$ the triple (F, σ, ι) is an associative coalgebra, that is, the following are morphisms of $\mathcal{C}$, satisfying the coalgebraic associativity and unit conditions:

$$\sigma \colon F \to F \otimes F \qquad\qquad \iota \colon F \to I$$

The triple (F, μ, ζ) is an associative algebra, that is, the following are morphisms of $\mathcal{C}$, satisfying the algebraic associativity and unit conditions:

$$\mu \colon F \otimes F \to F \qquad\qquad \zeta \colon I \to F$$

Moreover, the above σ and μ morphisms satisfy the following *Frobenius* condition:

$$(\mu \otimes 1_F) \circ (1_F \otimes \sigma) \;=\; \sigma \circ \mu \;=\; (1_F \otimes \mu) \circ (\sigma \otimes 1_F)$$

A Frobenius Algebra is *commutative* if it satisfies the following two conditions for $\delta : X \otimes Y \to Y \otimes X$, the symmetry morphism of $(\mathcal{C}, \otimes, I)$:

$$\delta \circ \sigma = \sigma \qquad\qquad \mu \circ \delta = \mu$$

Finally, a Frobenius Algebra is *isometric* or *special* if it satisfies the following condition:

$$\mu \circ \sigma = Id$$

In the category **FVect**, any vector space V with a fixed basis $\{\overrightarrow{v_i}\}_i$ has a commutative special Frobenius algebra over it, explicitly given as follows:

$$\sigma :: \overrightarrow{v_i} \mapsto \overrightarrow{v_i} \otimes \overrightarrow{v_i} \qquad\qquad \iota :: \overrightarrow{v_i} \mapsto 1$$
$$\mu :: \overrightarrow{v_i} \otimes \overrightarrow{v_i} \mapsto \overrightarrow{v_i} \qquad\qquad \zeta :: 1 \mapsto \overrightarrow{v_i}$$

In a Frobenius algebra over an orthonormal vector space, the coalgebra and algebra operations relate to each other via the equation $\sigma^\dagger = \mu$, where $\dagger$ is the adjoint, equal to the transpose for vector spaces over reals.

In such Frobenius algebras, the operation σ corresponds to copying and its unit ι corresponds to deleting of the vectors. They enable one to faithfully encode vectors of W into spaces with higher tensor ranks, such as $W \otimes W$, $W \otimes W \otimes W, \cdots$. In linear algebraic terms, for $v \in W$, we have that $\sigma(v)$ is a diagonal matrix whose diagonal elements are weights of v. The operation μ is referred to as uncopying; it loses some information when encoding a higher rank tensor into a lower rank space. In linear algebraic terms, for $z \in W \otimes W$, we have that $\mu(z)$ is a vector consisting only of the diagonal elements of z, hence losing the information encoded in the non-diagonal part.

§5. Pictorial calculi. The framework of compact closed categories comes with a complete diagrammatic calculus that allows convenient graphical representations of the derivations. We briefly introduce the fragment of this calculus that we are going to use in this paper. The objects of this fragment are the *tensors* of multi-linear algebra; that is, a vector is a rank-1 tensor, a matrix is a rank-2 tensor, and a 3d-array is a rank-3 tensor. Each tensor is represented by a triangle, whose rank can be determined by its wires. Words are represented by tensors that correspond to their meaning: subjects and objects are rank-1 tensors (vectors), adjectives and intransitive verbs are rank-2 tensors, and transitive verbs are rank-3 tensors. The ε maps are depicted as cups, whereas the identity morphism is a vertical straight line. The tensor products of vectors are represented by juxtaposing their corresponding triangles. For example, the meaning of a transitive sentence, following Definition(*), is depicted as follows:

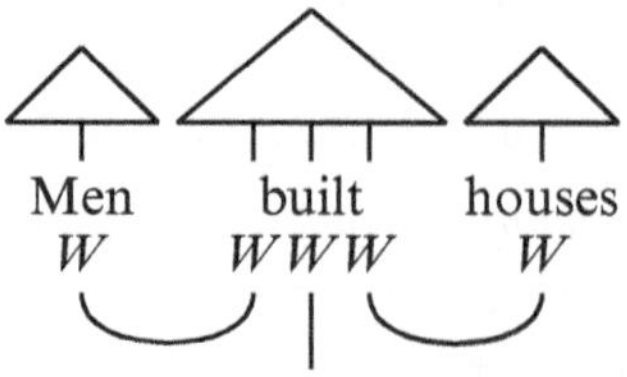

Computations with Frobenius algebras can also be represented within the more general diagrammatic calculus of symmetric monoidal categories, referred to as *string* diagrams, first formalized in [18]. Specifically, the linear maps of the coalgebra and algebra are depicted by:

$(\sigma, \iota) =$ 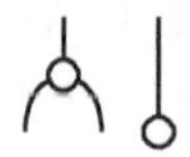 $(\mu, \zeta) =$ 

The *Frobenius* condition is depicted by:

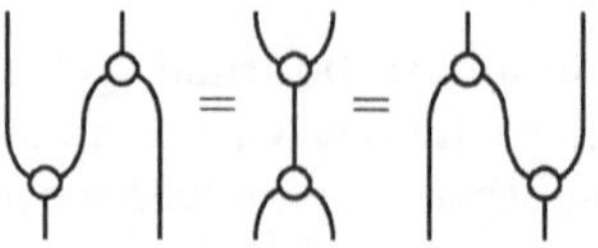

The commutativity conditions are shown as:

The isometry condition is depicted by:

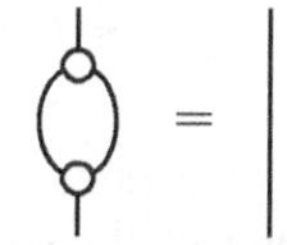

Finally, the Frobenius conditions guarantee that any diagram depicting a Frobenius algebraic computation can be reduced to a normal form that only depends on the number of input and output wires of the nodes, provided that the diagram of computation is connected. This justifies depicting computations with Frobenius algebras as *spiders*, referring to the right hand side diagram below:

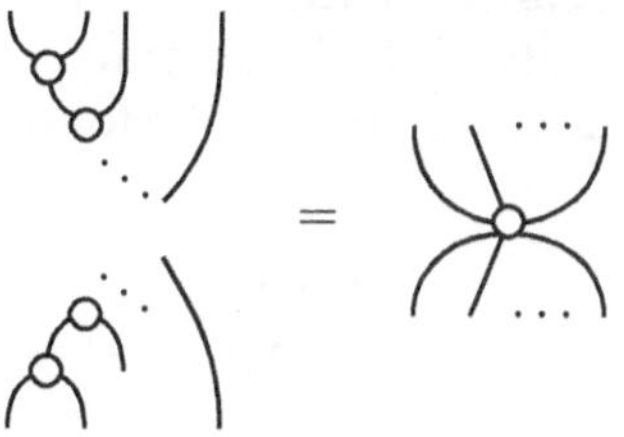

For an informal introduction to compact closed categories, Frobenius algebras, and their diagrammatic calculi, see [7].

§6. Building tensors for words with complex types. The type-logical models of meaning treat words with complex types as predicates. In a matrix calculus, predicates can be modeled as matrices (or equivalently, linear maps), over the semiring of booleans. In vector spaces over reals, one can extend these 0/1 entries to real numbers and model words with complex types as weighted predicates. These are predicates that not only tell us which instantiations of their arguments are related to each other, but also that to what extent these are related to each other. For instance, a transitive verb is a binary predicate that, in the type-logical model, tells us which noun phrases are related to other noun phrases. In a vector space model, the verb becomes a linear map that moreover tells us to what extent these noun phrases are related to each other.

Building such linear maps from a corpus turns out to be a non-trivial task. In previous work [17, 15] we argue that such a linear map can be constructed by taking the sum of the tensor products of the vectors/linear maps of its arguments. For instance, the linear map representing an n-ary predicate p with arguments a_1 to a_n is $\sum_i \overline{a}_1 \otimes \cdots \otimes \overline{a}_n$, where $\overline{a}_j$ is the vector/linear map associated to the argument a_j and the index i counts the number of times each word a_j has appeared in the corpus as the argument of p. Following this method, the linear maps corresponding to the predicates of our simple grammar are as follows:

intransitive verb	transitive verb	adjective
$\sum_i \overrightarrow{sbj}_i$	$\sum_i (\overrightarrow{sbj}_i \otimes \overrightarrow{obj}_i)$	$\sum_i \overrightarrow{noun}_i$

There is a problem: this method provides us with a linear map in a space whose tensor rank is one less than the rank of the space needed by Definition(*). For instance, the linear map of the transitive verb ends up being in $W \otimes W$, but we need a linear map in $W \otimes W \otimes W$. This problem is overcome by using the operations of a Frobenius algebra over vector spaces. We use the pictorial calculi of the compact closed categories and Frobenius algebras to depict the resulting linear maps and sentence vectors.

6.1. Adjectives and intransitive verbs. The linear maps of adjectives and intransitive verbs are elements of W. In order to encode them in $W \otimes W$, we use the Frobenius σ operation and obtain the following linear map:

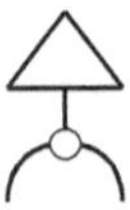

For the intransitive verb, when substituted in Definition(*), that is, when applied to its subject, the above will result in the left hand side vector below, which is then normalized to the right hand side vector.

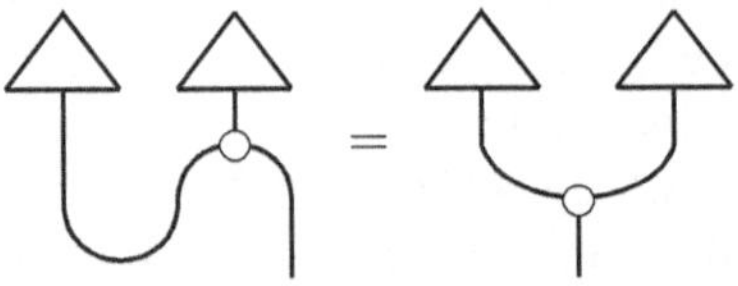

When an adjective is applied to a noun, the order of the above application is swapped: the triangle of the adjective will change place with the triangle of the subject of the intransitive verb.

6.2. Transitive verbs. The linear map of a transitive verb is an element of $W \otimes W$; this has to be encoded in $W \otimes W \otimes W$. We face a few options here, which geometrically speaking provide us with different ways of "diagonally" placing a plane into a cube.

CPSBJ. The first option is to copy the "row" dimension of the linear map corresponding to the verb; this dimension encodes the information of the subjects of the verb from the corpus. In the left hand side diagram below we see how σ transforms the verb in this way. Once substituted in Definition(*), we obtain the diagram in the right hand side:

Verb:  **Sentence**: 

In this case, the σ map transforms the matrix of the verb as follows:

$$\sigma :: \sum_{ij} c_{ij}(\overrightarrow{n_i} \otimes \overrightarrow{n_j}) \quad \mapsto \quad \sum_{iij} c_{ii}(\overrightarrow{n_i} \otimes \overrightarrow{n_i} \otimes \overrightarrow{n_j})$$

CPOBJ. Our other option is to copy the "column" dimension of the matrix, which encodes the information about the objects of the verb from the corpus:

Verb:  **Sentence**: 

Now the σ-map does the following transformation:

$$\sigma :: \sum_{ij} c_{ij}(\overrightarrow{n_i} \otimes \overrightarrow{n_j}) \quad \mapsto \quad \sum_{ijj} c_{jj}(\overrightarrow{n_i} \otimes \overrightarrow{n_j} \otimes \overrightarrow{n_j})$$

The diagrams above simplify the calculations involved, since they suggest a closed form formula for each case. Taking as an example the diagram of the copy-subject method, we see that: (a) the object interacts with the verb; (b) the result of this interaction serves as input for the σ map; (c) one wire of the output of σ interacts with the subject, while the other branch delivers the result. Linear algebraically, this corresponds to the computation $\sigma(\overline{verb} \times \overrightarrow{obj})^{\mathsf{T}} \times \overrightarrow{sbj}$, which expresses the fact that the meaning of a sentence is obtained by first applying the meaning of the verb to the meaning of the object, then applying the (σ version of the) result to the meaning of the subject. This computation results in the Equation 1 below:

$$\overrightarrow{sbj\ verb\ obj} = \overrightarrow{sbj} \odot (\overline{verb} \times \overrightarrow{obj}) \tag{1}$$

This order of application is the exact same way formalized in the generative rules of the language. On the contrary, the meaning of a transitive sentence for the copy-object results is given by Equation 2 below, which expresses the fact that the meaning of a sentence is obtained by first applying the (transposed) meaning of the verb to the meaning of the subject and then applying the result to the meaning of the object:

$$\overrightarrow{sbj\ verb\ obj} = \overrightarrow{obj} \odot (\overline{verb}^{\mathsf{T}} \times \overrightarrow{sbj}) \tag{2}$$

Note that equipped with the above closed forms we do not need to create or manipulate rank-3 tensors at any point of the computation, something that would cause high computational overhead.

Purely syntactically speaking, in a pregroup grammar the order of application of a transitive verb does not matter: it is applied to its subject and object in parallel. Semantically, as originated in the work of Montague [28], a transitive verb is first applied to its object and then to its subject. In the more modern approaches to semantics via logical grammars, this order is some times based on the choice of the specific verb [12]. Our work in this paper is more inline with the latter approach, where for the specific task of disambiguating the verbs of our dataset, first applying the verb to the subject then to the object seems to provide better experimental results. According to our general theoretical setting, the linear map corresponding to the transitive verb should be a rank-3 tensor, but at the moment, apart from work in progress which tries to conjoin efforts with Machine Learning to directly build these as rank-3 tensors, we do not have the technology to do other than described in this paper. However, in the ideal case that the linear maps of words are already in the spaces allocated to them by the theory, these choices will not arise, as the compact nature of the matrix calculus implies that the application can be done in parallel in all the cases that parallel applications are prescribed by the syntax. From a linear-algebra perspective, fully populated rank-3 tensors for verbs satisfy the following equality:

$$\overline{subj\ verb\ obj} = (\overline{verb} \times \overrightarrow{obj})^{\mathsf{T}} \times \overrightarrow{subj} = (\overline{verb}^{\mathsf{T}} \times \overrightarrow{subj}) \times \overrightarrow{obj}$$

which shows that the order of application does not actually play a role.

MIXCPDL. We can also use a mixture of σ and μ maps. There are three reasonable options here, all of which start by applying two σ's to the two wires of the linear map of the verb (that is, one for each of the dimensions). Then one can either apply a ι to one of the copies of the first wire, or a ι to one of the copies of the second wire. These two options are depicted as follows:

The first diagram has the same normal form as the copy-subject option, and the second one has the same normal form as the copy-object option.

Finally, one can apply a μ to one wire from each of the copied wires of the verb, the result of which is depicted in the following left hand side diagram. When substituted in Definition(*), we obtain the following right hand side diagram for the meaning of the transitive sentence:

Verb: 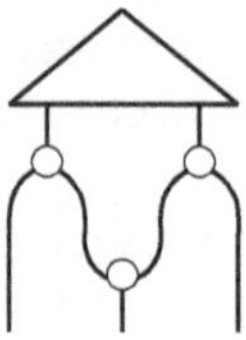 **Sentence:** 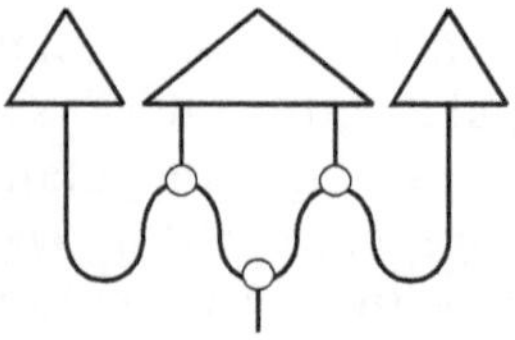

The normal form of the diagram of the sentence is obtained by collapsing the three dots and yanking the corresponding wires, resulting in the following diagram:

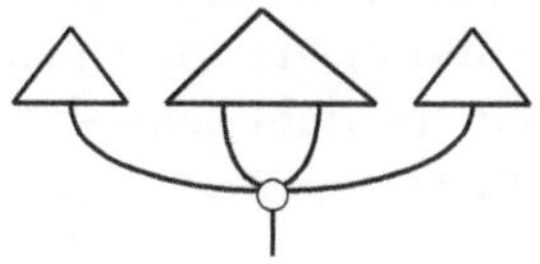

Linear algebraically, the spider form of the verb is equivalent to $\sum_i (\overrightarrow{sbj_i} \odot \overrightarrow{obj_i})$. A verb obtained in this way will only relate the properties of its subjects and objects on identical bases and there will be no interaction of properties across bases. For instance, for a certain verb v, this construction will result in a vector that only encodes to what extent v has related subjects and objects with property $\overrightarrow{w_1}$, and has no information about to what extent v has related subjects with property $\overrightarrow{w_1}$ to objects with property $\overrightarrow{w_2}$. The closed form of the above diagram is:

$$\overrightarrow{sbj} \odot \left(\sum_i (\overrightarrow{sbj_i} \odot \overrightarrow{obj_i})\right) \odot \overrightarrow{obj}$$

6.3. Encoding the existing non-predicative models. Apart from the predicative way of encoding meanings of words with complex types, there exists two other approaches in the literature, who simply work with the context vectors of such words [27, 16]. These two approaches are representable in our setting using the Frobenius operations.

MULTP. To represent the model of [27] in our setting, in which the meaning of a sentence is simply the point-wise multiplication of the context vectors of the words, we start from the context vector of the verb, denoted by $\overrightarrow{verb}$, and apply three σ's and then one μ to it. The result is depicted in the left hand side diagram below; once this verb is substituted in Definition(*), we obtain the right hand side diagram below as the meaning of a transitive sentence:

Verb: 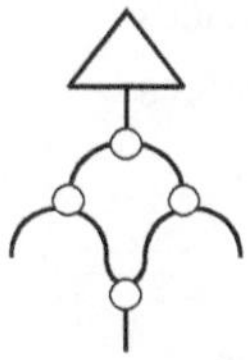 **Sentence:** 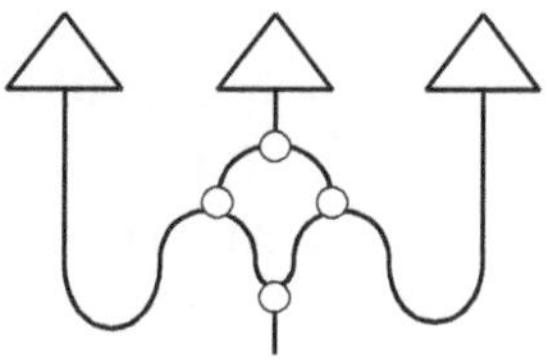

The normal form of the diagram of the sentence and its closed linear algebraic form are as follows:

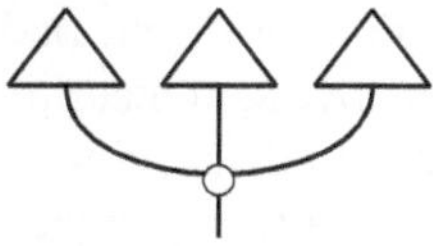 $=$ $\overrightarrow{sbj} \odot \overrightarrow{verb} \odot \overrightarrow{obj}$

KRON. In the model of [16], the tensor of a transitive sentence is calculated as the Kronecker product of the context vector of the verb with itself, so we have $\overline{verb} = \overrightarrow{verb} \otimes \overrightarrow{verb}$. To encode this, we start from the Kronecker product of the context vector of the verb with itself, apply one σ to each one of the vectors and then a μ to both of them jointly. The result is the following left hand side verb, which when substituted in the equation of Definition(*) results in a normal form (depicted in the right hand side below) very similar to the normal form of the MULTP model:

Verb: 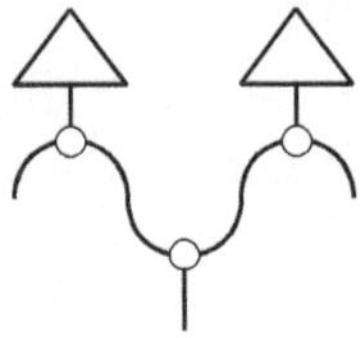 **Sentence**: 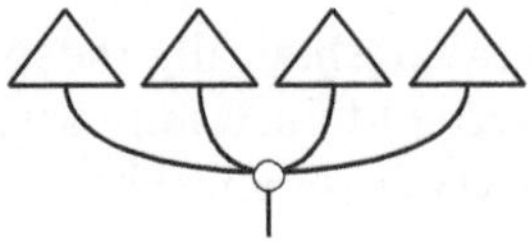

Linear algebraically, the above normal form is equivalent to:

$$\overrightarrow{sbj} \odot \overrightarrow{verb} \odot \overrightarrow{verb} \odot \overrightarrow{obj} \tag{3}$$

§7. **Experiments.** The different options presented in Section 6 and summarized in Table 1 provide us a number of models for testing our setting. We train our vectors on the British National Corpus (BNC), which has about six million sentences and one million words, classified into a hundred million different lexical tokens. We use the set of its 2000 most frequent lemmas as a basis of our basic distributional vector space W. The weights of each vector are set to the ratio of the probability of the context word given the target word to the probability of the context word overall. As our similarity measure we use the cosine distance between the vectors.

TABLE 1. The models.

Model	**Description**
CPSBJ	Copy subject on relational matrices
CPOBJ	Copy object on relational matrices
MIXCPDL	Diagonalize on relational matrices
KRON	Diagonalize on direct matrices
MULTP	Multiplicative model

7.1. **Disambiguation.** We first test our models on a disambiguation task, which is an extension of Schütze's original disambiguation task from words to sentences. A dataset for this task was originally developed in [27] for intransitive sentences, and later extended to transitive sentences in [15]; we use the latter. The goal is to assess how well a model can discriminate between the different senses of an ambiguous verb, given the context (subject and object) of that verb. The entries of this dataset consist of a target verb, a subject, an object, and a landmark verb used for the comparison. One such entry for example is, "write, pupil, name, spell". A good model should be able to understand that the sentence "pupil write name" is closer to the sentence "pupil spell name" than, for example, to "pupil publish name". On the other hand, given the context "writer, book" these results should be reversed. The evaluation of this experiment is performed by calculating Spearman's ρ, which measures the degree of correlation of the cosine distance with the judgements of 25 human evaluators, who has been asked to assess the similarity of each pair of sentences using a scale from 1 to 7. As our baseline we use a simple additive model (ADDTV), where the meaning of a transitive sentence is computed as the addition of the relevant context vectors.

Results. The results of this experiment are shown in Table 2, indicating that the most successful model for this task is the copy-object model. The better performance of this model against the copy-subject approach provides us some insights about the role of subjects and objects in disambiguating our verbs. By copying the dimension associated with the object, the compression of the original sentence matrix, as this was calculated in [15], takes place along the dimension of subjects (rows), meaning that the resulting vector will bring much more information from the objects than the subjects (this is also suggested by Equation 2). Hence, the fact that this vector performs better than the one of the copy-subject method provides an indication that the object of some ambiguous verbs (which turns out to be the case for our dataset) can be more important for disambiguating that verb than the subject. Intuitively, we can imagine that the crucial factor to disambiguate between the verbs "write", "publish" and "spell" is more the object than the subject: a book or a paper can be both published and written, but a letter or a shopping list can only be written. Similarly, a word can be spelled and written, but a book can only be written. The subject in all these cases is not so important.

TABLE 2. Disambiguation results. High and Low refer to average scores for high and low landmarks, respectively. UpperBound refers to agreement between annotators.

Model	**High**	**Low**	ρ
ADDTV	0.90	0.90	0.050
MULTP	0.67	0.60	0.163
MIXCPDL	0.75	0.77	0.000
KRON	0.31	0.21	0.168
CPSBJ	0.95	0.95	0.143
CPOBJ	0.89	0.90	0.172
UpperBound	4.80	2.49	0.620

The copy-object model is followed closely by the (KRON) and the MULTP models. The similar performance of these two is not a surprise, given their almost identical nature. Finally, the bad performance of the model (MIXCPDL) that is obtained by the application of the uncopying μ map conforms to the predictions of the theory, as these were expressed in Section 6.

7.2. Comparing transitive and intransitive sentences. In this section we will examine the potential of the above approach in practice, in the context of an experiment aiming to compare transitive and intransitive sentences. In order to do that, we use the dataset of the previous verb disambiguation task (see detailed description in Section 7.1) to conduct the following simple experiment: We create intransitive versions of all the transitive sentences from target verbs and their high and low landmarks by dropping the object; then, we compare each transitive sentence coming from the target verbs with all the other intransitive sentences, expecting that the highest similarity would come from its own intransitive version, the next higher similarity would come from the intransitive version that uses the corresponding high landmark verb, and so on. To present a concrete example, consider the entry "write, pupil, name, spell, publish". Our transitive sentence here is s_{tr} = "pupil write name"; the intransitive version of this is s_{in} = "pupil write". We also create intransitive versions using the high and the low landmarks, getting s_{hi} = "pupil spell" and s_{lo} = "pupil publish". If the similarity between two sentences s_1 and s_2 is given by $sim(s_1, s_2)$, we would expect that:

$$sim(s_{tr}, s_{it}) > sim(s_{tr}, s_{hi}) > sim(s_{tr}, s_{lo}) > sim(s_{tr}, s_u)$$

where s_u represents an unrelated intransitive version coming from a target verb different than the one of s_{tr}. The results of this experiment are shown in Table 3 below, for 100 target verbs.

TABLE 3. Results of the comparison between transitive and intransitive sentences.

Case	**Errors**	%
$sim(s_{tr}, s_{hi}) > sim(s_{tr}, s_{it})$	7 of 93	7.5
$sim(s_{tr}, s_{lo}) > sim(s_{tr}, s_{it})$	6 of 93	5.6
$sim(s_{tr}, s_u) > sim(s_{tr}, s_{it})$	36 of 9900	0.4

The outcome follows indeed our expectations for this task. We see, for example, that the highest error rate comes from cases where the intransitive sentence of the high landmark verb is closer to a transitive sentence than the intransitive version coming from the sentence itself (first row of the table). Since the meaning of a target verb and the high landmark verb were specifically selected to be very similar given the context (subject and object), this is naturally the most error-prone category. The seven misclassified cases are presented in Table 4, where the similarity of the involved intransitive versions is apparent.

TABLE 4. Errors in the first category of comparisons.

s_{tr}	s_{in}	s_{hi}
people run round	people run	people move
boy meet girl	boy meet	boy visit
management accept responsibility	management accept	management bear
patient accept treatment	patient accept	patient bear
table draw eye	table draw	table attract
claimant draw benefit	claimant draw	claimant attract
tribunal try crime	tribunal try	tribunal judge

The six cases of the second category (where an intransitive sentence from a low-landmark gives higher similarity than the normal intransitive version) are quite similar, since in many cases dropping the object leads to semantically identical expressions. For the transitive sentence "tribunal try crime", for example, the low-landmark intransitive version "tribunal test" has almost identical meaning with the normal intransitive version "tribunal try", so it is easier to be "mistakenly" selected by the model as the one closest to the original transitive sentence.

Finally, the model performs really well for cases when an unrelated intransitive sentence is compared with a transitive one, with only a 0.4% error rate. Here many of the misclassifications can also be attributed to the increased ambiguity of the involved verbs when the object is absent. For example, the similarity between "man draw sword" and "man draw" is considered smaller than the similarity of the first sentence with "man write". Although this is an obvious error, we should acknowledge that the two intransitive sentences, "man draw" and "man write", are not so different semantically, so the error was not completely unjustified.

7.3. Definition classification. The ability of reliably comparing the meaning of single words with larger textual fragments, e.g. phrases or even sentences, can be an invaluable tool for many challenging NLP tasks, such as definition classification, sentiment analysis, or even the simple everyday search on the internet. In this task we examine the extend to which our models can correctly match a number of terms (single words) with a number of definitions. Our dataset consists of 112 terms (72 nouns and 40 verbs) extracted from a junior dictionary together with their main definition. For each term we added two more definitions, either by using entries of WordNet for the term or by simple paraphrase of the main definition using a thesaurus, getting a total of three definitions per term. In all cases a definition for a noun-term is a noun phrase, whereas the definitions for the verb-terms consist of verb phrases. A sample of the dataset entries can be found in Table 5.

TABLE 5. Sample of the dataset for the term/definition comparison task (noun-terms in the top part, verb-terms in the bottom part).

Term	**Main definition**	**Alternative def. 1**	**Alternative def. 2**
blaze	large strong fire	huge potent flame	substantial heat
husband	married man	partner of a woman	male spouse
foal	young horse	adolescent stallion	juvenile mare
horror	great fear	intense fright	disturbing feeling
apologise	say sorry	express regret or sadness	acknowledge shortcoming or failing
embark	get on a ship	enter boat or vessel	commence trip
vandalize	break things	cause damage	produce destruction
regret	be sad or sorry	feel remorse	express dissatisfaction

We approach this evaluation problem as a classification task, where terms have the role of classes. First, we calculate the distance between each definition and every term in the dataset. The definition is "classified" to the term that gives the higher similarity. Due to the nature the dataset, this task did not require human annotation and we evaluate the results by calculating separate F1-scores for each term, and getting their average as an overall score for the whole model. The results are presented in Table 6.

TABLE 6. Results of the term/definition comparison task.

	Nouns			**Verbs**		
Model	P	R	F1	P	R	F1
ADDTV	0.21	0.17	0.16	0.28	0.25	0.23
MULTP	0.21	0.22	0.19	0.31	0.30	0.26
RELTN	0.22	0.24	0.21	0.32	0.28	0.27

Results. Since this experiment includes verb phrases, where the subject is missing, we construct our verb vectors by summing over all context vectors of objects with which the verb appears in the corpus; that is, we use $\overline{verb} = \sum_i \overrightarrow{obj_i}$.

This is referred to as the relational model (RELTN), and is compared with the multiplicative model. Additive model serves again as our baseline. We evaluate separately the performance on the noun terms and the performance on the verb terms, since a mixing of the two sets would be inconsistent.

The relational model delivers again the best performance, although the difference from the multiplicative model is small. All models perform better on the verb terms than the noun part of the dataset, yet in general F-scores tend to be low. This is natural, since the challenge that this task poses to a machine is great, and F-score considers anything but the perfect result (every definition assigned to the correct term) as unacceptable.

An error analysis shows that for the noun-term set the relational model returns the correct main definition in 25 of the 72 cases, whereas in 47 cases (65%) the correct definition is in the top-five list for that term (Table 7). The multiplicative model performs similarly, and better for the verb-term set. For this experiment we also calculated Mean Reciprocal Ranks values, which again were very close for the two models. Furthermore, some of the "misclassified" cases can also be considered as somehow "correct". For example, the definition we originally assigned to the term 'jacket' was 'short coat'; however, the system "preferred" the definition 'waterproof cover', which is also correct. Some interesting other cases are presented in Table 8.

TABLE 7. Results of the term/definition comparison task based on the rank of the main definition.

		MULTP		RELTN	
	Rank	Count	%	Count	%
Nouns	1	26	36.1	25	34.7
	2-5	20	27.8	22	30.6
	6-10	11	15.3	5	6.9
	11-72	15	20.8	20	27.8
Verbs	1	15	37.5	8	20.0
	2-5	10	25.0	13	32.5
	6-10	6	15.0	4	10.0
	11-40	9	22.5	15	37.5

§8. Conclusion and future work. In summary, after a brief review of the definitions of compact closed categories and monoidal functors, we ramified their applications to natural language syntax and semantics. We recasted the categorical setting of [9], which was based on the product of the category of a pregroup type-logic with the category of finite dimensional vector spaces,

TABLE 8. A sample of ambiguous cases where the model assigned a different definition than the original.

Term	Original definition	Assigned definition
rod	fishing stick	round handle
jacket	short coat	waterproof cover
mud	wet soil	wet ground
labyrinth	complicated maze	burial chamber

in terms of a monoidal functor from the former to the latter. This passage is similar to the vector space representation of category of manifolds in a topological quantum field theory. We showed how the operations of Frobenius algebras over vector spaces provide a concrete instantiation of this setting and used their pictorial calculus to simplify the multi-linear algebraic computations. This instantiation resulted in meanings of all sentences living in a basic vector space W, hence we became able to compare their meanings with one another and also with meanings of single words and phrases. We developed experiments based on this instantiation and evaluated the predictions of our model in a number of sentence and phrase synonymy tasks.

We conclude that the concrete setting of this paper provides a robust and scalable base for an implementation of the framework of [9], ready for further experimentation and applications. It overcomes the shortcomings of our first implementation [15], whose main problem was that the vector space representation of the atomic type s was taken to be the tensor space $W \otimes W$ (for a transitive sentence), hence the logical and the concrete types did not match. As a consequence, sentences with nested structures such as 'Mary saw John reading a book' could not be assigned a meaning; furthermore, meanings of phrases and sentences with different grammatical structure lived in different vector spaces, so a direct comparison of them was not possible.

An experimental future direction is a higher order evaluation of the definition classification task using an unambiguous vector space along the lines of [34], where each word is associated with one or more *sense* vectors. A model like this will avoid encoding different meanings of words in one vector, and will help us separate the two distinct tasks of composition and disambiguation that currently are interwoven in a single step.

From the theoretical perspective, one direction is to start from type-logical grammars that are more expressive than pregroups. In recent work [33] we have shown how the functorial passage to **FVect** can be extended from a pregroup algebra to the Lambek Calculus [22], which has a monoidal (rather than compact) structure. It remains to show how this passage can be extended to more expressive versions of Lambek calculi, such as Lambek-Grishin algebras

[29], calculi of discontinuity [30], or abstract categorial grammars [11]. More specifically, one wants to investigate if the new operations and axioms of these extensions are preserved by the vector space semantic functor.

Another venue to explore is meanings of logical words, where any context-dependant method fails to succeed. The abstract operations of Frobenius algebras have been used to model a limited class of logical operations on vector spaces [8], hence these might turn out to be promising in this aspect too.

REFERENCES

[1] S. Abramsky and B. Coecke, *A categorical semantics of quantum protocols*, ***Proceedings of the 19th Annual IEEE Symposium on Logic in Computer Science***, 2004, arXiv:quant-ph/0402130, pp. 415–425.

[2] M. Atiyah, *Topological quantum field theories*, ***Publications Mathématique de l'Institut des Hautes Études Scientifiques***, vol. 68 (1989), pp. 175–186.

[3] J. C. Baez and J. Dolan, *Higher-dimensional algebra and topological quantum field theory*, ***Journal of Mathematical Physics***, vol. 36 (1995), pp. 6073D0–6105.

[4] G. Birkhoff and J. von Neumann, *The logic of quantum mechanics*, ***Annals of Mathematics***, vol. 35 (1936), pp. 823–843.

[5] A. Carboni and R. F. C. Walters, *Cartesian bicategories I*, ***Journal of Pure and Applied Algebra***, vol. 49 (1987).

[6] S. Clark and S. Pulman, *Combining symbolic and distributional models of meaning*, ***Proceedings of AAAI Spring Symposium on Quantum Interaction***, AAAI Press, 2007.

[7] B. Coecke and E. Paquette, *Categories for the practicing physicist*, ***New Structures for Physics*** (B. Coecke, editor), Lecture Notes in Physics, vol. 813, Springer, 2011, pp. 167–271.

[8] B. Coecke, D. Pavlovic, and J. Vicary, *A new description of orthogonal bases*, ***Mathematical Structures in Computer Science***, vol. 1 (2008).

[9] B. Coecke, M. Sadrzadeh, and S. Clark, *Mathematical foundations for distributed compositional model of meaning*, ***Lambek Festschrift*** (J. van Benthem, M. Moortgat, and W. Buszkowski, editors), ***Linguistic Analysis***, vol. 36, 2010, pp. 345–384.

[10] J. Curran, ***From Distributional to Semantic Similarity***, Ph.D. thesis, University of Edinburgh, 2004.

[11] P. de Groote, *Towards abstract categorial grammars*, ***Proceedings of Association for Computational Linguistic***, 2001, pp. 148–155.

[12] P. de Groote and F. Lamarche, *Classical non-associative Lambek calculus*, ***Studia Logica***, vol. 71 (2002), pp. 355–388.

[13] J. R. Firth, *A synopsis of linguistic theory* 1930-1955, ***Studies in Linguistic Analysis***, (1957).

[14] F. G. Frobenius, *Theorie der hyperkomplexen Größen*, ***Sitzungsberichte der Preußischen Akademie der Wissenschaften zu Berlin***, (1903).

[15] E. Grefenstette and M. Sadrzadeh, *Experimental support for a categorical compositional distributional model of meaning*, ***Proceedings of Conference on Empirical Methods in Natural Language Processing (EMNLP)***, 2011.

[16] ———, *Experimenting with Transitive Verbs in a DisCoCat*, ***Proceedings of Workshop on Geometrical Models of Natural Language Semantics (GEMS)***, 2011.

[17] E. Grefenstette, M. Sadrzadeh, S. Clark, B. Coecke, and S. Pulman, *Concrete compositional sentence spaces for a compositional distributional model of meaning*, ***International Conference on Computational Semantics (IWCS'11)***, 2011.

[18] A. JOYAL and R. STREET, *The geometry of tensor calculus I*, ***Advances in Mathematics***, vol. 88 (1991).

[19] G. M. KELLY, *Many-variable functorial calculus* (*I*), ***Coherence in Categories*** (G. M. Kelly, M. Laplaza, G. Lewis, and S. MacLane, editors), Lecture Notes in Mathematics, vol. 281, Springer, 1972, pp. 66–105.

[20] A. KOCK, *Strong functors and monoidal monads*, ***Archiv der Mathematik***, vol. 23 (1972), pp. 113–120.

[21] J. KOCK, ***Frobenius Algebras and 2D Topological Quantum Field Theories***, London Mathematical Society Student Texts, Cambridge University Press, 2003.

[22] J. LAMBEK, *The mathematics of sentence structure*, ***American Mathematics Monthly***, vol. 65 (1958), pp. 154–170.

[23] ———, ***From Word to Sentence***, Polimetrica, Milan, 2008.

[24] J. LAMBEK and C. CASADIO, ***Computational Algebraic Approaches to Natural Language***, Polimetrica, Milan, 2006.

[25] T. LANDAUER and S. DUMAIS, *A solution to Plato's problem: The latent semantic analysis theory of acquisition, induction, and representation of knowledge*, ***Psychological Review***, (1997).

[26] C. D. MANNING, P. RAGHAVAN, and H. SCHÜTZE, ***Introduction to Information Retrieval***, Cambridge University Press, 2008.

[27] J. MITCHELL and M. LAPATA, *Vector-based models of semantic composition*, ***Proceedings of the 46th Annual Meeting of the Association for Computational Linguistics***, 2008, pp. 236–244.

[28] R. MONTAGUE, *English as a formal language*, ***Formal Philosophy***, Yale University Press, New Haven, 1974, pp. 189–223.

[29] M. MOORTGAT, *Symmetric categorial grammar*, ***Journal of Philosophical Logic***, (2009), pp. 681–710.

[30] G. MORRILL, *Discontinuity in categorial grammar*, ***Linguistics and Philosophy***, vol. 18 (1995), pp. 175–219.

[31] A. PRELLER and J. LAMBEK, *Free compact 2-categories*, ***Mathematical Structures in Computer Science***, vol. 17 (2007), pp. 309–340.

[32] A. PRELLER and M. SADRZADEH, *Bell states and negative sentences in the distributed model of meaning*, ***Proceedings of the 6th QPL Workshop on Quantum Physics and Logic***, Electronic Notes in Theoretical Computer Science, vol. 270, 2011, pp. 141–153.

[33] M. SADRZADEH, E. GREFENSTETTE, and B. COECKE, *Lambek vs. Lambek: Vector Space Semantics and String Diagrams for Lambek Calculus*, ***Annals of Pure and Applied Logic***, vol. 164 (2013), pp. 1079–1100.

[34] H. SCHÜTZE, *Automatic word sense discrimination*, ***Computational Linguistics***, vol. 24 (1998), pp. 97–123.

[35] P. SELINGER, *A survey of graphical languages for monoidal categories*, ***New Structures for Physics*** (B. Coecke, editor), Springer-Verlag, 2011, pp. 275–337.

SCHOOL OF ELECTRONIC ENGINEERING AND COMPUTER SCIENCE
QUEEN MARY UNIVERSITY OF LONDON
E-mail: d.kartsaklis@qmul.ac.uk
E-mail: m.sadrzadeh@qmul.ac.uk

DEPARTMENT OF COMPUTER SCIENCE
UNIVERSITY OF OXFORD
WOLFSON BUILDING, PARKS ROAD
OXFORD, OX1 3QD
E-mail: stephen.pulman@cs.ox.ac.uk
E-mail: coecke@cs.ox.ac.uk

KNOT LOGIC AND TOPOLOGICAL QUANTUM COMPUTING WITH MAJORANA FERMIONS

LOUIS H. KAUFFMAN

Abstract. This paper is an introduction to relationships between quantum topology and quantum computing. We show how knots are related not just to braiding and quantum operators, but to quantum set theoretical foundations, algebras of fermions, and we show how the operation of negation in logic, seen as both a value and an operator, can generate the fusion algebra for a Majorana fermion. We call negation in this mode the mark, as it operates on itself to change from marked to unmarked states. The mark viewed recursively as a simplest discrete dynamical system naturally generates the fermion algebra, the quaternions and the braid group representations related to Majorana fermions. The paper begins with these fundamentals. It then discusses unitary solutions to the Yang-Baxter equation that are universal quantum gates, quantum entanglement and topological entanglement, and gives an exposition of knot-theoretic recoupling theory, its relationship with topological quantum field theory and applies these methods to produce unitary representations of the braid groups that are dense in the unitary groups. These methods are rooted in the bracket state sum model for the Jones polynomial. A self-contained study of the quantum universal Fibonacci model is given. Results are applied to give quantum algorithms for the computation of the colored Jones polynomials for knots and links, and the Witten-Reshetikhin-Turaev invariant of three manifolds. Two constructions are given for the Fibonacci model, one based in Temperley-Lieb recoupling theory, the other quite elementary and also based on the Temperley-Lieb algebra. This paper begins an exploration of quantum epistemology in relation to the structure of discrimination as the underpinning of basic logic, perception and measurement.

§1. Introduction. This paper is an introduction to relationships between quantum topology and quantum computing. We take a foundational approach, showing how knots are related not just to braiding and quantum operators, but to quantum set theoretical foundations and algebras of fermions. We show how the operation of negation in logic, seen as both a value and an operator, can generate the fusion algebra for a Majorana fermion, a particle that is its own anti-particle and interacts with itself either to annihilate itself or to produce itself. We call negation in this mode *the mark*, as it operates on itself to change from marked to unmarked states. The mark viewed recursively as a simplest discrete dynamical system naturally generates the fermion algebra, the quaternions and the braid group representations related to Majorana fermions. The paper begins with these fundmentals. They provide a conceptual key to

Logic and Algebraic Structures in Quantum Computing
Edited by J. Chubb, A. Eskandarian and V. Harizanov
Lecture Notes in Logic, 45
© 2016, Association for Symbolic Logic

many of the models that appear later in the paper. In particular, the Fibonacci model for topological quantum computing is seen to be based on the fusion rules for a Majorana fermion and these in turn are the interaction rules for the mark seen as a logical particle. It requires a shift in viewpoint to see that the operator of negation can also be seen as a logical value. This is explained in Sections 3, 4 and 5. The quaternions emerge naturally from the reentering mark. All these models have their roots in unitary representations of the Artin braid group to the quaternions.

An outline of the parts of this paper is given below.

§1. Introduction
§2. Knots and braids
§3. Knot logic
§4. Fermions, Majorana fermions and algebraic knot sets
§5. Laws of Form
§6. Quantum mechanics and quantum computation
§7. Braiding operators and universal quantum gates
§8. A remark about *EPR* entanglement and Bell's inequality
§9. The Aravind hypothesis
§10. $SU(2)$ representations of the Artin braid group
§11. The bracket polynomial and the Jones polynomial
§12. Quantum topology, cobordism categories, Temperley-Lieb algebra and topological quantum field theory
§13. Braiding and topological quantum field theory
§14. Spin networks and Temperley-Lieb recoupling theory
§15. Fibonacci particles
§16. The Fibonacci recoupling model
§17. Quantum computation of colored Jones polynomials and the Witten-Reshetikhin-Turaev invariant
§18. A direct construction of the Fibonacci model

Much of what is new in this paper proceeds from thinking about knots and sets and distinctions. The Sections 3, 4 and 5 are self-contained and self-explanatory. These sections show how a formal system discovered by Spencer-Brown [92], underlying Boolean logic, is composed of a "logical particle", the mark ⏋, that interacts with itself to either produce itself or to cancel itself.

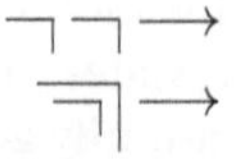

In this sense the mark is a formal model of a Majorana fermion. The oirginal formal structure of the mark gives the fusion algebra for the Majorana fermion. In Section 5 we show that this iconic representation of the particle is directly related to modeling with surface cobordisms and this theme occurs throughout the paper. In Section 5 we also show that the mark, viewed as a generator of a discrete dynamical system, generates the Clifford algebra associated with a Majorana fermion and we end this section by showing how this iterant viewpoint leads naturally to the Dirac equation using the approach of [86]. This is part of the contents of the Sections 3, 4, 5. In these sections we examine relationships with knots as models of non-standard set theory. The algebra of fermions is directly relevant to this knot set theory and can be formulated in terms of the Clifford algebra of Majorana fermions.

We weave this material with the emergence of unitary braid group representations that are significant for quantum information theory. In particular we weave the topology with the algebra of fermions and in order to clarify this development, we give a quick summary of that algebra and a quick summary of topological quantum computing in the rest of this introduction.

Fermion algebra. Recall fermion algebra. One has fermion annihilation operators ψ and their conjugate creation operators $\psi^\dagger$. One has $\psi^2 = 0 = (\psi^\dagger)^2$. There is a fundamental commutation relation

$$\psi\psi^\dagger + \psi^\dagger\psi = 1.$$

If you have more than one of them say ψ and ϕ, then they anti-commute:

$$\psi\phi = -\phi\psi.$$

The Majorana fermions c satisfy $c^\dagger = c$ so that they are their own anti-particles. There is a lot of interest in these as quasi-particles and they are related to braiding and to topological quantum computing. A group of researchers [78] claims, at this writing, to have found quasiparticle Majorana fermions in edge effects in nano-wires. (A line of fermions could have a Majorana fermion happen non-locally from one end of the line to the other.) The Fibonacci model that we discuss is also based on Majorana particles, possibly related to collective electronic excitations. If P is a Majorana fermion particle, then P can interact with itself to either produce itself or to annihilate itself. This is the simple "fusion algebra" for this particle. One can write $P^2 = P + 1$ to denote the two possible self-interactions the particle P. The patterns of interaction and braiding of such a particle P give rise to the Fibonacci model.

Majoranas are related to standard fermions as follows: The algebra for Majoranas is $c = c^\dagger$ and $cc' = -c'c$ if c and c' are distinct Majorana fermions with $c^2 = 1$ and $c'^2 = 1$. One can make a standard fermion from two Majoranas via

$$\psi = (c + ic')/\sqrt{2},$$
$$\psi^\dagger = (c - ic')/\sqrt{2}.$$

Similarly one can mathematically make two Majoranas from any single fermion. Now if you take a set of Majoranas

$$\{c_1, c_2, c_3, \ldots, c_n\}$$

then there are natural braiding operators that act on the vector space with these c_k as the basis. The operators are mediated by algebra elements

$$\tau_k = (1 + c_{k+1}c_k)/\sqrt{2},$$
$$\tau_k^{-1} = (1 - c_{k+1}c_k)/\sqrt{2}.$$

Then the braiding operators are

$$T_k : Span\{c_1, c_2, \ldots, , c_n\} \longrightarrow Span\{c_1, c_2, \ldots, , c_n\}$$

via

$$T_k(x) = \tau_k x \tau_k^{-1}.$$

The braiding is simply:

$$T_k(c_k) = c_{k+1},$$
$$T_k(c_{k+1}) = -c_k,$$

and T_k is the identity otherwise. This gives a very nice unitary representaton of the Artin braid group and it deserves better understanding.

It is worth noting that a triple of Majorana fermions say a, b, c gives rise to a representation of the quaternion group. This is a generalization of the well-known association of Pauli matrices and quaternions. We have $a^2 = b^2 = c^2 = 1$ and they anticommute. Let $I = ba, J = cb, K = ac$. Then

$$I^2 = J^2 = K^2 = IJK = -1,$$

giving the quaternions. The operators

$$A = (1/\sqrt{2})(1 + I)$$
$$B = (1/\sqrt{2})(1 + J)$$
$$C = (1/\sqrt{2})(1 + K)$$

braid one another:

$$ABA = BAB, \quad BCB = CBC, \quad ACA = CAC.$$

This is a special case of the braid group representation described above for an arbitrary list of Majorana fermions. These braiding operators are entangling and so can be used for universal quantum computation, but they give only partial topological quantum computation due to the interaction with single qubit operators not generated by them.

In Section 5 we show how the dynamics of the reentering mark leads to two (algebraic) Majorana fermions e and η that correspond to the spatial and temporal aspects of this recursive process. The corresponding standard fermion operators are then given by the formulas below.

$$\psi = (e + i\eta)/\sqrt{2} \quad \text{and} \quad \psi^{\dagger} = (e - i\eta)/\sqrt{2}.$$

This gives a model of a fermion creation operator as a point in a non-commutative spacetime. This suggestive point of view, based on knot logic and Laws of Form, will be explored in subsequent publications.

Topological quantum computing. This paper describes relationships between quantum topology and quantum computing as a modified version of Chapter 14 of the book [12] and an expanded version of [60] and an expanded version of a chapter in [62]. Quantum topology is, roughly speaking, that part of low-dimensional topology that interacts with statistical and quantum physics. Many invariants of knots, links and three dimensional manifolds have been born of this interaction, and the form of the invariants is closely related to the form of the computation of amplitudes in quantum mechanics. Consequently, it is fruitful to move back and forth between quantum topological methods and the techniques of quantum information theory.

We sketch the background topology, discuss analogies (such as topological entanglement and quantum entanglement), show direct correspondences between certain topological operators (solutions to the Yang-Baxter equation) and universal quantum gates. We describe the background for topological quantum computing in terms of Temperley-Lieb (we will sometimes abbreviate this to TL) recoupling theory. This is a recoupling theory that generalizes standard angular momentum recoupling theory, generalizes the Penrose theory of spin networks and is inherently topological. Temperley-Lieb recoupling Theory is based on the bracket polynomial model [35, 39] for the Jones polynomial. It is built in terms of diagrammatic combinatorial topology. The same structure can be explained in terms of the $SU(2)_q$ quantum group, and has relationships with functional integration and Witten's approach to topological quantum field theory. Nevertheless, the approach given here is elementary. The structure is built from simple beginnings and this structure and its recoupling language can be applied to many things including colored Jones polynomials, Witten-Reshetikhin-Turaev invariants of three manifolds, topological quantum field theory and quantum computing.

In quantum computing, the simplest non-trivial example of the Temperley-Lieb recoupling Theory gives the so-called Fibonacci model. The recoupling theory yields representations of the Artin braid group into unitary groups $U(n)$ where n is a Fibonacci number. These representations are *dense* in the unitary group, and can be used to model quantum computation universally in terms of representations of the braid group. Hence the term: topological quantum computation. In this paper, we outline the basics of the Temperley-Lieb Recoupling Theory, and show explicitly how the Fibonacci model arises from it. The diagrammatic computations in the sections 12 to 18 are completely self-contained and can be used by a reader who has just learned the bracket polynomial, and wants to see how these dense unitary braid group representations arise from it. In the final section of the paper we give a separate construction for the Fibnacci model that is based on 2×2 complex matrix representations of the Temperley-Lieb algebra. This is a completely elementary construction independent of the recoupling theory of the previous sections. We studied this construction in [61] and a version of it has been used in [89].

The relationship of the work here with the mathematics of Chern-Simons theory and conformal field theory occurs through the work of Witten, Moore and Seiberg and Moore and Read [76]. One can compare the mathematical techniques of the present paper with the physics of the quantum Hall effect and its possibilities for topological quantum computing. This part of the story will await a sequel to the present exposition.

Here is a very condensed presentation of how unitary representations of the braid group are constructed via topological quantum field theoretic methods, leading to the Fibonacci model and its generalizations. These representations are more powerful, in principle, than the representations we have just given, because they encompass a dense collection of all unitary transformations, including single qubit transformations needed for universal quantum computing. One has a mathematical particle with label P that can interact with itself to produce either itself labeled P or itself with the null label $*$. We shall denote the interaction of two particles P and Q by the expression PQ, but it is understood that the "value" of PQ is the result of the interaction, and this may partake of a number of possibilities. Thus for our particle P, we have that PP may be equal to P or to $*$ in a given situation. When $*$ interacts with P the result is always P. When $*$ interacts with $*$ the result is always $*$. One considers process spaces where a row of particles labeled P can successively interact, subject to the restriction that the end result is P. For example the space $V[(ab)c]$ denotes the space of interactions of three particles labeled P. The particles are placed in the positions a, b, c. Thus we begin with $(PP)P$. In a typical sequence of interactions, the first two P's interact to produce a $*$, and the $*$ interacts with P to produce P.

$$(PP)P \longrightarrow (*)P \longrightarrow P.$$

In another possibility, the first two P's interact to produce a P, and the P interacts with P to produce P.

$$(PP)P \longrightarrow (P)P \longrightarrow P.$$

It follows from this analysis that the space of linear combinations of processes $V[(ab)c]$ is two dimensional. The two processes we have just described can be taken to be the qubit basis for this space. One obtains a representation of the three strand Artin braid group on $V[(ab)c]$ by assigning appropriate phase changes to each of the generating processes. One can think of these phases as corresponding to the interchange of the particles labeled a and b in the association $(ab)c$. The other operator for this representation corresponds to the interchange of b and c. This interchange is accomplished by a *unitary change of basis mapping*

$$F : V[(ab)c] \longrightarrow V[a(bc)].$$

If

$$A : V[(ab)c] \longrightarrow V[(ba)c]$$

is the first braiding operator (corresponding to an interchange of the first two particles in the association) then the second operator

$$B : V[(ab)c] \longrightarrow V[(ac)b]$$

is accomplished via the formula $B = F^{-1}RF$ where the R in this formula acts in the second vector space $V[a(bc)]$ to apply the phases for the interchange of b and c. These issues are illustrated in Figure 1, where the parenthesization of the particles is indicated by circles and by also by trees. The trees can be taken to indicate patterns of particle interaction, where two particles interact at the branch of a binary tree to produce the particle product at the root. See also Figure 50 for an illustration of the braiding $B = F^{-1}RF$.

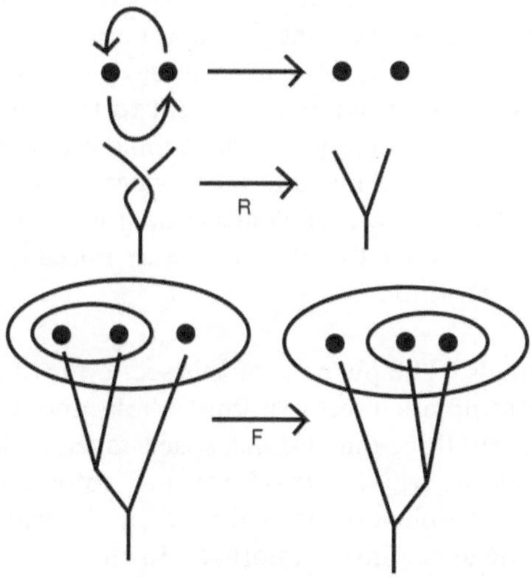

FIGURE 1. Braiding anyons.

In this scheme, vector spaces corresponding to associated strings of particle interactions are interrelated by *recoupling transformations* that generalize the mapping F indicated above. A full representation of the Artin braid group on each space is defined in terms of the local interchange phase gates and the recoupling transformations. These gates and transformations have to satisfy a number of identities in order to produce a well-defined representation of the braid group. These identities were discovered originally in relation to topological quantum field theory. In our approach the structure of phase gates and recoupling transformations arise naturally from the structure of the bracket model for the Jones polynomial. Thus we obtain a knot-theoretic basis for topological quantum computing.

In modeling the quantum Hall effect [94, 19, 10, 90], the braiding of quasi-particles (collective excitations) leads to non-trival representations of the Artin braid group. Such particles are called *Anyons*. The braiding in these models is related to topological quantum field theory. It is hoped that the mathematics we explain here will form a bridge between theoretical models of anyons and their applications to quantum computing.

Acknowledgements. Much of this paper is based upon joint work with Samuel J. Lomonaco in the papers [56, 54, 58, 71, 72, 73, 63, 57, 59, 56, 61, 62]. I have woven this work into the present paper in a form that is coupled with recent and previous work on relations with logic and with Majorana fermions. The relations with logic stem from the following previous papers of the author [67, 33, 45, 34, 41, 47, 51, 31, 32, 65, 64, 46, 50, 52]. These previous papers are an exploration of the foundations of knot theory in relation to Laws of Form, non-standard set theory, recursion and discrete dynamical systems. At the level of discrete dynamical systems the papers are related to foundations of physics. More work needs to be done in all these domains.

Two recent books contain material relevant to the context of this paper. They are [87] and [86]. The interested reader should examine these approaches to fundamental physics. It is planned to use this paper and other joint work as a springboard for a book [55] on topological quantum information theory and for a book that expands on the foundational issues raised in this paper and the previous papers of the author.

§2. Knots and braids. The purpose of this section is to give a quick introduction to the diagrammatic theory of knots, links and braids. A *knot* is an embedding of a circle in three-dimensional space, taken up to ambient isotopy. The problem of deciding whether two knots are isotopic is an example of a *placement problem*, a problem of studying the topological forms that can be made by placing one space inside another. In the case of knot theory we consider the placements of a circle inside three dimensional space. There are many applications of the theory of knots. Topology is a background for the physical structure of real knots made from rope of cable. As a result, the field of practical knot tying is a field of applied topology that existed well before the mathematical discipline of topology arose. Then again long molecules such as rubber molecules and DNA molecules can be knotted and linked.

There have been a number of intense applications of knot theory to the study of *DNA* [17] and to polymer physics [42]. Knot theory is closely related to theoretical physics as well with applications in quantum gravity [91, 85, 53] and many applications of ideas in physics to the topological structure of knots themselves [39].

Quantum topology is the study and invention of topological invariants via the use of analogies and techniques from mathematical physics. Many invariants such as the Jones polynomial are constructed via partition functions and generalized quantum amplitudes. As a result, one expects to see relationships between knot theory and physics. In this paper we will study how knot theory can be used to produce unitary representations of the braid group. Such representations can play a fundamental role in quantum computing.

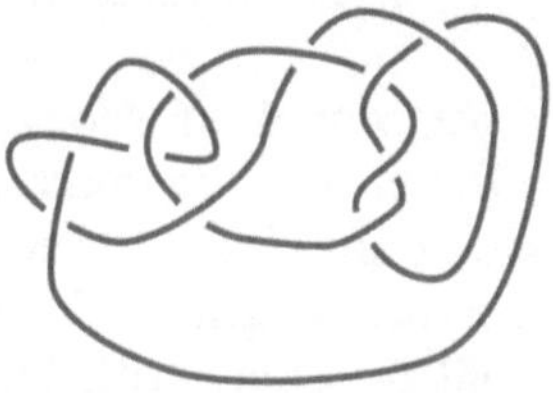

FIGURE 2. A knot diagram.

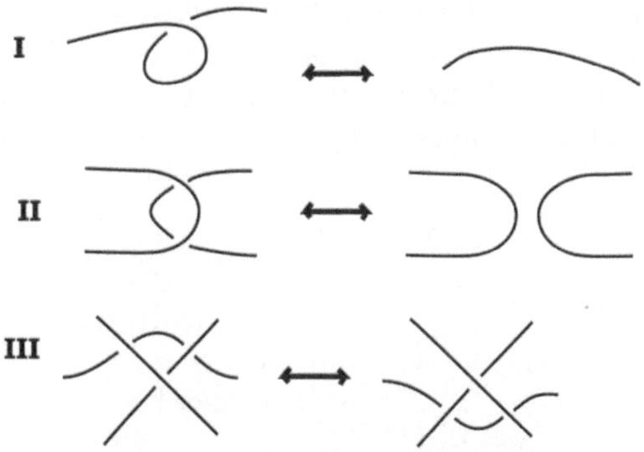

FIGURE 3. The Reidemeister moves.

That is, Two knots are regarded as equivalent if one embedding can be obtained from the other through a continuous family of embeddings of circles in three-space. A *link* is an embedding of a disjoint collection of circles, taken up to ambient isotopy. Figure 2 illustrates a diagram for a knot. The diagram is regarded both as a schematic picture of the knot, and as a plane graph with extra structure at the nodes (indicating how the curve of the knot passes over or under itself by standard pictorial conventions).

Ambient isotopy is mathematically the same as the equivalence relation generated on diagrams by the *Reidemeister moves*. These moves are illustrated

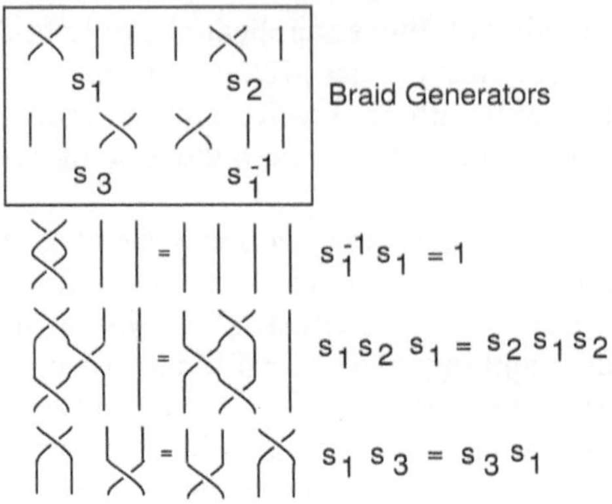

FIGURE 4. Braid generators.

in Figure 3. Each move is performed on a local part of the diagram that is topologically identical to the part of the diagram illustrated in this figure (these figures are representative examples of the types of Reidemeister moves) without changing the rest of the diagram. The Reidemeister moves are useful in doing combinatorial topology with knots and links, notably in working out the behaviour of knot invariants. A *knot invariant* is a function defined from knots and links to some other mathematical object (such as groups or polynomials or numbers) such that equivalent diagrams are mapped to equivalent objects (isomorphic groups, identical polynomials, identical numbers). The Reidemeister moves are of great use for analyzing the structure of knot invariants and they are closely related to the *Artin braid group*, which we discuss below.

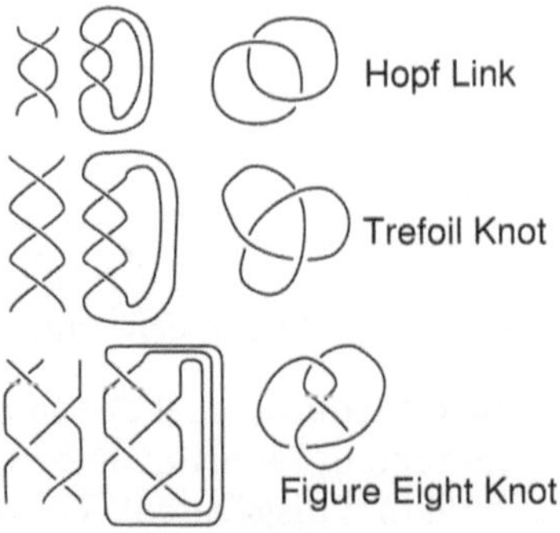

FIGURE 5. Closing braids to form knots and links.

A *braid* is an embedding of a collection of strands that have their ends in two rows of points that are set one above the other with respect to a choice of vertical. The strands are not individually knotted and they are disjoint from one another. See Figure 4, Figure 5 and Figure 6 for illustrations of braids and

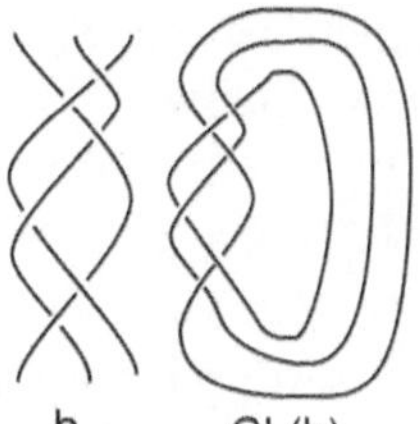

FIGURE 6. Borromean rings as a braid closure.

moves on braids. Braids can be multiplied by attaching the bottom row of one braid to the top row of the other braid. Taken up to ambient isotopy, fixing the endpoints, the braids form a group under this notion of multiplication. In Figure 4 we illustrate the form of the basic generators of the braid group, and the form of the relations among these generators. Figure 5 illustrates how to close a braid by attaching the top strands to the bottom strands by a collection of parallel arcs. A key theorem of Alexander states that every knot or link can be represented as a closed braid. Thus the theory of braids is critical to the theory of knots and links. Figure 6 illustrates the famous Borromean Rings (a link of three unknotted loops such that any two of the loops are unlinked) as the closure of a braid.

Let B_n denote the Artin braid group on n strands. We recall here that B_n is generated by elementary braids $\{s_1, \ldots, s_{n-1}\}$ with relations

1. $s_i s_j = s_j s_i$ for $|i - j| > 1$,
2. $s_i s_{i+1} s_i = s_{i+1} s_i s_{i+1}$ for $i = 1, \ldots n - 2$.

See Figure 4 for an illustration of the elementary braids and their relations. Note that the braid group has a diagrammatic topological interpretation, where a braid is an intertwining of strands that lead from one set of n points to another set of n points. The braid generators s_i are represented by diagrams where the i-th and $(i + 1)$-th strands wind around one another by a single half-twist (the sense of this turn is shown in Figure 4) and all other strands drop straight to the bottom. Braids are diagrammed vertically as in Figure 4, and the products are taken in order from top to bottom. The product of two braid diagrams is accomplished by adjoining the top strands of one braid to the bottom strands of the other braid.

In Figure 4 we have restricted the illustration to the four-stranded braid group B_4. In that figure the three braid generators of B_4 are shown, and then the inverse of the first generator is drawn. Following this, one sees the identities $s_1 s_1^{-1} = 1$ (where the identity element in B_4 consists in four vertical strands), $s_1 s_2 s_1 = s_2 s_1 s_2$, and finally $s_1 s_3 = s_3 s_1$.

Braids are a key structure in mathematics. It is not just that they are a collection of groups with a vivid topological interpretation. From the algebraic point of view the braid groups B_n are important extensions of the symmetric groups S_n. Recall that the symmetric group S_n of all permutations of n distinct objects has presentation as shown below.

1. $s_i^2 = 1$ for $i = 1, \ldots n-1$,
2. $s_i s_j = s_j s_i$ for $|i-j| > 1$,
3. $s_i s_{i+1} s_i = s_{i+1} s_i s_{i+1}$ for $i = 1, \ldots n-2$.

Thus S_n is obtained from B_n by setting the square of each braiding generator equal to one. We have an exact sequence of groups

$$1 \longrightarrow P_n \longrightarrow B_n \longrightarrow S_n \longrightarrow 1$$

exhibiting the Artin braid group as an extension of the pure braids P_n (inducing the identity permutation), by the symmetric group.

In the next sections we shall show how representations of the Artin braid group are rich enough to provide a dense set of transformations in the unitary groups. Thus the braid groups are *in principle* fundamental to quantum computation and quantum information theory.

§3. **Knot logic.** We shall use knot and link diagrams to represent sets. More about this point of view can be found in the author's paper "Knot Logic" [40].

Set theory is about an asymmetric relation called membership. We write $a \in S$ to say that a is a member of the set S. In this section we shall diagram the membership relation as in Figure 7.

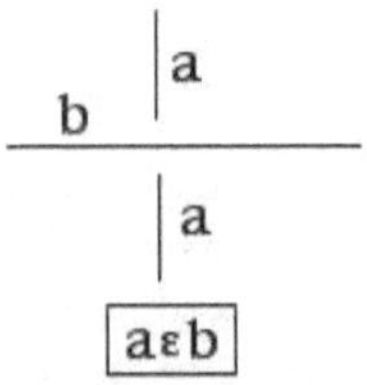

FIGURE 7. Membership.

The entities a and b that are in the relation $a \in b$ are diagrammed as segments of lines or curves, with the a-curve passing underneath the b-curve. Membership is represented by under-passage of curve segments. A curve or segment with no curves passing underneath it is the empty set.

In the Figure 8, we indicate two sets. The first (looking like a right-angle bracket that we refer to as the *mark*) is the empty set. The second, consisting of a mark crossing over another mark, is the set whose only member is the empty set. We can continue this construction, building the von Neumann construction of the natural numbers in this notation as in Figure 9.

This notation allows us to also have sets that are members of themselves as in Figure 10, and and sets can be members of each other as in Figure 11. This mutuality is diagrammed as topological linking. This leads to the question

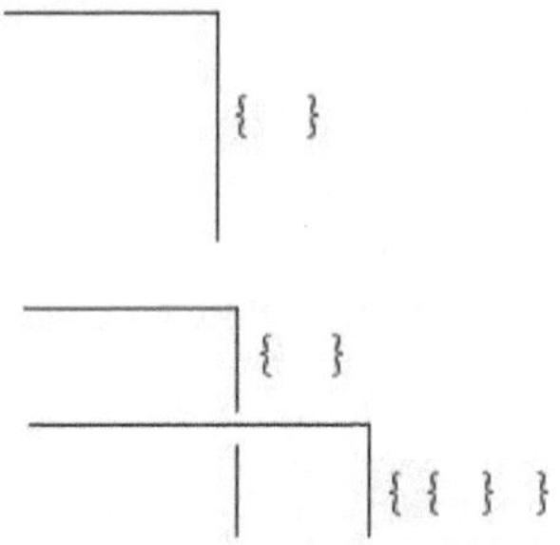

FIGURE 8. Von Neumann 1.

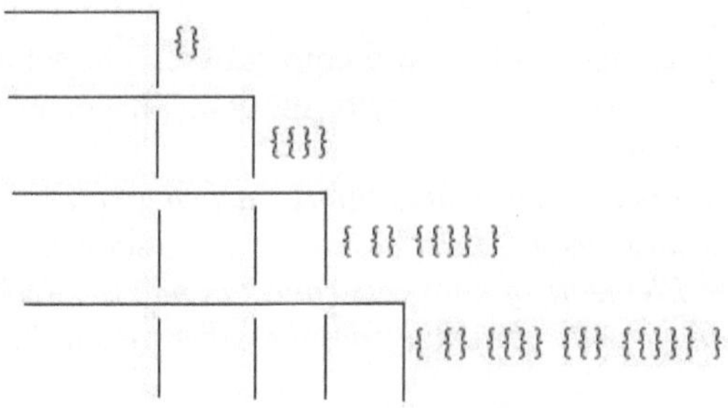

FIGURE 9. Von Neumann 2.

beyond flatland: Is there a topological interpretation for this way of looking at set-membership?

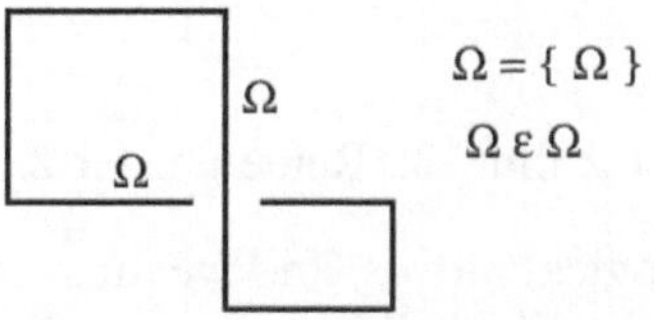

FIGURE 10. Omega is a member of Omega.

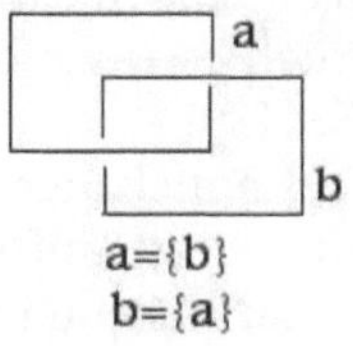

FIGURE 11. Mutual membership.

Consider the example in Figure 12, modified from the previous one. The link consisting of a and b in this example is not topologically linked. The two

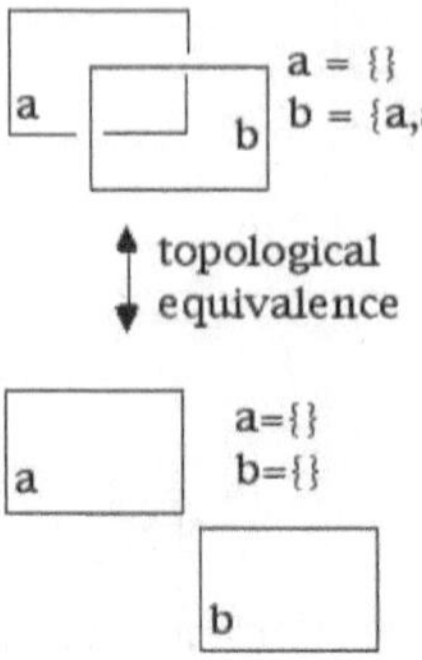

FIGURE 12. Cancellation.

components slide over one another and come apart. The set a remains empty, but the set b changes from $b = \{a, a\}$ to empty. This example suggests the following interpretation.

Regard each diagram as specifying a multi-set (where more than one instance of an element can occur), and the rule for reducing to a set with one representative for each element is: Elements of knot sets cancel in pairs. Two knot sets are said to be equivalent if one can be obtained from the other by a finite sequence of pair cancellations.

This equivalence relation on knot sets is in exact accord with the first Reidemeister move as shown in Figure 13.

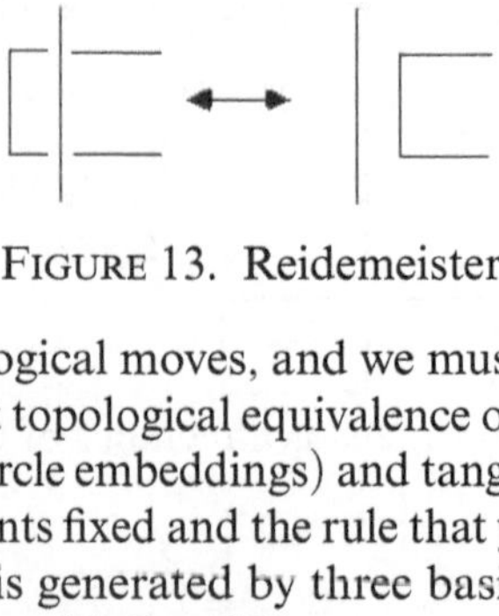

FIGURE 13. Reidemeister 2.

There are other topological moves, and we must examine them as well. In fact, it is well-known that topological equivalence of knots (single circle embeddings), links (mutltiple circle embeddings) and tangles (arbitrary diagrammatic embeddings with end points fixed and the rule that you are not allowed to move strings over endpoints) is generated by three basic moves (the Reidemeister moves) as shown in Figure 14. See [39].

It is apparent that move III does not change any of the relationships in the knot multi-sets. The line that moves just shifts and remains underneath the other two lines. On the other hand move number one can change the self-referential nature of the corresponding knot-set. One goes, in the first move, between a set that indicates self-membership to a set that does not indicate self-membership (at the site in question). See Figure 15 This means that in knot-set theory every set has representatives (the diagrams are the representatives of

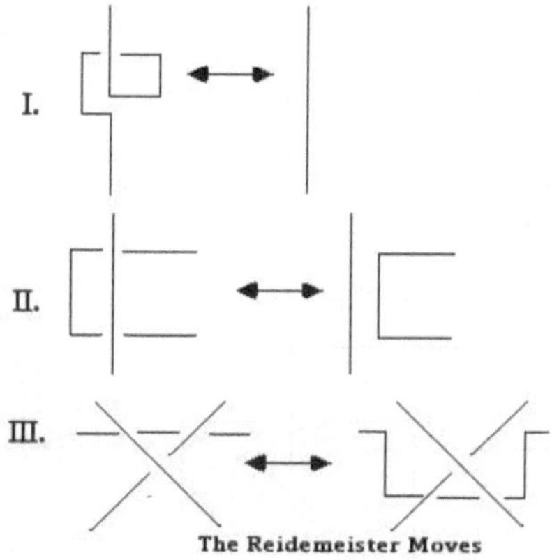

FIGURE 14. Reidemeister moves.

the sets) that are members of themselves, and it has representatives that are not members of themselves. In this domain, self-membership does not mean infinite descent. We do not insist that

$$a = \{a\}$$

implies that

$$a = \{\{\{\{\ldots\}\}\}\}.$$

Rather, $a = \{a\}$ just means that a has a little curl in its diagram. The Russell set of all sets that are not members of themselves is meaningless in this domain.

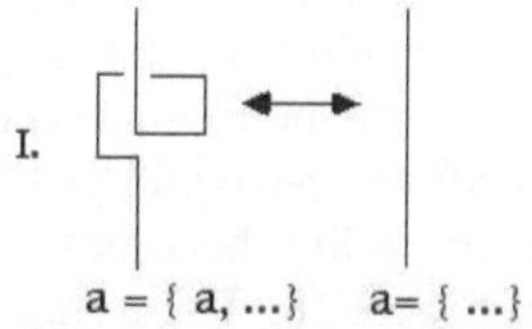

FIGURE 15. Reidemeister I: Replacing self-membership with no self-membership.

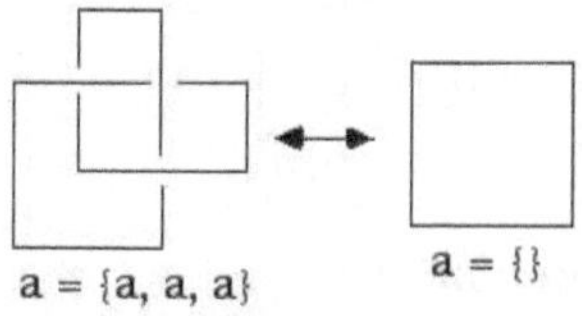

FIGURE 16. Trefoil is an empty knotset.

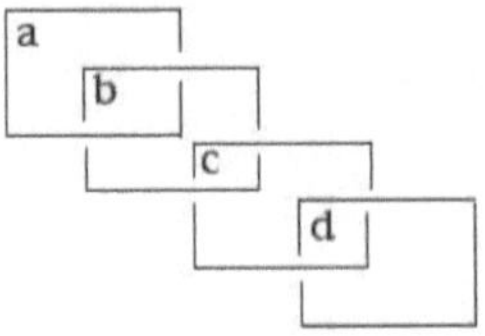

FIGURE 17. Chain.

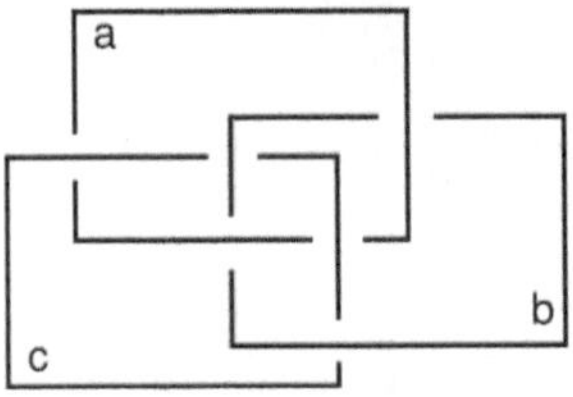

FIGURE 18. Borromean rings.

We can summarize this first level of knot-set theory in the following two equivalences:

1. Self-Reference: $a = \{b, c, \ldots\} \Longleftrightarrow a = \{a, b, c, \ldots\}$
2. Pair Cancellation: $S = \{a, a, b, c, \ldots\} \Longleftrightarrow S = \{b, c, \ldots\}$

With this mode of dealing with self-reference and multiplicity, knot-set theory has the interpretation in terms of topological classes of diagrams. We could imagine that the flatlanders felt the need to invent three dimensional space and topology, just so their set theory would have such an elegant interpretation.

But how elegant is this interpretation, from the point of view of topology? Are we happy that knots are equivalent to the empty knot-set as shown in Figure 16? For this, an extension of the theory is clearly in the waiting. We are happy that many topologically non-trivial links correspond to non-trivial knot-sets. In the Figure 17 , a chain link becomes a linked chain of knot-sets. But consider the link shown in Figure 18. These rings are commonly called the Borromean Rings. The Rings have the property that if you remove any one of them, then the other two are topologically unlinked. They form a topological tripartite relation. Their knot-set is described by the three equations

$$a = \{b, b\}$$
$$b = \{c, c\}$$
$$c = \{a, a\}.$$

Thus we see that this representative knot-set is a "scissors-paper-stone" pattern. Each component of the Rings lies over one other component, in a cyclic pattern. But in terms of the equivalence relation on knot sets that we have used, the knot set for the Rings is empty (by pair cancellation).

In order to go further in the direction of topological invariants for knots and links it is necessary to use more structure than the simple membership relation that motivates the knots-sets. Viewed from the point of view of the diagrams for knots and links there are a number of possible directions. For example, one can label all the arcs of the diagram and introduce algebraic relations at each crossing. This leads to the fundamental group and the quandle [39]. One can also label all the arcs of the diagram from an index set and view this labeling as a state in analogous to a state of a physical system in statistical mechanics. Then evaluations of these states and summations of the evaluations over all the states give the class of knot invariants called *quantum invariants* for knots and links [39]. These include the Jones polynomial and its generalizations. In this paper we will explain and use the Jones polynomial and the so-called colored Jones polynomials. See Section 17 for this development. The purpose of this section has been to introduce the subject of knot and link diagrams in the context of thinking about foundations of mathematics. However, it is worthwhile adding structure to the knot set theory so that it can at least see the higher order linking of the Borommean rings. We do this in the next subsection by keeping track of the order in which sets are encountered along the arc of a given component, and by keeping track of both membership and *co-membership* where we shall say that A is co-member of B if B is a member of A. As one moves along an arc one sequentially encounters members and co-members.

3.1. Ordered knot sets. Take a walk along a given component. Write down the sequence of memberships and belongings that you encounter on the walk as shown in Figure 19.

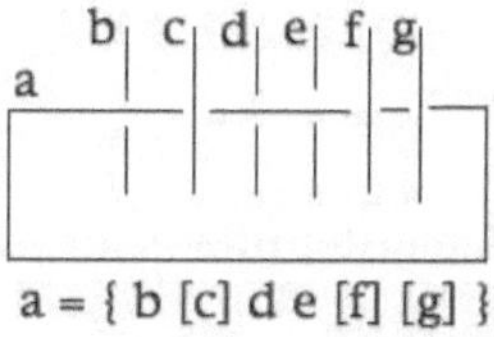

FIGURE 19. An ordered knot set.

In this notation, we record the order in which memberships and "co-memberships" (a is a co-member of b if and only if b is a member of a) occur along the strand of a given component of the knot-set. We do not choose a direction of traverse, so it is ok to reverse the total order of the contents of a given component and to take this order up to cyclic permutation. Thus we now have the representation of the Borromean Rings as shown in Figure 20.

With this extra information in front of us, it is clear that we should not allow the pair cancellations unless they occur in direct order, with no intervening co-memberships. Lets look at the revised Reidemeister moves as in Figure 21.

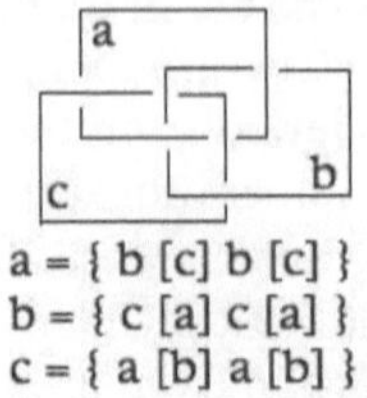

FIGURE 20. Borromean rings as ordered knot set.

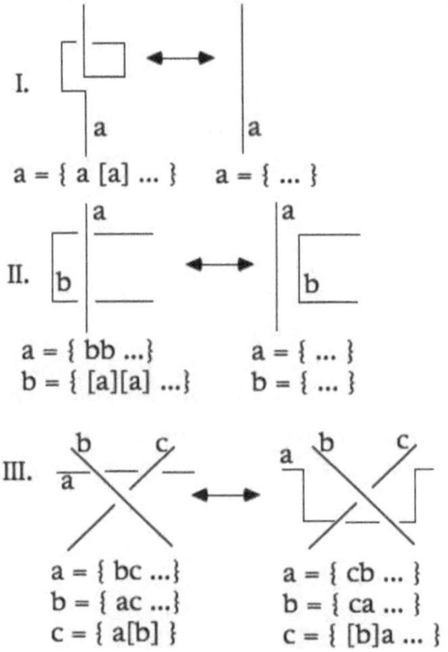

FIGURE 21. Reidemeister moves for ordered knot sets.

As is clear from the above diagrams, the Reidemeister moves tell us that we should impose some specific equivalences on these ordered knot sets:

1. We can erase any appearance of $a[a]$ or of $[a]a$ inside the set for a.
2. If bb occurs in a and $[a][a]$ occurs in b, then they can both be erased.
3. If bc is in a, ac is in b and $a[b]$ is in c, then we can reverse the order of each of these two element strings.

We take these three rules (and a couple of variants suggested by the diagrams) as the notion of equivalence of ordered knot-sets. One can see that the ordered knot-set for the Borromean rings is non-trivial in this equivalence relation. In this sense we have a a proof that the Borromean rings are linked, based on their scissors, paper, stone structure. The only proof that I know for the non-triviality of the Borommean ordered knot set uses the concept of coloring discussed in the next subsection.

Knots and links are represented by the diagrams themselves, taken up the equivalence relation generated by the Reidemeister moves. This calculus of diagrams is quite complex and it is remarkable, the number and depth of different mathematical approaches that are used to study this calculus and its properties. Studying knots and links is rather like studying number theory. The objects of study themselves can be constructed directly, and form a countable set. The problems that seem to emanate naturally from these objects are challenging and fascinating. For more about knot-sets, see [41]

3.2. Quandles and colorings of knot diagrams. There is an approach to studying knots and links that is very close to our ordered knot sets, but starts from a rather different premise. In this approach each arc of the diagram receives a label or "color". An arc of the diagram is a continuous curve in the diagram that starts at one under crossing and ends at another under crossing. For example, the trefoil diagram is related to this algebra as shown in Figure 22.

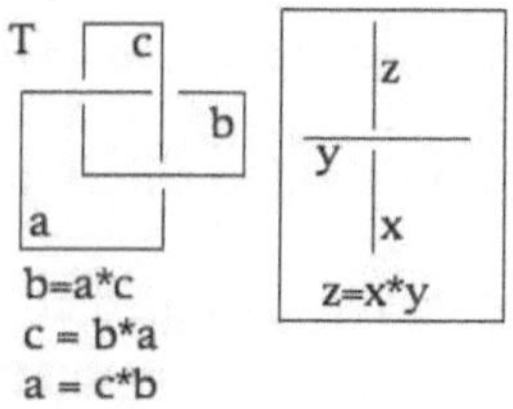

FIGURE 22. The quandle for the trefoil knot.

Each arc corresponds to an element of a "color algebra" $IQ(T)$ where T denotes the trefoil knot. We have that$IQ(T)$ is generated by colors a,b and c with the relations $c * b = a, a * c = b, b * a = c, a * a = a$. Each of these relations is a description of one of the crossings in T. These relations are specific to the trefoil knot. If we take on an algebra of this sort, we want its coloring structure to be invariant under the Reidemeister moves. This implies the following global relations:

$$x * x = x$$

$$(x * y) * y = x$$

$$(x * y) * z = (x * z) * (y * z)$$

for any x, y and z in the algebra (set of colors) $IQ(T)$. An algebra that satisfies these rules is called an Involutory Quandle (See [39]), hence the initials IQ. These global relations are really expressions of the concept of self-crossing and iterated crossing in the multiplicity of crossings that are available in a calculus of boundaries where the notation indicates the choice of interpretation, where one boundary is seen to cross (over) the other boundary. If we adopt these global relations for the algebra $IQ(K)$ for any knot or link diagram K, then two diagrams that are related by the Reidemeister moves will have isomorphic algebras. They will also inherit colorings of their arcs from one another. Thus the calculation of the algebra $IQ(K)$ for a knot or link K has the potentiality for bringing forth deep topological structure from the diagram.

In the case of the trefoil, one can see that the algebra actually closes at the set of elements a, b, c. We have the complete set of relations

$$c * b = a,\ a * c = b,\ b * a = c,\ a * a = a,\ a * a = a,\ b * b = b,\ c * c = c,$$

forming the *three-color quandle*. Three-coloring turns out to be quite useful for many knots and links. Thus we have seen that the trefoil knot is knotted due to its having a non-trivial three-coloring. By the same token, one can see that the Borommean rings are linked by checking that they do *not* have a non-trivial three-coloring! This fact is easy to check by directly trying to color the rings. That uncolorability implies that the rings are linked follows from the fact that there is a non-trivial coloring of three unlinked rings (color each ring by a separate color). This coloring of the unlinked rings would then induce a coloring of the Borommean rings. Since there is no such coloring, the Borommean rings must be linked.

The ordered knot set corresponding to a link can be colored or not colored in the same manner as a link diagram. The spaces between the letters in the ordered code of the knot set can be assigned colors in the same way as the arcs of a link diagram. In this way, the coloring proofs can be transferred to ordered knot sets in the case of links. We leave the details of this analysis of link sets to another paper.

Knot theory can be seen as a natural articulation not of three dimensional space (a perfectly good interpretation) but of the properties of interactions of boundaries. Each boundary can be regarded as that boundary transgressed by another boundary. The choice of who is the transgressed and who transgresses is the choice of a crossing, the choice of membership in the context of knot-set theory.

§4. Fermions, Majorana fermions and algebraic knot sets. In the last part of our discussion of knot sets we added order and co-membership to the structure. In this way of thinking, the knot set is an ordered sequence of memberships and co-memberships that are encountered as one moves along the strand of that part of the weave. Lets take this view, but go back to the ordinary knot sets that just catalog memberships. Then the knot set is a ordered list of the memberships that are encountered along the weave. For example, in Figure 17 we have $a = \{b\}, b = \{a, c\}, c = \{b, d\}, c = \{c\}$, and this would become the algebraic statements $a = \{b\}, b = \{ac\}, c = \{bd\}, c = \{c\}$, where we remove the parentheses and write the contents of each set as a algebraic product. We retain the brackets in order to continue to differentiate the set from its contents. Then we would have that $\{bccd\} = \{bd\}$ since repetitions are eliminated, and we see that the rule $x^2 = 1$ should be obeyed by this algebra of products of set members.

What shall we do about $a = \{bcdc\}$? We could decide that $xy = yx$ for all x and y in a given knot set. This commutative law would disregard the ordering, and we would have $\{bcdc\} = \{bccd\} = \{bd\}$. The simplest algebraic version of the knot sets is to have a commutative algebra with $x^2 = 1$ for all members. Then we can define $X \Delta Y$ for sets $X = \{\alpha\}$ and $Y = \{\beta\}$ by the equation

$$X \Delta Y = \{\alpha\beta\}$$

where $\alpha\beta$ represents the product of the members of X and Y taken together. The operation $X \Delta Y$ represents the union of knot sets and corresponds to *exclusive or* in standard set theory.

For example, suppose

$$A = \{yx\}, B = \{zy\}, C = \{xz\}.$$

Then we have $A^2 = B^2 = C^2 = \{1\}$ where it is understood that $\{1\} = \{\}$ represents the empty set. (That is, in the algebra 1 represents the empty word.) Furthermore we have $AB = C, BC = A, CA = B$. The relations in this example are very close to the quaternions. This example suggests that we could change the algebraic structure so that members satisfy $xy = -yx$, adding a notion of *sign* to the algebraic representation of the knot sets. We then get the pattern of the quaternion group: $A^2 = B^2 = C^2 = ABC = -1$ where -1 denotes the "negative" empty set.

By introducing the *Clifford algebra* with $x^2 = 1$ and $xy = -yx$ for generators, we bring the knot sets into direct correspondence with an algebra of *Majorana fermions*. The generators of this Clifford algebra represent fermions that are their own anti-particles. For a long time it has been conjectured that neutrinos may be Majorana fermions. More recently, it has been suggested that Majorana fermions may occur in collective electronic phenomena [74, 28, 8, 26, 70].

There is a natural association of fermion algebra to knot sets. In order to explain this association, we first give a short exposition of the algebra of fermion operators. In a standard collection of fermion operators $m_1, \ldots, m_k$ one has that each m_i is a linear operator on a Hilbert space with an adjoint operator $m_i^\dagger$ (corresponding to the anti-particle for the particle created by m_i) and relations

$$\begin{aligned} m_i^2 &= 0, \\ m_i m_i^\dagger + m_i^\dagger m_i &= 1, \\ m_i m_j + m_j m_i &= 0 \end{aligned}$$

when $i \neq j$.

There is another brand of Fermion algebra where we have generators $c_1, \ldots c_k$ and $c_i^2 = 1$ while $c_i c_j = -c_j c_i$ for all $i \neq j$. These are the *Majorana fermions*. There is a algebraic translation between the fermion algebra and Majorana fermion algebra. Given two Majorana fermions a and b with $a^2 = b^2 = 1$ and $ab = -ba$, define

$$m = (a + ib)/\sqrt{2}$$

and

$$m^\dagger = (a - ib)/\sqrt{2}.$$

It is then easy to see that $a^2 = b^2 = 1$ and $ab = -ba$ imply that m and $m^\dagger$ form a fermion in the sense that $m^2 = (m^\dagger)^2 = 0$ and $mm^\dagger + m^\dagger m = 1$. Thus pairs of Majorana fermions can be construed as ordinary fermions. Conversely, if m is an ordinary fermion, then formal real and imaginary parts of m yield a mathematical pair of Majorana fermions. A chain of electrons in a nano-wire, conceived in this way can give rise to a chain of Majorana fermions with a non-localized pair corresponding to the distant ends of the chain. The non-local nature of this pair is promising for creating topologically protected qubits, and there is at this writing an experimental search for evidence for the existence of such end-effect Majorana fermions.

We now see that it is exactly the Majorana fermion algebra that matches the properties of the knot sets. Here is an example that shows how the topology comes in. Let x, y, z be three Majorana fermions. Let $A = yx, B = zy, C = xz$. We have already seen that A, B, C represent the quaternions. Now define

$$s_1 = (1 + A)/\sqrt{2}, s_2 = (1 + B)/\sqrt{2}, s_3 = (1 + C)/\sqrt{2}.$$

It is easy to see that s_i and s_j satisfy the braiding relation for any $i \neq j$. For example, here is the verification for $i = 1, j = 2$.

$$\begin{aligned} s_1 s_2 s_1 &= (1/2\sqrt{2})(1 + A)(1 + B)(1 + A) \\ &= (1/2\sqrt{2})(1 + A + B + AB)(1 + A) \\ &= (1/2\sqrt{2})(1 + A + B + AB + A + A^2 + BA + ABA) \\ &= (1/2\sqrt{2})(1 + A + B + AB + A - 1 - AB + B) \\ &= (1/\sqrt{2})(A + B). \end{aligned}$$

Similarly,

$$\begin{aligned} s_2 s_1 s_2 &= (1/2\sqrt{2})(1 + B)(1 + A)(1 + B) \\ &= (1/2\sqrt{2})(1 + A + B + BA)(1 + B) \\ &= (1/2\sqrt{2})(1 + A + B + BA + B + AB + B^2 + BAB) \\ &= (1/2\sqrt{2})(1 + A + B + BA + B - BA - 1 + A) \\ &= (1/\sqrt{2})(A + B). \end{aligned}$$

Thus

$$s_1 s_2 s_1 = s_2 s_1 s_2,$$

and so a natural braid group representation arises from the Majorana fermions. This braid group representation is significant for quantum computing as we shall see in Section 7. For the purpose of this discussion, the braid group representation shows that the Clifford algebraic representation for knot sets is related to topology at more than one level. The relation $x^2 = 1$ for generators makes the individual sets, taken as products of generators, invariant under the Reidemeister moves (up to a global sign). But braiding invariance of certain linear combinations of sets is a relationship with knotting at a second level. This multiple relationship certainly deserves more thought. We will make one more remark here, and reserve further analysis for a subsequent paper.

These braiding operators can be seen to act on the vector space over the complex numbers that is spanned by the fermions x, y, z. To see how this works, consider

$$s = \frac{1 + yx}{\sqrt{2}},$$

$$T(p) = sps^{-1} = (\frac{1 + yx}{\sqrt{2}})p(\frac{1 - yx}{\sqrt{2}}),$$

and verify that $T(x) = y$ and $T(y) = -x$. Now view Figure 23 where we have illustrated a topological interpretation for the braiding of two fermions. In the topological interpretation the two fermions are connected by a flexible belt. On interchange, the belt becomes twisted by 2π. In the topological interpretation a twist of 2π corresponds to a phase change of -1. (For more information on this topological interpretation of 2π rotation for fermions, see [39].) Without a further choice it is not evident which particle of the pair should receive the phase change. The topology alone tells us only the relative change of phase between the two particles. The Clifford algebra for Majorana fermions makes a specific choice in the matter and in this way fixes the representation of the braiding.

Finally, we remark that linear combinations of products in the Clifford algebra can be regarded as superpositions of the knot sets. Thus $xy + xz$ is a superposition of the sets with members $\{x, y\}$ and$\{x, z\}$. Superposition of sets suggests that we are creating a species of *quantum set theory* and indeed Clifford algebra based quantum set theories have been suggested (see [18]) by David Finkelstein and others. It may come as a surprise to a quantum set theorist to find that knot theoretic topology is directly related to this subject. It is also clear that this Clifford algebraic quantum set theory should be related to our previous constructions for quantum knots [58, 71, 72, 73, 63]. This requires more investigation, and it suggests that knot theory and the theory of braids occupy a fundamental place in the foundations of quantum mechanics.

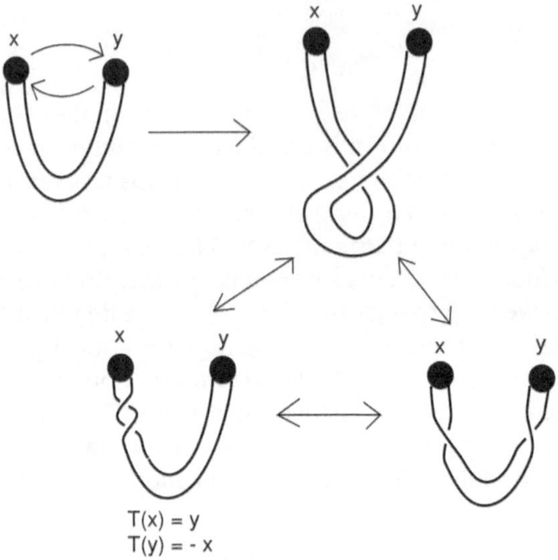

FIGURE 23. Braiding action on a pair of fermions.

§5. Laws of Form. In this section we discuss a formalism due the G. Spencer-Brown [92] that is often called the "calculus of indications". This calculus is a study of mathematical foundations with a topological notation based on one symbol, the mark:

$$\urcorner .$$

This single symbol represents a distinction between its own inside and outside. As is evident from Fgure 24, the mark is regarded as a shorthand for a rectangle drawn in the plane and dividing the plane into the regions inside and outside the rectangle.

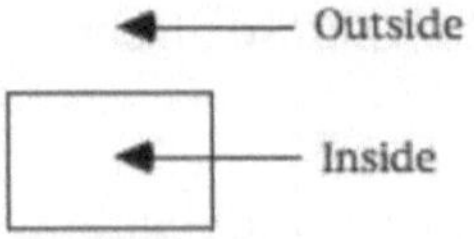

FIGURE 24. Inside and outside.

The reason we introduce this notation is that in the calculus of indications the mark can interact with itself in two possible ways. The resulting formalism becomes a version of Boolean arithmetic, but fundamentally simpler than the usual Boolean arithmetic of 0 and 1 with its two binary operations and one unary operation (negation). In the calculus of indications one takes a step in the direction of simplicity, and also a step in the direction of physics. The patterns of this mark and its self-interaction match those of a *Majorana fermion* as discussed in the previous section. A Majorana fermion is a particle that is its own anti-particle. [74]. We will later see, in this paper, that by adding braiding to the calculus of indications we arrive at the Fibonacci model, that can in principle support quantum computing.

In the previous section we described Majorana fermions in terms of their algebra of creation and annihilation operators. Here we describe the particle directly in terms of its interactions. This is part of a general scheme called "fusion rules" [76] that can be applied to discrete particle interacations. A fusion rule represents all of the different particle interactions in the form of a set of equations. The bare bones of the Majorana fermion consist in a particle P such that P can interact with itself to produce a neutral particle $*$ or produce itself P. Thus the possible interactions are

$$PP \longrightarrow *$$

and

$$PP \longrightarrow P.$$

This is the bare minimum that we shall need. The fusion rule is

$$P^2 = 1 + P.$$

This represents the fact that P can interact with itself to produce the neutral particle (represented as 1 in the fusion rule) or itself (represented by P in the fusion rule). We shall come back to the combinatorics related to this fusion equation.

Is there a *linguistic* particle that is its own anti-particle? Certainly we have

$$\sim\sim Q = Q$$

for any proposition Q (in Boolean logic). And so we might write

$$\sim\sim \longrightarrow *$$

where $*$ is a neutral linguistic particle, an identity operator so that

$$*Q = Q$$

for any proposition Q. But in the normal use of negation there is no way that the negation sign combines with itself to produce itself. This appears to ruin the analogy between negation and the Majorana fermion. Remarkably, the calculus of indications provides a context in which we can say exactly that a certain logical particle, the mark, can act as negation *and* can interact with itself to produce itself.

In the calculus of indications patterns of non-intersecting marks (i.e. non-intersecting rectangles) are called *expressions.* For example in Figure 25 we see how patterns of boxes correspond to patterns of marks.

In Figure 25, we have illustrated both the rectangle and the marked version of the expression. In an expression you can say definitively of any two marks whether one is or is not inside the other. The relationship between two marks is

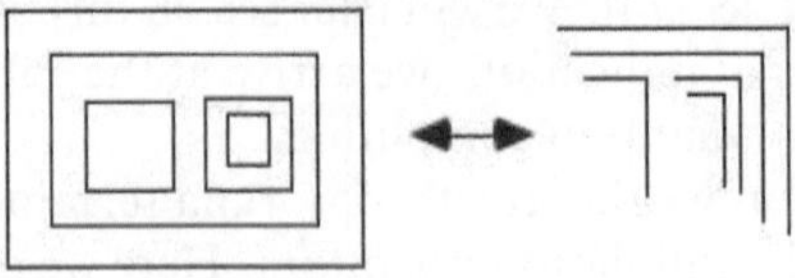

FIGURE 25. Boxes and marks.

either that one is inside the other, or that neither is inside the other. These two conditions correspond to the two elementary expressions shown in Figure 26.

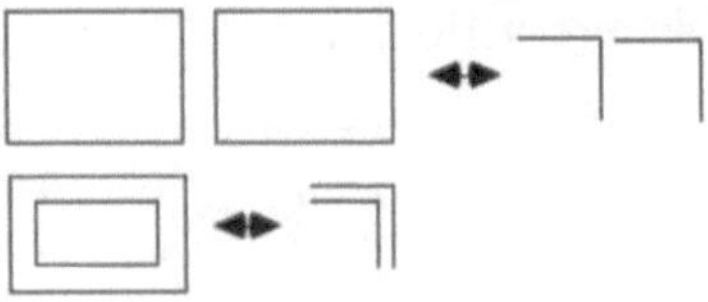

FIGURE 26. Translation between boxes and marks.

The mathematics in Laws of Form begins with two laws of transformation about these two basic expressions. Symbolically, these laws are:

1. Calling: 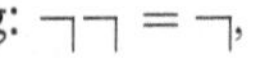
2. Crossing: 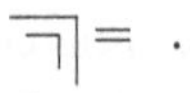

The equals sign denotes a replacement step that can be performed on instances of these patterns (two empty marks that are adjacent or one mark surrounding an empty mark). In the first of these equations two adjacent marks condense to a single mark, or a single mark expands to form two adjacent marks. In the second equation two marks, one inside the other, disappear to form the unmarked state indicated by nothing at all. That is, two nested marks can be replaced by an empty word in this formal system. Alternatively, the unmarked state can be replaced by two nested marks. These equations give rise to a natural calculus, and the mathematics can begin. For example, *any expression can be reduced uniquely to either the marked or the unmarked state.* The following example illustrates the method:

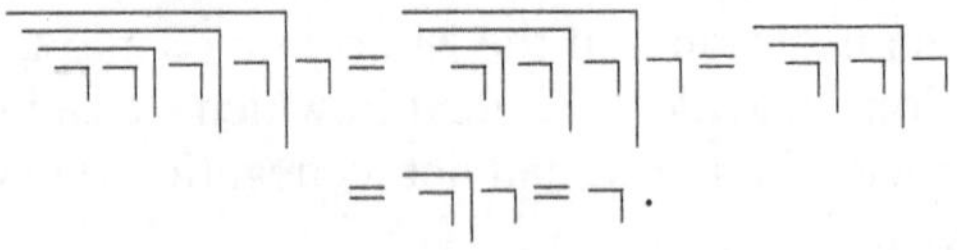

The general method for reduction is to locate marks that are at the deepest places in the expression (depth is defined by counting the number of inward crossings of boundaries needed to reach the given mark). Such a deepest mark must be empty and it is either surrounded by another mark, or it is adjacent to an empty mark. In either case a reduction can be performed by either calling or crossing.

Laws of Form begins with the following statement. "We take as given the idea of a distinction and the idea of an indication, and that it is not possible to make an indication without drawing a distinction. We take therefore the form of distinction for the form." Then the author makes the following two statements (laws):

1. *The value of a call made again is the value of the call.*
2. *The value of a crossing made again is not the value of the crossing.*

The two symbolic equations above correspond to these statements. First examine the law of calling. It says that the value of a repeated name is the value of the name. In the equation

$$\urcorner\,\urcorner = \urcorner$$

one can view either mark as the name of the state indicated by the outside of the other mark. In the other equation

$$\overline{\urcorner}\big| = \qquad .$$

the state indicated by the outside of a mark is the state obtained by crossing from the state indicated on the inside of the mark. Since the marked state is indicated on the inside, the outside must indicate the unmarked state. The Law of Crossing indicates how opposite forms can fit into one another and vanish into nothing, or how nothing can produce opposite and distinct forms that fit one another, hand in glove. The same interpretation yields the equation

$$\urcorner = \urcorner$$

where the left-hand side is seen as an instruction to cross from the unmarked state, and the right hand side is seen as an indicator of the marked state. The mark has a double carry of meaning. It can be seen as an operator, transforming the state on its inside to a different state on its outside, and it can be seen as the name of the marked state. That combination of meanings is compatible in this interpretation.

From the calculus of indications, one moves to algebra. Thus

$$\overline{\overline{A}\big|}\Big|$$

stands for the two possibilities

$$\overline{\overline{\urcorner}\big|}\Big| = \urcorner \longleftrightarrow A = \urcorner$$

$$\overline{\urcorner}\big| = \qquad \longleftrightarrow A = $$

In all cases we have

$$\overline{\overline{A}\big|}\Big| = A.$$

By the time we articulate the algebra, the mark can take the role of a unary operator

$$A \longrightarrow \overline{A}|.$$

But it retains its role as an element in the algebra. Thus begins algebra with respect to this non-numerical arithmetic of forms. The primary algebra that emerges is a subtle precursor to Boolean algebra. One can translate back and forth between elementary logic and primary algebra:

1. $\overline{\ \ }| \longleftrightarrow T$
2. $\overline{\overline{\ \ }|}| \longleftrightarrow F$
3. $\overline{A}| \longleftrightarrow \sim A$
4. $AB \longleftrightarrow A \vee B$
5. $\overline{\overline{A}|\ \overline{B}|}| \longleftrightarrow A \wedge B$
6. $\overline{A}|B \longleftrightarrow A \Rightarrow B$

The calculus of indications and the primary algebra form an efficient system for working with basic symbolic logic.

By reformulating basic symbolic logic in terms of the calculus of indications, we have a ground in which negation is represented by the mark *and* the mark is also interpreted as a value (a truth value for logic) and these two intepretations are compatible with one another in the formalism. The key to this compatibility is the choice to represent the value "false" by a literally unmarked state in the notational plane. With this the empty mark (a mark with nothing on its inside) can be interpreted as the negation of "false" and hence represents "true". The mark interacts with itself to produce itself (calling) and the mark interacts with itself to produce nothing (crossing). We have expanded the conceptual domain of negation so that it satisfies the mathematical pattern of an abstract Majorana fermion.

Another way to indicate these two interactions symbolically is to use a box,for the marked state and a blank space for the unmarked state. Then one has two modes of interaction of a box with itself:

1. Adjacency: □ □
   and
2. Nesting: 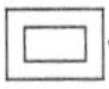.

With this convention we take the adjacency interaction to yield a single box, and the nesting interaction to produce nothing:

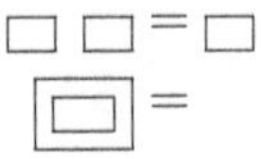

We take the notational opportunity to denote nothing by an asterisk (*). The syntatical rules for operating the asterisk are Thus the asterisk is a stand-in for no mark at all and it can be erased or placed wherever it is convenient to do so. Thus

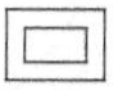 $= *.$

At this point the reader can appreciate what has been done if he returns to the usual form of symbolic logic. In that form we that

$$\sim\sim X = X$$

for all logical objects (propositions or elements of the logical algebra) X. We can summarize this by writing

$$\sim\sim \quad =$$

as a symbolic statement that is outside the logical formalism. Furthermore, one is committed to the interpretation of negation as an operator and not as an operand. The calculus of indications provides a formalism where the mark (the analog of negation in that domain) is both a value and an object, and so can act on itself in more than one way.

The Majorana particle is its own anti-particle. It is exactly at this point that physics meets logical epistemology. Negation as logical entity is its own anti-particle. Wittgenstein says (Tractatus [96] 4.0621) "... the sign '∼' corresponds to nothing in reality." And he goes on to say (Tractatus 5.511) "How can all-embracing logic which mirrors the world use such special catches and manipulations? Only because all these are connected into an infinitely fine network, the great mirror." For Wittgenstein in the Tractatus, the negation sign is part of the mirror making it possible for thought to reflect reality through combinations of signs. These remarks of Wittgenstein are part of his early picture theory of the relationship of formalism and the world. In our view, the world and the formalism we use to represent the world are not separate. The observer and the mark are (formally) identical. A path is opened between logic and physics.

The visual iconics that create via the boxes or half-boxes of the calculus of indications a model for a logical Majorana fermion can also be seen in terms of cobordisms of surfaces. View Figure 27. There the boxes have become circles and the interactions of the circles have been displayed as evolutions in an extra dimension, tracing out surfaces in three dimensions. The condensation of two circles to one is a simple cobordism betweem two circles and a single circle. The cancellation of two circles that are concentric can be seen as the right-hand lower cobordism in this figure with a level having a continuum of critical points where the two circles cancel. A simpler cobordism is illustrated above on the right where the two circles are not concentric, but nevertheless are cobordant to the empty circle. Another way of putting this is that two topological closed strings can interact by cobordism to produce a single string or to cancel one another. Thus a simple circle can be a topological model for a Majorana fermion.

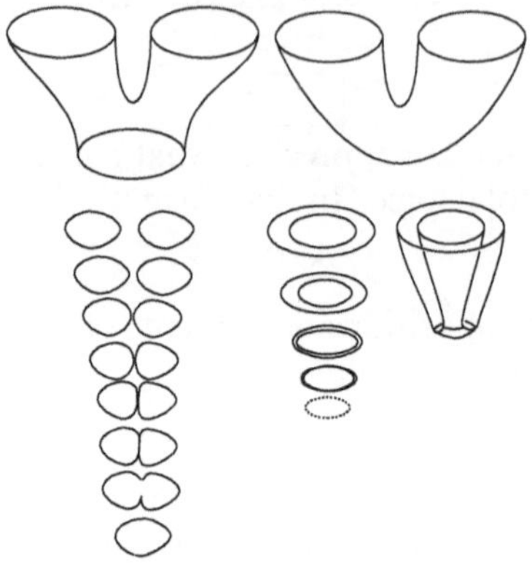

FIGURE 27. Calling, crossing and cobordism.

In Sections 15 and 16 we detail how the Fibonacci model for anyonic quantum computing [68, 81] can be constructed by using a version of the two-stranded bracket polynomial and a generalization of Penrose spin networks. This is a fragment of the Temperley-Lieb recoupling theory [40].

5.1. The square root of minus one is an eigenform and a clock. So far we have seen that the mark can represent the fusion rules for a Majorana fermion since it can interact with itself to produce either itself or nothing. But we have not yet seen the anti-commuting fermion algebra emerge from this context of making a distinction. Remarkably, this algebra does emerge when one looks at the mark recursively.

Consider the transformation

$$F(X) = \overline{X}\rceil.$$

If we iterate it and take the limit we find

$$G = F(F(F(F(\ldots)))) = \overline{\overline{\overline{\ldots}\rceil}\rceil}\rceil$$

an infinite nest of marks satisfying the equation

$$G = \overline{G}\rceil.$$

With $G = F(G)$, I say that G is an *eigenform* for the transformation F. See [47] for more about this point of view. See Figure 28 for an illustration of this nesting with boxes and an arrow that points inside the reentering mark to indicate its appearance inside itself. If one thinks of the mark itself as a Boolean logical value, then extending the language to include the reentering mark G goes beyond the boolean. We will not detail here how this extension can be related to non-standard logics, but refer the reader to [41]. Taken at face value the reentering mark cannot be just marked or just unmarked, for by its very definition, if it is marked then it is unmarked and if it is unmarked then it is marked. In this sense the reentering mark has the form of a self-contradicting paradox. There is no paradox since we do not have to permanently assign it to either value. The simplest interpretation of the reentering mark is that it is temporal and that it represents an oscillation between markedness and unmarkedness. In numerical terms it is a discrete dynamical system oscillating between +1 (marked) and −1 (not marked).

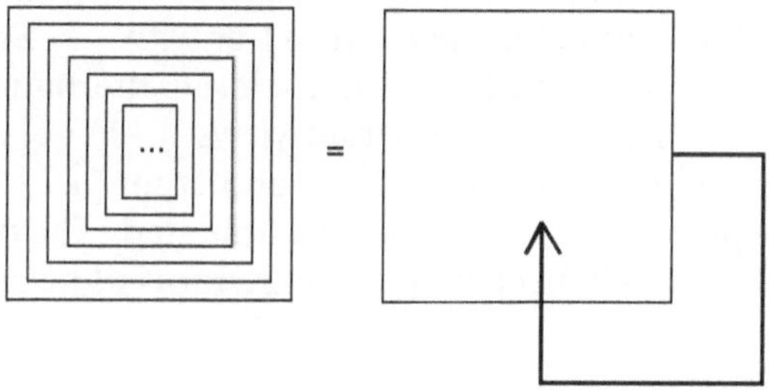

FIGURE 28.

With the reentering mark in mind consider now the transformation on real numbers given by

$$T(x) = -1/x.$$

This has the fixed points i and $-i$, the complex numbers whose squares are negative unity. But lets take a point of view more directly associated with the analogy of the recursive mark. Begin by starting with a simple periodic process that is associated directly with the classical attempt to solve for i as a solution to a quadratic equation. We take the point of view that solving $x^2 = ax + b$ is the same (when $x \neq 0$) as solving

$$x = a + b/x,$$

and hence is a matter of finding a fixed point. In the case of i we have

$$x^2 = -1$$

and so desire a fixed point

$$x = -1/x.$$

There are no real numbers that are fixed points for this operator and so we consider the oscillatory process generated by

$$T(x) = -1/x.$$

The fixed point would satisfy

$$i = -1/i$$

and multiplying, we get that

$$ii = -1.$$

On the other hand the iteration of T yields

$$1,\ T(1) = -1,\ T(T(1)) = +1,\ T(T(T(1))) = -1, +1, -1, +1, -1, \ldots.$$

The square root of minus one is a perfect example of an eigenform that occurs in a new and wider domain than the original context in which its recursive process arose. The process has no fixed point in the original domain.

Looking at the oscillation between $+1$ and -1, we see that there are naturally two phase-shifted viewpoints. We denote these two views of the oscillation by $[+1, -1]$ and$[-1, +1]$. These viewpoints correspond to whether one regards the oscillation at time zero as starting with $+1$ or with -1. See Figure 29. We shall let the word *iterant* stand for an undisclosed alternation or ambiguity between $+1$ and -1. There are two iterant views: $[+1, -1]$ and $[-1, +1]$ for the basic process we are examining. Given an iterant $[a, b]$, we can think of $[b, a]$ as the same process with a shift of one time step. The two iterant views, $[+1, -1]$ and $[-1, +1]$, will become the square roots of negative unity, i and $-i$.

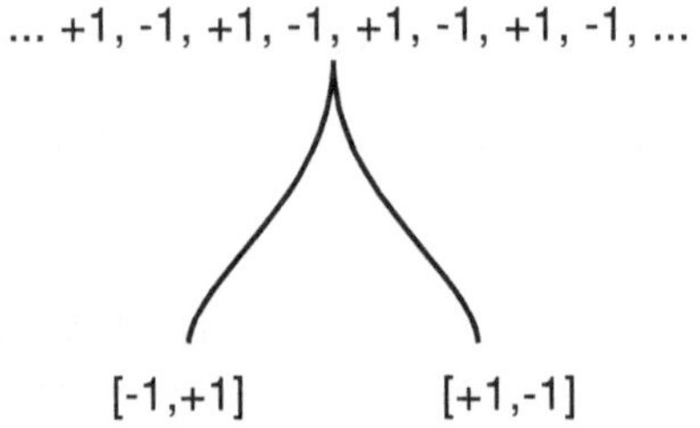

FIGURE 29.

We introduce a *temporal shift operator* η such that

$$[a, b]\eta = \eta[b, a]$$

and

$$\eta\eta = 1$$

for any iterant $[a, b]$, so that concatenated observations can include a time step of one-half period of the process

$$\ldots ababab ab \ldots.$$

We combine iterant views term-by-term as in

$$[a, b][c, d] = [ac, bd].$$

We now define i by the equation

$$i = [1, -1]\eta.$$

This makes i both a value and an operator that takes into account a step in time.

We calculate

$$ii = [1, -1]\eta[1, -1]\eta = [1, -1][-1, 1]\eta\eta = [-1, -1] = -1.$$

Thus we have constructed a square root of minus one by using an iterant viewpoint. In this view i represents a discrete oscillating temporal process and it is an eigenform for $T(x) = -1/x$, participating in the algebraic structure of the complex numbers. In fact the corresponding algebra structure of linear combinations $[a, b] + [c, d]\eta$ is isomorphic with 2×2 matrix algebra and iterants can be used to construct $n \times n$ matrix algebra. We treat this generalization elsewhere [46, 50].

Now we can make contact with the algebra of the Majorana fermions. Let $e = [1, -1]$. Then we have $e^2 = [1, 1] = 1$ and $e\eta = [1, -1]\eta = \eta[-1, 1] = -\eta e$. Thus we have

$$e^2 = 1, \quad \eta^2 = 1, \quad \text{and} \quad e\eta = -\eta e.$$

We can regard e and η as a fundamental pair of Majorana fermions. This is a formal correspondence, but it is striking how this Marjorana fermion algebra emerges from an analysis of the recursive nature of the reentering mark, while the fusion algebra for the Majorana fermion emerges from the distinctive properties of the mark itself. We see how the seeds of the fermion algebra live in this extended logical context.

Note how the development of the algebra works at this point. We have that

$$(e\eta)^2 = -1$$

and so regard this as a natural construction of the square root of minus one in terms of the phase synchronization of the clock that is the iteration of the reentering mark. Once we have the square root of minus one it is natural to introduce *another* one and call this one i, letting it commute with the other operators. Then we have the $(ie\eta)^2 = +1$ and so we have a triple of Majorana fermions:

$$a = e, b = \eta, c = ie\eta$$

and we can construct the quaternions

$$I = ba = \eta e, J = cb = ie, K = ac = i\eta.$$

With the quaternions in place, we have the braiding operators

$$A = \frac{1}{\sqrt{2}}(1 + I),\ B = \frac{1}{\sqrt{2}}(1 + J),\ C = \frac{1}{\sqrt{2}}(1 + K),$$

and can continue as we did in Section 4.

There is one more comment that is appropriate for this section. Recall from Section 4 that a pair of Majorana fermions can be assembled to form a single standard fermion. In our case we have the two Marjorana fermions e and η

and the corresponding standard fermion annihilation and creation operators are then given by the formulas below.

$$\psi = (e + i\eta)/\sqrt{2} \quad \text{and} \quad \psi^{\dagger} = (e - i\eta)/\sqrt{2}.$$

Since e represents a spatial view of the basic discrete oscillation and η is the time-shift operator for this oscillation it is of interest to note that the standard fermion built by these two can be regarded as a quantum of spacetime, retrieved from the way that we decomposed the process into space and time. Since all this is initially built in relation to extending the Boolean logic of the mark to a non-boolean recursive context, there is further analysis needed of the relation of the physics and the logic. This will be taken up in a separate paper.

5.2. Relativity and the Dirac equation. Starting with the algebra structure of e and η and adding a commuting square root of -1, i, we have constructed fermion algebra and quaternion algebra. We can now go further and construct the Dirac equation. This may sound circular, in that the fermions arise from solving the Dirac equation, but in fact the algebra underlying this equation has the same properties as the creation and annihilation algebra for fermions, so it is by way of this algebra that we will come to the Dirac equation. If the speed of light is equal to 1 (by convention), then energy E, momentum p and mass m are related by the (Einstein) equation

$$E^2 = p^2 + m^2.$$

Dirac constructed his equation by looking for an algebraic square root of $p^2 + m^2$ so that he could have a linear operator for E that would take the same role as the Hamiltonian in the Schrodinger equation. We will get to this operator by first taking the case where p is a scalar (we use one dimension of space and one dimension of time. Let $E = \alpha p + \beta m$ where α and β are elements of a a possibly non-commutative, associative algebra. Then

$$E^2 = \alpha^2 p^2 + \beta^2 m^2 + pm(\alpha\beta + \beta\alpha).$$

Hence we will satisfiy $E^2 = p^2 + m^2$ if $\alpha^2 = \beta^2 = 1$ and $\alpha\beta + \beta\alpha = 0$. This is our familiar Clifford algebra pattern and we can use the iterant algebra generated by e and η if we wish. Then, because the quantum operator for momentum is $-i\partial/\partial x$ and the operator for energy is $i\partial/\partial t$, we have the Dirac equation

$$i\partial\psi/\partial t = -i\alpha\partial\psi/\partial x + \beta m\psi.$$

Let

$$\mathcal{O} = i\partial/\partial t + i\alpha\partial/\partial x - \beta m$$

so that the Dirac equation takes the form

$$\mathcal{O}\psi(x,t) = 0.$$

Now note that

$$\mathcal{O}e^{i(px-Et)} = (E - \alpha p + \beta m)e^{i(px-Et)}$$

and that if

$$U = (E - \alpha p + \beta m)\beta\alpha = \beta\alpha E + \beta p + \alpha m,$$

then

$$U^2 = -E^2 + p^2 + m^2 = 0,$$

from which it follows that

$$\psi = Ue^{i(px - Et)}$$

is a (plane wave) solution to the Dirac equation.

In fact, this calculation suggests that we should multiply the operator $\mathcal{O}$ by $\beta\alpha$ on the right, obtaining the operator

$$\mathcal{D} = \mathcal{O}\beta\alpha = i\beta\alpha\partial/\partial t + i\beta\partial/\partial x + \alpha m,$$

and the equivalent Dirac equation

$$\mathcal{D}\psi = 0.$$

In fact for the specific ψ above we will now have $\mathcal{D}(Ue^{i(px-Et)}) = U^2 e^{i(px-Et)} = 0$. This way of reconfiguring the Dirac equation in relation to nilpotent algebra elements U is due to Peter Rowlands [86]. We will explore this relationship with the Rowlands formulation in a separate paper.

Return now to the original version of the Dirac equation.

$$i\partial\psi/\partial t = -i\alpha\partial\psi/\partial x + \beta m\psi.$$

We can rewrite this as

$$\partial\psi/\partial t = \alpha\partial\psi/\partial x + i\beta m\psi.$$

We see that if $i\beta$ is real, then we can write a fully real version of the Dirac equation. For example, we can take the equation

$$\partial\psi/\partial t = e\partial\psi/\partial x + e\eta m\psi.$$

where we represent

$$e = \begin{pmatrix} -1 & 0 \\ 0 & 1 \end{pmatrix}$$

and

$$\eta = \begin{pmatrix} 0 & 1 \\ 1 & 0 \end{pmatrix}$$

as matrix versions of the iterants associated with the reentering mark. For the case of one dimension of space and one dimension of time, this is the Majorana representation for the Dirac equation (compare [65]). Since the equation can have real solutions, these are their own complex conjugates and correspond to particles that are their own anti-particles. As the reader can check, the corresponding Rowland nilpotent U is given by the formula

$$U = -i\eta E + ie\eta p + em.$$

For effective application to the topics in this paper, one needs to use two dimensions of space and one dimension of time. This will be explored in another paper. In the present paper we have given a picture of how, *starting with the mark as a logical and recursive particle, one can tell a story that reaches the Dirac equation and its algebra.*

§6. Quantum mechanics and quantum computation. We shall quickly indicate the basic principles of quantum mechanics. The quantum information context encapsulates a concise model of quantum theory:

The initial state of a quantum process is a vector $|v\rangle$ in a complex vector space H. Measurement returns basis elements β of H with probability

$$|\langle\beta\,|v\rangle|^2/\langle v\,|v\rangle$$

where $\langle v\,|w\rangle = v^\dagger w$ with $v^\dagger$ the conjugate transpose of v. A physical process occurs in steps $|v\rangle \longrightarrow U\,|v\rangle = |\,Uv\rangle$ where U is a unitary linear transformation.

Note that since $\langle Uv\,|Uw\rangle = \langle v\,|U^\dagger U|w\rangle = \langle v\,|w\rangle =$ when U is unitary, it follows that probability is preserved in the course of a quantum process.

One of the details required for any specific quantum problem is the nature of the unitary evolution. This is specified by knowing appropriate information about the classical physics that supports the phenomena. This information is used to choose an appropriate Hamiltonian through which the unitary operator is constructed via a correspondence principle that replaces classical variables with appropriate quantum operators. (In the path integral approach one needs a Langrangian to construct the action on which the path integral is based.) One needs to know certain aspects of classical physics to solve any specific quantum problem.

A key concept in the quantum information viewpoint is the notion of the superposition of states. If a quantum system has two distinct states $|v\rangle$ and $|w\rangle$, then it has infinitely many states of the form $a|v\rangle + b|w\rangle$ where a and b are complex numbers taken up to a common multiple. States are "really" in the projective space associated with H. There is only one superposition of a single state $|v\rangle$ with itself. On the other hand, it is most convenient to regard the states $|v\rangle$ and $|w\rangle$ as vectors in a vector space. We than take it as part of the procedure of dealing with states to normalize them to unit length. Once again, the superposition of a state with itself is again itself.

Dirac [15] introduced the "bra-(c)-ket" notation $\langle A\,|B\rangle = A^\dagger B$ for the inner product of complex vectors $A, B \in H$. He also separated the parts of the bracket into the *bra* $\langle A\,|$ and the *ket* $|B\rangle$. Thus

$$\langle A\,|B\rangle = \langle A\,|\;|B\rangle$$

In this interpretation, the ket $|B\rangle$ is identified with the vector $B \in H$, while the bra $< A\,|$ is regarded as the element dual to A in the dual space H^*. The dual element to A corresponds to the conjugate transpose $A^\dagger$ of the vector A, and the inner product is expressed in conventional language by the matrix product $A^\dagger B$ (which is a scalar since B is a column vector). Having separated the bra and the ket, Dirac can write the "ket-bra" $|A\rangle\langle B| = AB^\dagger$. In conventional notation, the ket-bra is a matrix, not a scalar, and we have the following formula for the square of $P = |A\rangle\langle B|$:

$$P^2 = |A\rangle\langle B||A\rangle\langle B| = A(B^\dagger A)B^\dagger = (B^\dagger A)AB^\dagger = \langle B|A\rangle P.$$

The standard example is a ket-bra $P = |A\rangle\langle A|$ where $\langle A|A\rangle = 1$ so that $P^2 = P$. Then P is a projection matrix, projecting to the subspace of H that is spanned by the vector $|A\rangle$. In fact, for any vector $|B\rangle$ we have

$$P|B\rangle = |A\rangle\langle A||B\rangle = |A\rangle\langle A|B\rangle = \langle A|B\rangle|A\rangle.$$

If $\{|C_1\rangle, |C_2\rangle, \ldots |C_n\rangle\}$ is an orthonormal basis for H, and

$$P_i = |C_i\rangle\langle C_i|,$$

then for any vector $|A\rangle$ we have

$$|A\rangle = \langle C_1|A\rangle|C_1\rangle + \cdots + \langle C_n|A\rangle|C_n\rangle.$$

Hence

$$\langle B|A\rangle = \langle B|C_1\rangle\langle C_1|A\rangle + \cdots + \langle B|C_n\rangle\langle C_n|A\rangle$$

One wants the probability of starting in state $|A\rangle$ and ending in state $|B\rangle$. The probability for this event is equal to $|\langle B|A\rangle|^2$. This can be refined if we have more knowledge. If the intermediate states $|C_i\rangle$ are a complete set of orthonormal alternatives then we can assume that $\langle C_i|C_i\rangle = 1$ for each i and that $\Sigma_i|C_i\rangle\langle C_i| = 1$. This identity now corresponds to the fact that 1 is the sum of the probabilities of an arbitrary state being projected into one of these intermediate states.

If there are intermediate states between the intermediate states this formulation can be continued until one is summing over all possible paths from A to B. This becomes the path integral expression for the amplitude $\langle B|A\rangle$.

6.1. What is a quantum computer? A *quantum computer* is, abstractly, a composition U of unitary transformations, together with an initial state and a choice of measurement basis. One runs the computer by repeatedly initializing it, and then measuring the result of applying the unitary transformation U to the initial state. The results of these measurements are then analyzed for the desired information that the computer was set to determine. The key to using the computer is the design of the initial state and the design of the composition of unitary transformations. The reader should consult [79] for more specific examples of quantum algorithms.

Let H be a given finite dimensional vector space over the complex numbers C. Let

$$\{W_0, W_1, \ldots, W_n\}$$

be an orthonormal basis for H so that with $|i\rangle := |W_i\rangle$ denoting W_i and $\langle i|$ denoting the conjugate transpose of $|i\rangle$, we have

$$\langle i|j\rangle = \delta_{ij}$$

where δ_{ij} denotes the Kronecker delta (equal to one when its indices are equal to one another, and equal to zero otherwise). Given a vector v in H let $|v|^2 := \langle v|v\rangle$. Note that $\langle i|v$ is the i-th coordinate of v.
An *measurement of* v returns one of the coordinates $|i\rangle$ of v with probability $|\langle i|v|^2$. This model of measurement is a simple instance of the situation with a quantum mechanical system that is in a mixed state until it is observed. The result of observation is to put the system into one of the basis states.

When the dimension of the space H is two ($n = 1$), a vector in the space is called a *qubit*. A qubit represents one quantum of binary information. On measurement, one obtains either the ket $|0\rangle$ or the ket $|1\rangle$. This constitutes the binary distinction that is inherent in a qubit. Note however that the information obtained is probabilistic. If the qubit is

$$|\psi\rangle = \alpha|0\rangle + \beta\,|1\rangle,$$

then the ket $|0\rangle$ is observed with probability $|\alpha|^2$, and the ket $|1\rangle$ is observed with probability $|\beta|^2$. In speaking of an idealized quantum computer, we do not specify the nature of measurement process beyond these probability postulates.

In the case of general dimension n of the space H, we will call the vectors in H *qunits*. It is quite common to use spaces H that are tensor products of two-dimensional spaces (so that all computations are expressed in terms of qubits) but this is not necessary in principle. One can start with a given space, and later work out factorizations into qubit transformations.

A *quantum computation* consists in the application of a unitary transformation U to an initial qunit $\psi = a_0|0\rangle + \cdots + a_n|n\rangle$ with $|\psi|^2 = 1$, plus an measurement of $U\psi$. A measurement of $U\psi$ returns the ket $|i\rangle$ with probability $|\langle i|U\psi|^2$. In particular, if we start the computer in the state $|i\rangle$, then the probability that it will return the state $|j\rangle$ is $|\langle j|U|i\rangle|^2$.

It is the necessity for writing a given computation in terms of unitary transformations, and the probabilistic nature of the result that characterizes quantum computation. Such computation could be carried out by an idealized quantum mechanical system. It is hoped that such systems can be physically realized.

§7. **Braiding operators and universal quantum gates.** A class of invariants of knots and links called quantum invariants can be constructed by using representations of the Artin braid group, and more specifically by using solutions to the Yang-Baxter equation [7], first discovered in relation to $1 + 1$ dimensional quantum field theory, and 2 dimensional statistical mechanics.

Braiding operators feature in constructing representations of the Artin braid group, and in the construction of invariants of knots and links.

A key concept in the construction of quantum link invariants is the association of a Yang-Baxter operator R to each elementary crossing in a link diagram. The operator R is a linear mapping

$$R\colon\ V \otimes V \longrightarrow V \otimes V$$

defined on the 2-fold tensor product of a vector space V, generalizing the permutation of the factors (i.e., generalizing a swap gate when V represents one qubit). Such transformations are not necessarily unitary in topological applications. It is useful to understand when they can be replaced by unitary transformations for the purpose of quantum computing. Such unitary R-matrices can be used to make unitary representations of the Artin braid group.

A solution to the Yang-Baxter equation, as described in the last paragraph is a matrix R, regarded as a mapping of a two-fold tensor product of a vector space $V \otimes V$ to itself that satisfies the equation

$$(R \otimes I)(I \otimes R)(R \otimes I) = (I \otimes R)(R \otimes I)(I \otimes R).$$

From the point of view of topology, the matrix R is regarded as representing an elementary bit of braiding represented by one string crossing over another. In Figure 30 we have illustrated the braiding identity that corresponds to the Yang-Baxter equation. Each braiding picture with its three input lines (below) and output lines (above) corresponds to a mapping of the three fold tensor product of the vector space V to itself, as required by the algebraic equation quoted above. The pattern of placement of the crossings in the diagram corresponds to the factors $R \otimes I$ and $I \otimes R$. This crucial topological move has an algebraic expression in terms of such a matrix R. Our approach in this section to relate topology, quantum computing, and quantum entanglement is through the use of the Yang-Baxter equation. In order to accomplish this aim, *we need to study solutions of the Yang-Baxter equation that are unitary.* Then the R matrix can be seen *either* as a braiding matrix *or* as a quantum gate in a quantum computer.

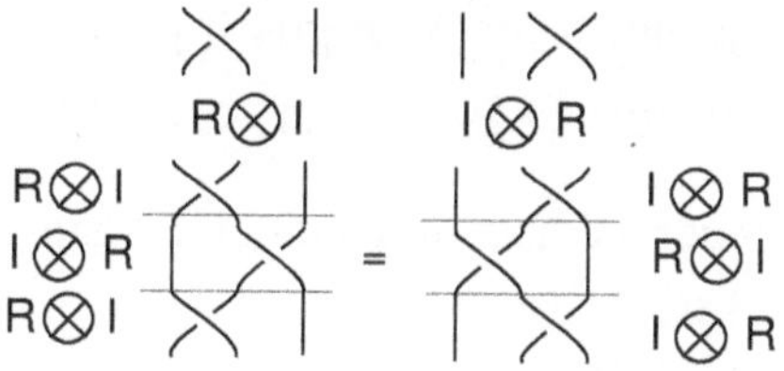

FIGURE 30. The Yang-Baxter equation.

The problem of finding solutions to the Yang-Baxter equation that are unitary turns out to be surprisingly difficult. Dye [16] has classified all such matrices of size 4×4. A rough summary of her classification is that all 4×4 unitary solutions to the Yang-Baxter equation are similar to one of the following types of matrix:

$$R = \begin{pmatrix} 1/\sqrt{2} & 0 & 0 & 1/\sqrt{2} \\ 0 & 1/\sqrt{2} & -1/\sqrt{2} & 0 \\ 0 & 1/\sqrt{2} & 1/\sqrt{2} & 0 \\ -1/\sqrt{2} & 0 & 0 & 1/\sqrt{2} \end{pmatrix}$$

$$R' = \begin{pmatrix} a & 0 & 0 & 0 \\ 0 & 0 & b & 0 \\ 0 & c & 0 & 0 \\ 0 & 0 & 0 & d \end{pmatrix}$$

$$R'' = \begin{pmatrix} 0 & 0 & 0 & a \\ 0 & b & 0 & 0 \\ 0 & 0 & c & 0 \\ d & 0 & 0 & 0 \end{pmatrix}$$

where a, b, c, d are unit complex numbers.

For the purpose of quantum computing, one should regard each matrix as acting on the stamdard basis $\{|00\rangle, |01\rangle, |10\rangle, |11\rangle\}$ of $H = V \otimes V$, where V is a two-dimensional complex vector space. Then, for example we have

$$R|00\rangle = (1/\sqrt{2})|00\rangle - (1/\sqrt{2})|11\rangle,$$
$$R|01\rangle = (1/\sqrt{2})|01\rangle + (1/\sqrt{2})|10\rangle,$$
$$R|10\rangle = -(1/\sqrt{2})|01\rangle + (1/\sqrt{2})|10\rangle,$$
$$R|11\rangle = (1/\sqrt{2})|00\rangle + (1/\sqrt{2})|11\rangle.$$

The reader should note that R is the familiar change-of-basis matrix from the standard basis to the Bell basis of entangled states.

In the case of R', we have

$$R'|00\rangle = a|00\rangle, \; R'|01\rangle = c|10\rangle,$$
$$R'|10\rangle = b|01\rangle, \; R'|11\rangle = d|11\rangle.$$

Note that R' can be regarded as a diagonal phase gate P, composed with a swap gate S.

$$P = \begin{pmatrix} a & 0 & 0 & 0 \\ 0 & b & 0 & 0 \\ 0 & 0 & c & 0 \\ 0 & 0 & 0 & d \end{pmatrix}$$

$$S = \begin{pmatrix} 1 & 0 & 0 & 0 \\ 0 & 0 & 1 & 0 \\ 0 & 1 & 0 & 0 \\ 0 & 0 & 0 & 1 \end{pmatrix}$$

Compositions of solutions of the (Braiding) Yang-Baxter equation with the swap gate S are called *solutions to the algebraic Yang-Baxter equation*. Thus the diagonal matrix P is a solution to the algebraic Yang-Baxter equation.

Remark 1. Another avenue related to unitary solutions to the Yang-Baxter equation as quantum gates comes from using extra physical parameters in this equation (the rapidity parameter) that are related to statistical physics. In [99] we discovered that solutions to the Yang-Baxter equation with the rapidity parameter allow many new unitary solutions. The significance of these gates for quatnum computing is still under investigation.

7.1. Universal gates. A *two-qubit gate* G is a unitary linear mapping $G : V \otimes V \longrightarrow V$ where V is a two complex dimensional vector space. We say that the gate G is *universal for quantum computation* (or just *universal*) if G together with local unitary transformations (unitary transformations from V to V) generates all unitary transformations of the complex vector space of dimension 2^n to itself. It is well-known [79] that *CNOT* is a universal gate. (On the standard basis, *CNOT* is the identity when the first qubit is $|0\rangle$, and it flips the second qbit, leaving the first alone, when the first qubit is $|1\rangle$.)
A gate G, as above, is said to be *entangling* if there is a vector

$$|\alpha\beta\rangle = |\alpha\rangle \otimes |\beta\rangle \in V \otimes V$$

such that $G|\alpha\beta\rangle$ is not decomposable as a tensor product of two qubits. Under these circumstances, one says that $G|\alpha\beta\rangle$ is *entangled.*
In [11], the Brylinskis give a general criterion of G to be universal. They prove that *a two-qubit gate G is universal if and only if it is entangling.*

Remark 2. A two-qubit pure state

$$|\phi\rangle = a|00\rangle + b|01\rangle + c|10\rangle + d|11\rangle$$

is entangled exactly when $(ad - bc) \neq 0$. It is easy to use this fact to check when a specific matrix is, or is not, entangling.

Remark 3. There are many gates other than *CNOT* that can be used as universal gates in the presence of local unitary transformations. Some of these are themselves topological (unitary solutions to the Yang-Baxter equation, see [57]) and themselves generate representations of the Artin braid group. Replacing *CNOT* by a solution to the Yang-Baxter equation does not place the local unitary transformations as part of the corresponding representation of the braid group. Thus such substitutions give only a partial solution to creating topological quantum computation. In this paper we are concerned with braid group representations that include all aspects of the unitary group. Accordingly, in the next section we shall first examine how the braid group on three strands can be represented as local unitary transformations.

THEOREM 1. *Let D denote the phase gate shown below. D is a solution to the algebraic Yang-Baxter equation (see the earlier discussion in this section). Then D is a universal gate.*

$$D = \begin{pmatrix} 1 & 0 & 0 & 0 \\ 0 & 1 & 0 & 0 \\ 0 & 0 & 1 & 0 \\ 0 & 0 & 0 & -1 \end{pmatrix}$$

PROOF. It follows at once from the Brylinski Theorem that D is universal. For a more specific proof, note that $CNOT = QDQ^{-1}$, where $Q = H \otimes I$, H is the 2×2 Hadamard matrix. The conclusion then follows at once from this identity and the discussion above. We illustrate the matrices involved in this proof below:

$$H = (1/\sqrt{2}) \begin{pmatrix} 1 & 1 \\ 1 & -1 \end{pmatrix}$$

$$Q = (1/\sqrt{2}) \begin{pmatrix} 1 & 1 & 0 & 0 \\ 1 & -1 & 0 & 0 \\ 0 & 0 & 1 & 1 \\ 0 & 0 & 1 & -1 \end{pmatrix}$$

$$D = \begin{pmatrix} 1 & 0 & 0 & 0 \\ 0 & 1 & 0 & 0 \\ 0 & 0 & 1 & 0 \\ 0 & 0 & 0 & -1 \end{pmatrix}$$

$$QDQ^{-1} = QDQ = \begin{pmatrix} 1 & 0 & 0 & 0 \\ 0 & 1 & 0 & 0 \\ 0 & 0 & 0 & 1 \\ 0 & 0 & 1 & 0 \end{pmatrix} = CNOT$$

⊣

Remark 4. We thank Martin Roetteles [84] for pointing out the specific factorization of $CNOT$ used in this proof.

THEOREM 2. *The matrix solutions R' and R'' to the Yang-Baxter equation, described above, are universal gates exactly when $ad - bc \neq 0$ for their internal parameters a, b, c, d. In particular, let R_0 denote the solution R' (above) to the Yang-Baxter equation with $a = b = c = 1$, $d = -1$.*

$$R' = \begin{pmatrix} a & 0 & 0 & 0 \\ 0 & 0 & b & 0 \\ 0 & c & 0 & 0 \\ 0 & 0 & 0 & d \end{pmatrix}$$

$$R_0 = \begin{pmatrix} 1 & 0 & 0 & 0 \\ 0 & 0 & 1 & 0 \\ 0 & 1 & 0 & 0 \\ 0 & 0 & 0 & -1 \end{pmatrix}$$

Then R_0 is a universal gate.

PROOF. The first part follows at once from the Brylinski Theorem. In fact, letting H be the Hadamard matrix as before, and

$$\sigma = \begin{pmatrix} 1/\sqrt{2} & i/\sqrt{2} \\ i/\sqrt{2} & 1/\sqrt{2} \end{pmatrix}, \quad \lambda = \begin{pmatrix} 1/\sqrt{2} & 1/\sqrt{2} \\ i/\sqrt{2} & -i/\sqrt{2} \end{pmatrix}$$

$$\mu = \begin{pmatrix} (1-i)/2 & (1+i)/2 \\ (1-i)/2 & (-1-i)/2 \end{pmatrix}.$$

Then

$$CNOT = (\lambda \otimes \mu)(R_0(I \otimes \sigma)R_0)(H \otimes H).$$

This gives an explicit expression for $CNOT$ in terms of R_0 and local unitary transformations (for which we thank Ben Reichardt). ⊣

Remark 5. Let $SWAP$ denote the Yang-Baxter Solution R' with $a = b = c = d = 1$.

$$SWAP = \begin{pmatrix} 1 & 0 & 0 & 0 \\ 0 & 0 & 1 & 0 \\ 0 & 1 & 0 & 0 \\ 0 & 0 & 0 & 1 \end{pmatrix}$$

$SWAP$ is the standard swap gate. Note that $SWAP$ is not a universal gate. This also follows from the Brylinski Theorem, since $SWAP$ is not entangling. Note also that R_0 is the composition of the phase gate D with this swap gate.

THEOREM 3. *Let*

$$R = \begin{pmatrix} 1/\sqrt{2} & 0 & 0 & 1/\sqrt{2} \\ 0 & 1/\sqrt{2} & -1/\sqrt{2} & 0 \\ 0 & 1/\sqrt{2} & 1/\sqrt{2} & 0 \\ -1/\sqrt{2} & 0 & 0 & 1/\sqrt{2} \end{pmatrix}$$

be the unitary solution to the Yang-Baxter equation discussed above. Then R is a universal gate. The proof below gives a specific expression for $CNOT$ in terms of R.

PROOF. This result follows at once from the Brylinksi Theorem, since R is highly entangling. For a direct computational proof, it suffices to show that $CNOT$ can be generated from R and local unitary transformations. Let

$$\alpha = \begin{pmatrix} 1/\sqrt{2} & 1/\sqrt{2} \\ 1/\sqrt{2} & -1/\sqrt{2} \end{pmatrix}$$

$$\beta = \begin{pmatrix} -1/\sqrt{2} & 1/\sqrt{2} \\ i/\sqrt{2} & i/\sqrt{2} \end{pmatrix}$$

$$\gamma = \begin{pmatrix} 1/\sqrt{2} & i/\sqrt{2} \\ 1/\sqrt{2} & -i/\sqrt{2} \end{pmatrix}$$

$$\delta = \begin{pmatrix} -1 & 0 \\ 0 & -i \end{pmatrix}$$

Let $M = \alpha \otimes \beta$ and $N = \gamma \otimes \delta$. Then it is straightforward to verify that

$$CNOT = MRN.$$

This completes the proof. ⊣

Remark 6. See [57] for more information about these calculations.

7.2. Majorana fermions generate universal braiding gates. Recall that in Section 4 we showed how to construct braid group representations by using Majorana fermions in the special case of three particles. Here we generalize this construction and show how the Marjorana fermions give rise to universal topological gates. Let $c_1, c_2, \ldots c_n$ denote n Majorana fermion creation operators. Thus we assume that

$$c_k^2 = 1$$

and

$$c_i c_j = -c_j c_i$$

for each $k = 1 \ldots n$ and whenever $i \neq j$. Then define operators

$$s_k = \frac{1}{\sqrt{2}}(1 + c_{k+1} c_k)$$

for $k = 1 \ldots n-1$. Then by the same algebra as we explored in Section 4 it is easy to verify that $s_{k+1} s_k s_{k+1} = s_k s_{k+1} s_k$ and that $s_i s_j = s_j s_i$ whenever $|i - j| > 1$. Thus the s_i give a representation of the n-strand braid group B_n. Furthermore, it is easy to see that a specific representation is given on the complex vector space V_n with basis $\{c_1, c_2, \ldots c_n\}$ via the linear transformations $T_k : V_n \longrightarrow V_n$ defined by

$$T_k(v) = s_k v s_k^{-1}.$$

Note that $s_k^{-1} = \frac{1}{\sqrt{2}}(1 - c_{k+1} c_k)$. It is then easy to verify that

$$T_k(c_k) = c_{k+1},$$
$$T_k(c_{k+1}) = -c_k$$

and that T_k is the identity otherwise.

For universality, take $n = 4$ and regard each T_k as operating on $V \otimes V$ where V is a single qubit space. Then the braiding operators T_k each satisfy the Yang-Baxter equation and so we have universal gates (in the presence of single qubit unitary operators) from Majorana fermions. If experimental work shows that Majorana fermions can be detected and controlled, then it is possible that quantum computers based on these topological unitary representations will be constructed.

In the later sections of this paper we will describe the Fibonacci model, which also uses Majorana fermions, and a different subtler representation of the braid groups that is also promising for topological quantum computing.

§8. A remark about *EPR*, entanglement and Bell's inequality. A state $|\psi\rangle \in H^{\otimes n}$, where H is the qubit space, is said to be *entangled* if it cannot be written as a tensor product of vectors from non-trivial factors of $H^{\otimes n}$. Such states turn out to be related to subtle nonlocality in quantum physics. It helps to place this algebraic structure in the context of a gedanken experiment to see where the physics comes in. Thought experiments of the sort we are about to describe were first devised by Einstein, Podolosky and Rosen, referred henceforth as *EPR*.

Consider the entangled state

$$S = (|0\rangle|1\rangle + |1\rangle|0\rangle)/\sqrt{2}.$$

In an EPR thought experiment, we think of two "parts" of this state that are separated in space. We want a notation for these parts and suggest the following:

$$L = (\{|0\rangle\}|1\rangle + \{|1\rangle\}|0\rangle)/\sqrt{2},$$
$$R = (|0\rangle\{|1\rangle\} + |1\rangle\{|0\rangle\})/\sqrt{2}.$$

In the left state L, an observer can only observe the left hand factor. In the right state R, an observer can only observe the right hand factor. These "states" L and R together comprise the EPR state S, but they are accessible individually just as are the two photons in the usual thought experiement. One can transport L and R individually and we shall write

$$S = L * R$$

to denote that they are the "parts" (but not tensor factors) of S.

The curious thing about this formalism is that it includes a little bit of macroscopic physics implicitly, and so it makes it a bit more apparent what EPR were concerned about. After all, lots of things that we can do to L or R do not affect S. For example, transporting L from one place to another, as in the original experiment where the photons separate. On the other hand, if Alice has L and Bob has R and Alice performs a local unitary transformation on "her" tensor factor, this applies to both L and R since the transformation is actually being applied to the state S. This is also a "spooky action at a distance" whose consequence does not appear until a measurement is made.

To go a bit deeper it is worthwhile seeing what entanglement, in the sense of tensor indecomposability, has to do with the structure of the *EPR* thought experiment. To this end, we look at the structure of the Bell inequalities using the Clauser, Horne, Shimony, Holt formalism (*CHSH*) as explained in the book by Nielsen and Chuang [79]. For this we use the following observables with eigenvalues ± 1.

$$Q = \begin{pmatrix} 1 & 0 \\ 0 & -1 \end{pmatrix}_1,$$

$$R = \begin{pmatrix} 0 & 1 \\ 1 & 0 \end{pmatrix}_1,$$

$$S = \begin{pmatrix} -1 & -1 \\ -1 & 1 \end{pmatrix}_2 /\sqrt{2},$$

$$T = \begin{pmatrix} 1 & -1 \\ -1 & -1 \end{pmatrix}_2 /\sqrt{2}.$$

The subscripts 1 and 2 on these matrices indicate that they are to operate on the first and second tensor factors, repsectively, of a quantum state of the form

$$\phi = a|00\rangle + b|01\rangle + c|10\rangle + d|11\rangle.$$

To simplify the results of this calculation we shall here assume that the coefficients a, b, c, d are real numbers. We calculate the quantity

$$\Delta = \langle\phi|QS|\phi\rangle + \langle\phi|RS|\phi\rangle + \langle\phi|RT|\phi\rangle - \langle\phi|QT|\phi\rangle,$$

finding that

$$\Delta = (2 - 4(a+d)^2 + 4(ad - bc))/\sqrt{2}.$$

Classical probability calculation with random variables of value ± 1 gives the value of $QS + RS + RT - QT = \pm 2$ (with each of Q, R, S and T equal to ± 1). Hence the classical expectation satisfies the Bell inequality

$$E(QS) + E(RS) + E(RT) - E(QT) \leq 2.$$

That quantum expectation is not classical is embodied in the fact that Δ can be greater than 2. The classic case is that of the Bell state

$$\phi = (|01\rangle - |10\rangle)/\sqrt{2}.$$

Here

$$\Delta = 6/\sqrt{2} > 2.$$

In general we see that the following inequality is needed in order to violate the Bell inequality

$$(2 - 4(a+d)^2 + 4(ad - bc))/\sqrt{2} > 2.$$

This is equivalent to

$$(\sqrt{2} - 1)/2 < (ad - bc) - (a+d)^2.$$

Since we know that ϕ is entangled exactly when $ad - bc$ is non-zero, this shows that an unentangled state cannot violate the Bell inequality. This formula *also* shows that it is possible for a state to be entangled and yet not violate the Bell inequality. For example, if

$$\phi = (|00\rangle - |01\rangle + |10\rangle + |11\rangle)/2,$$

then $\Delta(\phi)$ satisfies Bell's inequality, but ϕ is an entangled state. We see from this calculation that entanglement in the sense of tensor indecomposability, and entanglement in the sense of Bell inequality violation for a given choice of Bell operators are not equivalent concepts. On the other hand, Benjamin Schumacher has pointed out [88] that any entangled two-qubit state will violate Bell inequalities for an appropriate choice of operators. This deepens the context for our question of the relationship between topological entanglement and quantum entanglement. The Bell inequality violation is an indication of quantum mechanical entanglement. One's intuition suggests that it is *this* sort of entanglement that should have a topological context.

§**9. The Aravind hypothesis.** Link diagrams can be used as graphical devices and holders of information. In this vein Aravind [5] proposed that the entanglement of a link should correspond to the entanglement of a state. *Measurement of a link would be modeled by deleting one component of the link.* A key example is the Borromean rings. See Figure 18. Deleting any component of the Boromean rings yields a remaining pair of unlinked rings. The Borromean rings are entangled, but any two of them are unentangled. In this sense the Borromean rings are analogous to the GHZ state $|GHZ\rangle = (1/\sqrt{2})(|000\rangle + |111\rangle)$. Measurement in any factor of the GHZ yields an unentangled state. Aravind points out that this property is basis dependent. *We point out that there are states whose entanglement after an measurement is a matter of probability (via quantum amplitudes).* Consider for example the state

$$|\psi\rangle = |001\rangle + |010\rangle + |100\rangle.$$

Measurement in any coordinate yields probabilistically an entangled or an unentangled state. For example

$$|\psi\rangle = |0\rangle(|01\rangle + |10\rangle) + |1\rangle|00\rangle.$$

so that projecting to $|1\rangle$ in the first coordinate yields an unentangled state, while projecting to $|0\rangle$ yields an entangled state.

New ways to use link diagrams must be invented to map the properties of such states. One direction is to consider appropriate notions of quantum knots so that one can formlate superpositions of topological types as in [58]. But one needs to go deeper in this consideration. The relationship of topology and physics needs to be examined carefully. We take the stance that topological properties of systems are properties that remain invariant under certain transformations that are identified as "topological equivalences". In making quantum physical models, these equivalences should correspond to unitary transformations of an appropriate Hilbert space. Accordingly, we have formulated a model for *quantum knots* [71, 72, 73, 63] that meets these requirements. A quantum knot system represents the "quantum embodiment" of a closed knotted physical piece of rope. A quantum knot (i.e., an element $|K\rangle$ lying in an appropriate Hilbert space H_n, as a state of this system, represents the state of such a knotted closed piece of rope, i.e., the particular spatial configuration of the knot tied in the rope. Associated with a quantum knot system is a group of unitary transformations A_n, called the *ambient group*, which represents all possible ways of moving the rope around (without cutting the rope, and without letting the rope pass through itself.) Of course, unlike a classical closed piece of rope, a quantum knot can exhibit non-classical behavior, such as quantum superposition and quantum entanglement. The *knot type* of a quantum knot $|K\rangle$ is simply the orbit of the quantum knot under the action of the ambient group A_n. This leads to new questions connecting quantum computing and knot theory.

§10. $SU(2)$ representations of the Artin braid group. The purpose of this section is to determine all the representations of the three strand Artin braid group B_3 to the special unitary group $SU(2)$ and concomitantly to the unitary group $U(2)$. One regards the groups $SU(2)$ and $U(2)$ as acting on a single qubit, and so $U(2)$ is usually regarded as the group of local unitary transformations in a quantum information setting. If one is looking for a coherent way to represent all unitary transformations by way of braids, then $U(2)$ is the place to start. Here we will show that there are many representations of the three-strand braid group that generate a dense subset of $U(2)$. Thus it is a fact that local unitary transformations can be "generated by braids" in many ways.

We begin with the structure of $SU(2)$. A matrix in $SU(2)$ has the form

$$M = \begin{pmatrix} z & w \\ -\bar{w} & \bar{z} \end{pmatrix},$$

where z and w are complex numbers, and $\bar{z}$ denotes the complex conjugate of z. To be in $SU(2)$ it is required that $Det(M) = 1$ and that $M^\dagger = M^{-1}$ where Det denotes determinant, and $M^\dagger$ is the conjugate transpose of M. Thus if $z = a + bi$ and $w = c + di$ where a, b, c, d are real numbers, and $i^2 = -1$, then

$$M = \begin{pmatrix} a + bi & c + di \\ -c + di & a - bi \end{pmatrix}$$

with $a^2 + b^2 + c^2 + d^2 = 1$. It is convenient to write

$$M = a\begin{pmatrix} 1 & 0 \\ 0 & 1 \end{pmatrix} + b\begin{pmatrix} i & 0 \\ 0 & -i \end{pmatrix} + c\begin{pmatrix} 0 & 1 \\ -1 & 0 \end{pmatrix} + d\begin{pmatrix} 0 & i \\ i & 0 \end{pmatrix},$$

and to abbreviate this decomposition as

$$M = a + bI + cJ + dK$$

where

$$1 \equiv \begin{pmatrix} 1 & 0 \\ 0 & 1 \end{pmatrix}, \; I \equiv \begin{pmatrix} i & 0 \\ 0 & -i \end{pmatrix}, \; J \equiv \begin{pmatrix} 0 & 1 \\ -1 & 0 \end{pmatrix}, \; K \equiv \begin{pmatrix} 0 & i \\ i & 0 \end{pmatrix}$$

so that

$$I^2 = J^2 = K^2 = IJK = -1$$

and

$$\begin{array}{lll} IJ = K & JK = I & KI = J \\ JI = -K & KJ = -I & IK = -J. \end{array}$$

The algebra of $1, I, J, K$ is called the *quaternions* after William Rowan Hamilton who discovered this algebra prior to the discovery of matrix algebra. Thus the unit quaternions are identified with $SU(2)$ in this way. We shall use this identification, and some facts about the quaternions to find the $SU(2)$ representations of braiding. First we recall some facts about the quaternions.

1. Note that if $q = a + bI + cJ + dK$ (as above), then $q^\dagger = a - bI - cJ - dK$ so that $qq^\dagger = a^2 + b^2 + c^2 + d^2 = 1$.
2. A general quaternion has the form $q = a + bI + cJ + dK$ where the value of $qq^\dagger = a^2 + b^2 + c^2 + d^2$, is not fixed to unity. The *length* of q is by definition $\sqrt{qq^\dagger}$.
3. A quaternion of the form $rI + sJ + tK$ for real numbers r, s, t is said to be a *pure* quaternion. We identify the set of pure quaternions with the vector space of triples (r, s, t) of real numbers R^3.
4. Thus a general quaternion has the form $q = a + bu$ where u is a pure quaternion of unit length and a and b are arbitrary real numbers. A unit quaternion (element of $SU(2)$) has the addition property that $a^2 + b^2 = 1$.
5. If u is a pure unit length quaternion, then $u^2 = -1$. Note that the set of pure unit quaternions forms the two-dimensional sphere $S^2 = \{(r, s, t) | r^2 + s^2 + t^2 = 1\}$ in R^3.
6. If u, v are pure quaternions, then

   $$uv = -u \cdot v + u \times v$$

   whre $u \cdot v$ is the dot product of the vectors u and v, and $u \times v$ is the vector cross product of u and v. In fact, one can take the definition of quaternion multiplication as

   $$(a + bu)(c + dv) = ac + bc(u) + ad(v) + bd(-u \cdot v + u \times v),$$

   and all the above properties are consequences of this definition. Note that quaternion multiplication is associative.

7. Let $g = a + bu$ be a unit length quaternion so that $u^2 = -1$ and $a = \cos(\theta/2), b = \sin(\theta/2)$ for a chosen angle θ. Define $\phi_g : R^3 \longrightarrow R^3$ by the equation $\phi_g(P) = gPg^\dagger$, for P any point in R^3, regarded as a pure quaternion. Then ϕ_g is an orientation preserving rotation of R^3 (hence an element of the rotation group $SO(3)$). Specifically, ϕ_g is a rotation about the axis u by the angle θ. The mapping

$$\phi : SU(2) \longrightarrow SO(3)$$

is a two-to-one surjective map from the special unitary group to the rotation group. In quaternionic form, this result was proved by Hamilton and by Rodrigues in the middle of the nineteeth century. The specific formula for $\phi_g(P)$ as shown below:

$$\phi_g(P) = gPg^{-1} = (a^2 - b^2)P + 2ab(P \times u) + 2(P \cdot u)b^2u.$$

We want a representation of the three-strand braid group in $SU(2)$. This means that we want a homomorphism $\rho : B_3 \longrightarrow SU(2)$, and hence we want elements $g = \rho(s_1)$ and $h = \rho(s_2)$ in $SU(2)$ representing the braid group generators s_1 and s_2. Since $s_1s_2s_1 = s_2s_1s_2$ is the generating relation for B_3, the only requirement on g and h is that $ghg = hgh$. We rewrite this relation as $h^{-1}gh = ghg^{-1}$, and analyze its meaning in the unit quaternions.

Suppose that $g = a + bu$ and $h = c + dv$ where u and v are unit pure quaternions so that $a^2 + b^2 = 1$ and $c^2 + d^2 = 1$. then $ghg^{-1} = c + d\phi_g(v)$ and $h^{-1}gh = a + b\phi_{h^{-1}}(u)$. Thus it follows from the braiding relation that $a = c, b = \pm d$, and that $\phi_g(v) = \pm\phi_{h^{-1}}(u)$. However, in the case where there is a minus sign we have $g = a + bu$ and $h = a - bv = a + b(-v)$. Thus we can now prove the following Theorem.

THEOREM 4. *Let u and v be pure unit quaternions and $g = a + bu$ and $h = c + dv$ have unit length. Then (without loss of generality), the braid relation $ghg = hgh$ is true if and only if $h = a + bv$, and $\phi_g(v) = \phi_{h^{-1}}(u)$. Furthermore, given that $g = a + bu$ and $h = a + bv$, the condition $\phi_g(v) = \phi_{h^{-1}}(u)$ is satisfied if and only if $u \cdot v = \frac{a^2-b^2}{2b^2}$ when $u \neq v$. If $u = v$ then $g = h$ and the braid relation is trivially satisfied.*

PROOF. We have proved the first sentence of the Theorem in the discussion prior to its statement. Therefore assume that $g = a + bu, h = a + bv$, and $\phi_g(v) = \phi_{h^{-1}}(u)$. We have already stated the formula for $\phi_g(v)$ in the discussion about quaternions:

$$\phi_g(v) = gvg^{-1} = (a^2 - b^2)v + 2ab(v \times u) + 2(v \cdot u)b^2u.$$

By the same token, we have

$$\begin{aligned}\phi_{h^{-1}}(u) = h^{-1}uh &= (a^2 - b^2)u + 2ab(u \times -v) + 2(u \cdot (-v))b^2(-v)\\ &= (a^2 - b^2)u + 2ab(v \times u) + 2(v \cdot u)b^2(v).\end{aligned}$$

Hence we require that

$$(a^2 - b^2)v + 2(v \cdot u)b^2 u = (a^2 - b^2)u + 2(v \cdot u)b^2(v).$$

This equation is equivalent to

$$2(u \cdot v)b^2(u - v) = (a^2 - b^2)(u - v).$$

If $u \neq v$, then this implies that

$$u \cdot v = \frac{a^2 - b^2}{2b^2}. \qquad \dashv$$

The Majorana fermion example. Note the case of the theorem where

$$g = a + bu, h = a + bv.$$

Suppose that $u \cdot v = 0$. Then the theorem tells us that we need $a^2 - b^2 = 0$ and since $a^2 + b^2 = 1$, we conclude that $a = 1/\sqrt{2}$ and b likewise. For definiteness, then we have for the braiding generators (since I, J and K are mutually orthogonal) the three operators

$$A = \frac{1}{\sqrt{2}}(1 + I),$$

$$B = \frac{1}{\sqrt{2}}(1 + J),$$

$$C = \frac{1}{\sqrt{2}}(1 + K).$$

Each pair satisfies the braiding relation so that $ABA = BAB$, $BCB = CBC$, $ACA = CAC$. We have already met this braiding triplet in our discussion of the construction of braiding operators from Majorana fermions in Section 4. This shows (again) how close Hamilton's quaternions are to topology and how braiding is fundamental to the structure of fermionic physics.

The Fibonacci example. Let

$$g = e^{I\theta} = a + bI$$

where $a = \cos(\theta)$ and $b = \sin(\theta)$. Let

$$h = a + b[(c^2 - s^2)I + 2csK]$$

where $c^2 + s^2 = 1$ and $c^2 - s^2 = \frac{a^2 - b^2}{2b^2}$. Then we can rewrite g and h in matrix form as the matrices G and H. Instead of writing the explicit form of H, we write $H = FGF^\dagger$ where F is an element of $SU(2)$ as shown below.

$$G = \begin{pmatrix} e^{i\theta} & 0 \\ 0 & e^{-i\theta} \end{pmatrix}$$

$$F = \begin{pmatrix} ic & is \\ is & -ic \end{pmatrix}$$

This representation of braiding where one generator G is a simple matrix of phases, while the other generator $H = FGF^{\dagger}$ is derived from G by conjugation by a unitary matrix, has the possibility for generalization to representations of braid groups (on greater than three strands) to $SU(n)$ or $U(n)$ for n greater than 2. In fact we shall see just such representations constructed later in this paper, by using a version of topological quantum field theory. The simplest example is given by

$$g = e^{7\pi I/10}$$
$$f = I\tau + K\sqrt{\tau}$$
$$h = fgf^{-1}$$

where $\tau^2 + \tau = 1$. Then g and h satisfy $ghg = hgh$ and generate a representation of the three-strand braid group that is dense in $SU(2)$. We shall call this the *Fibonacci* representation of B_3 to $SU(2)$.

Density. Consider representations of B_3 into $SU(2)$ produced by the method of this section. That is consider the subgroup $SU[G, H]$ of $SU(2)$ generated by a pair of elements $\{g, h\}$ such that $ghg = hgh$. We wish to understand when such a representation will be dense in $SU(2)$. We need the following lemma.

LEMMA 1. $e^{aI}e^{bJ}e^{cI} = \cos(b)e^{I(a+c)} + \sin(b)e^{I(a-c)}J$. *Hence any element of* $SU(2)$ *can be written in the form* $e^{aI}e^{bJ}e^{cI}$ *for appropriate choices of angles* a, b, c. *In fact, if* u *and* v *are linearly independent unit vectors in* R^3, *then any element of* $SU(2)$ *can be written in the form*

$$e^{au}e^{bv}e^{cu}$$

for appropriate choices of the real numbers a, b, c.

PROOF. See [59] for the details of this proof. ⊣

This Lemma can be used to verify the density of a representation, by finding two elements A and B in the representation such that the powers of A are dense in the rotations about its axis, and the powers of B are dense in the rotations about its axis, and such that the axes of A and B are linearly independent in R^3. Then by the Lemma the set of elements $A^{a+c}B^{b}A^{a-c}$ are dense in $SU(2)$. It follows for example, that the Fibonacci representation described above is dense in $SU(2)$, and indeed the generic representation of B_3 into $SU(2)$ will be dense in $SU(2)$. Our next task is to describe representations of the higher braid groups that will extend some of these unitary representations of the three-strand braid group. For this we need more topology.

§11. The bracket polynomial and the Jones polynomial. We now discuss the Jones polynomial. We shall construct the Jones polynomial by using the bracket state summation model [35]. The bracket polynomial, invariant under Reidmeister moves II and III, can be normalized to give an invariant of all three

Reidemeister moves. This normalized invariant, with a change of variable, is the Jones polynomial [29, 30]. The Jones polynomial was originally discovered by a different method than the one given here.

The *bracket polynomial* , $< K > = < K > (A)$, assigns to each unoriented link diagram K a Laurent polynomial in the variable A, such that

1. If K and K' are regularly isotopic diagrams, then $< K > = < K' >$.
2. If $K \sqcup O$ denotes the disjoint union of K with an extra unknotted and unlinked component O (also called 'loop' or 'simple closed curve' or 'Jordan curve'), then

$$< K \sqcup O > = \delta < K >,$$

   where

$$\delta = -A^2 - A^{-2}.$$

3. $< K >$ satisfies the following formulas

$$< \chi > = A < \asymp > + A^{-1} <)(>$$
$$< \overline{\chi} > = A^{-1} < \asymp > + A <)(>,$$

where the small diagrams represent parts of larger diagrams that are identical except at the site indicated in the bracket. We take the convention that the letter chi, χ, denotes a crossing where *the curved line is crossing over the straight segment*. The barred letter denotes the switch of this crossing, where *the curved line is undercrossing the straight segment*. See Figure 31 for a graphic illustration of this relation, and an indication of the convention for choosing the labels A and A^{-1} at a given crossing.

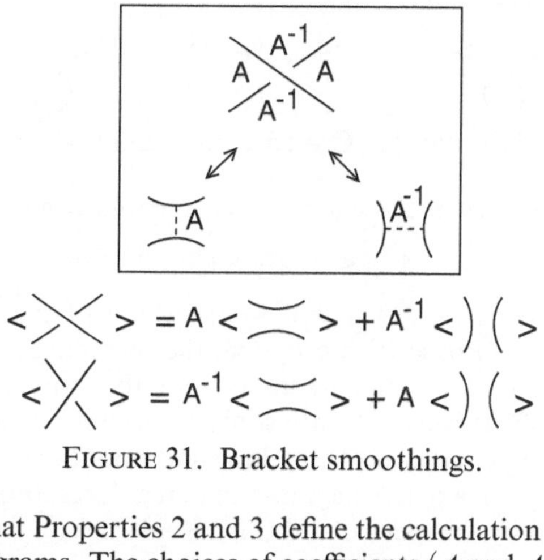

FIGURE 31. Bracket smoothings.

It is easy to see that Properties 2 and 3 define the calculation of the bracket on arbitrary link diagrams. The choices of coefficients (A and A^{-1}) and the value of δ make the bracket invariant under the Reidemeister moves II and III. Thus Property 1 is a consequence of the other two properties.

In computing the bracket, one finds the following behaviour under Reidemeister move I:

$$<\gamma>= -A^3 <\smile>$$

and

$$<\overline{\gamma}>= -A^{-3} <\smile>$$

where γ denotes a curl of positive type as indicated in Figure 32, and $\overline{\gamma}$ indicates a curl of negative type, as also seen in this figure. The type of a curl is the sign of the crossing when we orient it locally. Our convention of signs is also given in Figure 32. Note that the type of a curl does not depend on the orientation we choose. The small arcs on the right hand side of these formulas indicate the removal of the curl from the corresponding diagram.
The bracket is invariant under regular isotopy and can be normalized to an invariant of ambient isotopy by the definition

$$f_K(A) = (-A^3)^{-w(K)} < K > (A),$$

where we chose an orientation for K, and where $w(K)$ is the sum of the crossing signs of the oriented link K. $w(K)$ is called the *writhe* of K. The convention for crossing signs is shown in Figure 32.

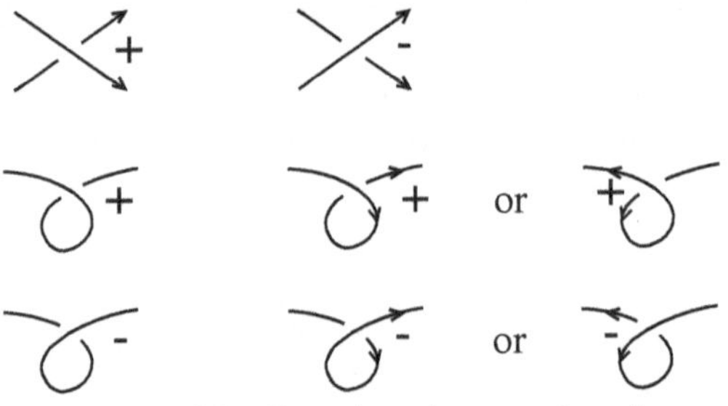

FIGURE 32. Crossing signs and curls.

One useful consequence of these formulas is the following *switching formula*

$$A < \chi > -A^{-1} < \overline{\chi} >= (A^2 - A^{-2}) < \asymp > .$$

Note that in these conventions the A-smoothing of χ is $\asymp$, while the A-smoothing of $\overline{\chi}$ is)(. Properly interpreted, the switching formula above says that you can switch a crossing and smooth it either way and obtain a three diagram relation. This is useful since some computations will simplify quite quickly with the proper choices of switching and smoothing. Remember that it is necessary to keep track of the diagrams up to regular isotopy (the equivalence relation generated by the second and third Reidemeister moves). Here is an example. View Figure 33.

FIGURE 33. Trefoil and two relatives.

Figure 33 shows a trefoil diagram K, an unknot diagram U and another unknot diagram U'. Applying the switching formula, we have

$$A^{-1} < K > -A < U >= (A^{-2} - A^{2}) < U' >$$

and $< U >= -A^3$ and $< U' >= (-A^{-3})^2 = A^{-6}$. Thus

$$A^{-1} < K > -A(-A^{3}) = (A^{-2} - A^{2})A^{-6}.$$

Hence

$$A^{-1} < K >= -A^{4} + A^{-8} - A^{-4}.$$

Thus

$$< K >= -A^{5} - A^{-3} + A^{-7}.$$

This is the bracket polynomial of the trefoil diagram K.
Since the trefoil diagram K has writhe $w(K) = 3$, we have the normalized polynomial

$$f_K(A) = (-A^3)^{-3} < K >= -A^{-9}(-A^5 - A^{-3} + A^{-7}) = A^{-4} + A^{-12} - A^{-16}.$$

The bracket model for the Jones polynomial is quite useful both theoretically and in terms of practical computations. One of the neatest applications is to simply compute, as we have done, $f_K(A)$ for the trefoil knot K and determine that $f_K(A)$ is not equal to $f_K(A^{-1}) = f_{-K}(A)$. This shows that the trefoil is not ambient isotopic to its mirror image, a fact that is much harder to prove by classical methods.

The state summation. In order to obtain a closed formula for the bracket, we now describe it as a state summation. Let K be any unoriented link diagram. Define a *state*, S, of K to be a choice of smoothing for each crossing of K. There are two choices for smoothing a given crossing, and thus there are 2^N states of a diagram with N crossings. In a state we label each smoothing with A or A^{-1} according to the left-right convention discussed in Property 3 (see Figure 31). The label is called a *vertex weight* of the state. There are two evaluations related to a state. The first one is the product of the vertex weights, denoted

$$< K|S > .$$

The second evaluation is the number of loops in the state S, denoted

$$||S||.$$

Define the *state summation*, $< K >$, by the formula

$$< K > = \sum_{S} < K|S > \delta^{||S||-1}.$$

It follows from this definition that $< K >$ satisfies the equations

$$< \chi > = A < \asymp > + A^{-1} <)(>,$$
$$< K \sqcup O > = \delta < K >,$$
$$< O > = 1.$$

The first equation expresses the fact that the entire set of states of a given diagram is the union, with respect to a given crossing, of those states with an A-type smoothing and those with an A^{-1}-type smoothing at that crossing. The second and the third equation are clear from the formula defining the state summation. Hence this state summation produces the bracket polynomial as we have described it at the beginning of the section.

Remark 7. By a change of variables one obtains the original Jones polynomial, $V_K(t)$, for oriented knots and links from the normalized bracket:

$$V_K(t) = f_K(t^{-\frac{1}{4}}).$$

Remark 8. The bracket polynomial provides a connection between knot theory and physics, in that the state summation expression for it exhibits it as a generalized partition function defined on the knot diagram. Partition functions are ubiquitous in statistical mechanics, where they express the summation over all states of the physical system of probability weighting functions for the individual states. Such physical partition functions contain large amounts of information about the corresponding physical system. Some of this information is directly present in the properties of the function, such as the location of critical points and phase transition. Some of the information can be obtained by differentiating the partition function, or performing other mathematical operations on it.

There is much more in this connection with statistical mechanics in that the local weights in a partition function are often expressed in terms of solutions to a matrix equation called the Yang-Baxter equation, that turns out to fit perfectly invariance under the third Reidemeister move. As a result, there are many ways to define partition functions of knot diagrams that give rise to invariants of knots and links. The subject is intertwined with the algebraic structure of Hopf algebras and quantum groups, useful for producing systematic solutions to the Yang-Baxter equation. In fact Hopf algebras are deeply connected with the problem of constructing invariants of three-dimensional manifolds in relation to invariants of knots. We have chosen, in this survey paper, to not discuss the details of these approaches, but rather to proceed to Vassiliev invariants and the relationships with Witten's functional integral. The reader is referred to [35, 37, 40, 36, 43, 39, 4, 29, 30, 66, 82, 83, 27, 93] for more information about relationships of knot theory with statistical mechanics, Hopf algebras and quantum groups. For topology, the key point is that Lie algebras can be used to construct invariants of knots and links.

11.1. Quantum computation of the Jones polynomial. *Can the invariants of knots and links such as the Jones polynomial be configured as quantum computers?* This is an important question because the algorithms to compute the Jones polynomial are known to be NP-hard, and so corresponding quantum algorithms may shed light on the relationship of this level of computational complexity with quantum computing (See [23]). Such models can be formulated in terms of the Yang-Baxter equation [35, 37, 39, 44, 56]. The next paragraph explains how this comes about.

In Figure 34, we indicate how topological braiding plus maxima (caps) and minima (cups) can be used to configure the diagram of a knot or link. This also can be translated into algebra by the association of a Yang-Baxter matrix R (not necessarily the R of the previous sections) to each crossing and other matrices to the maxima and minima. There are models of very effective invariants of knots and links such as the Jones polynomial that can be put into this form [44]. In this way of looking at things, the knot diagram can be viewed as a picture, with time as the vertical dimension, of particles arising from the vacuum, interacting (in a two-dimensional space) and finally annihilating one another. The invariant takes the form of an amplitude for this process that is computed through the association of the Yang-Baxter solution R as the scattering matrix at the crossings and the minima and maxima as creation and annihilation operators. Thus we can write the amplitude in the form

$$Z_K = \langle CUP|M|CAP\rangle$$

where $\langle CUP|$ denotes the composition of cups, M is the composition of elementary braiding matrices, and $|CAP\rangle$ is the composition of caps. We regard $\langle CUP|$ as the preparation of this state, and $|CAP\rangle$ as the measurement of this state. In order to view Z_K as a quantum computation, M must be a unitary operator. This is the case when the R-matrices (the solutions to the Yang-Baxter equation used in the model) are unitary. Each R-matrix is viewed as a a quantum gate (or possibly a composition of quantum gates), and the vacuum-vacuum diagram for the knot is interpreted as a quantum computer. This quantum computer will probabilistically (via quantum amplitudes) compute the values of the states in the state sum for Z_K.

We should remark, however, that it is not necessary that the invariant be modeled via solutions to the Yang-Baxter equation. One can use unitary representations of the braid group that are constructed in other ways. In fact, the presently successful quantum algorithms for computing knot invariants indeed use such representations of the braid group, and we shall see this below.

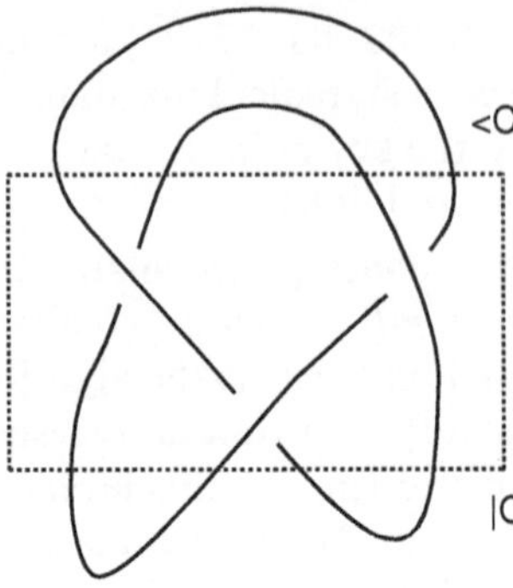

FIGURE 34. A knot quantum computer.

Nevertheless, it is useful to point out this analogy between the structure of the knot invariants and quantum computation.

Quantum algorithms for computing the Jones polynomial have been discussed elsewhere. See [44, 57, 3, 56, 2, 97]. Here, as an example, we give a local unitary representation that can be used to compute the Jones polynomial for closures of 3-braids. We analyze this representation by making explicit how the bracket polynomial is computed from it, and showing how the quantum computation devolves to finding the trace of a unitary transformation.

The idea behind the construction of this representation depends upon the algebra generated by two single qubit density matrices (ket-bras). Let $|v\rangle$ and $|w\rangle$ be two qubits in V, a complex vector space of dimension two over the complex numbers. Let $P = |v\rangle\langle v|$ and $Q = |w\rangle\langle w|$ be the corresponding ket-bras. Note that

$$P^2 = |v|^2 P,$$
$$Q^2 = |w|^2 Q,$$
$$PQP = |\langle v|w\rangle|^2 P,$$
$$QPQ = |\langle v|w\rangle|^2 Q.$$

P and Q generate a representation of the Temperley-Lieb algebra (See Section 12 of the present paper). One can adjust parameters to make a representation of the three-strand braid group in the form

$$s_1 \longmapsto rP + sI,$$
$$s_2 \longmapsto tQ + uI,$$

where I is the identity mapping on V and r, s, t, u are suitably chosen scalars. In the following we use this method to adjust such a representation so that it is unitary. Note also that this is a local unitary representation of B_3 to $U(2)$.

We leave it as an exersise for the reader to verify that it fits into our general classification of such representations as given in section 10 of the present paper.

Here is a specific representation depending on two symmetric matrices U_1 and U_2 with

$$U_1 = \begin{bmatrix} d & 0 \\ 0 & 0 \end{bmatrix} = d\,|w\rangle\langle w|$$

and

$$U_2 = \begin{bmatrix} d^{-1} & \sqrt{1-d^{-2}} \\ \sqrt{1-d^{-2}} & d - d^{-1} \end{bmatrix} = d\,|v\rangle\langle v|$$

where $w = (1,0)$, and $v = (d^{-1}, \sqrt{1-d^{-2}})$, assuming the entries of v are real. Note that $U_1^2 = dU_1$ and $U_2^2 = dU_1$. Moreover, $U_1U_2U_1 = U_1$ and $U_2U_1U_2 = U_1$. This is an example of a specific representation of the Temperley-Lieb algebra [35, 44]. The desired representation of the Artin braid group is given on the two braid generators for the three strand braid group by the equations:

$$\Phi(s_1) = AI + A^{-1}U_1,$$
$$\Phi(s_2) = AI + A^{-1}U_2.$$

Here I denotes the 2×2 identity matrix.

For any A with $d = -A^2 - A^{-2}$ these formulas define a representation of the braid group. With $A = e^{i\theta}$, we have $d = -2\cos(2\theta)$. We find a specific range of angles θ in the following disjoint union of angular intervals

$$\theta \in [0, \pi/6] \sqcup [\pi/3, 2\pi/3] \sqcup [5\pi/6, 7\pi/6] \sqcup [4\pi/3, 5\pi/3] \sqcup [11\pi/6, 2\pi]$$

that give unitary representations of the three-strand braid group. Thus a specialization of a more general represention of the braid group gives rise to a continuous family of unitary representations of the braid group.

LEMMA 2. *Note that the traces of these matrices are given by the formulas* $tr(U_1) = tr(U_2) = d$ *while* $tr(U_1U_2) = tr(U_2U_1) = 1$. *If* b *is any braid, let* $I(b)$ *denote the sum of the exponents in the braid word that expresses* b. *For* b *a three-strand braid, it follows that*

$$\Phi(b) = A^{I(b)}I + \Pi(b)$$

where I *is the* 2×2 *identity matrix and* $\Pi(b)$ *is a sum of products in the Temperley-Lieb algebra involving* U_1 *and* U_2.

We omit the proof of this Lemma. It is a calculation. To see it, consider an example. Suppose that $b = s_1 s_2^{-1} s_1$. Then

$$\begin{aligned}\Phi(b) = \Phi(s_1 s_2^{-1} s_1) &= \Phi(s_1)\Phi(s_2^{-1})\Phi(s_1) \\ &= (AI + A^{-1}U_1)(A^{-1}I + AU_2)(AI + A^{-1}U_1).\end{aligned}$$

The sum of products over the generators U_1 and U_2 of the Temperley-Lieb algebra comes from expanding this expression.

Since the Temperley-Lieb algebra in this dimension is generated by I, U_1, U_2, U_1U_2 and U_2U_1, it follows that the value of the bracket polynomial of the closure of the braid b, denoted $< \overline{b} >$, can be calculated directly from the trace of this representation, except for the part involving the identity matrix. The result is the equation

$$< \overline{b} >= A^{I(b)}d^2 + tr(\Pi(b))$$

where $\overline{b}$ denotes the standard braid closure of b, and the sharp brackets denote the bracket polynomial. From this we see at once that

$$< \overline{b} >= tr(\Phi(b)) + A^{I(b)}(d^2 - 2).$$

It follows from this calculation that the question of computing the bracket polynomial for the closure of the three-strand braid b is mathematically equivalent to the problem of computing the trace of the unitary matrix $\Phi(b)$.

The Hadamard test. In order to (quantum) compute the trace of a unitary matrix U, one can use the *Hadamard test* to obtain the diagonal matrix elements $\langle\psi|U|\psi\rangle$ of U. The trace is then the sum of these matrix elements as $|\psi\rangle$ runs over an orthonormal basis for the vector space. We first obtain

$$\frac{1}{2} + \frac{1}{2}Re\langle\psi|U|\psi\rangle$$

as an expectation by applying the Hadamard gate H

$$H|0\rangle = \frac{1}{\sqrt{2}}(|0\rangle + |1\rangle)$$
$$H|1\rangle = \frac{1}{\sqrt{2}}(|0\rangle - |1\rangle)$$

to the first qubit of

$$C_U \circ (H \otimes 1)|0\rangle|\psi\rangle = \frac{1}{\sqrt{2}}(|0\rangle \otimes |\psi\rangle + |1\rangle \otimes U|\psi\rangle.$$

Here C_U denotes controlled U, acting as U when the control bit is $|1\rangle$ and the identity mapping when the control bit is $|0\rangle$. We measure the expectation for the first qubit $|0\rangle$ of the resulting state

$$\begin{aligned}&\frac{1}{2}(H|0\rangle \otimes |\psi\rangle + H|1\rangle \otimes U|\psi\rangle)\\ &\qquad = \frac{1}{2}((|0\rangle + |1\rangle) \otimes |\psi\rangle + (|0\rangle - |1\rangle) \otimes U|\psi\rangle)\\ &\qquad = \frac{1}{2}(|0\rangle \otimes (|\psi\rangle + U|\psi\rangle) + |1\rangle \otimes (|\psi\rangle - U|\psi\rangle)).\end{aligned}$$

This expectation is

$$\frac{1}{2}(\langle\psi| + \langle\psi|U^{\dagger})(|\psi\rangle + U|\psi\rangle) = \frac{1}{2} + \frac{1}{2}Re\langle\psi|U|\psi\rangle.$$

The imaginary part is obtained by applying the same procedure to

$$\frac{1}{\sqrt{2}}(|0\rangle \otimes |\psi\rangle - i|1\rangle \otimes U|\psi\rangle)$$

This is the method used in [3], and the reader may wish to contemplate its efficiency in the context of this simple model. Note that the Hadamard test enables this quantum computation to estimate the trace of any unitary matrix U by repeated trials that estimate individual matrix entries $\langle\psi|U|\psi\rangle$. We shall return to quantum algorithms for the Jones polynomial and other knot polynomials in a subsequent paper.

§12. Quantum topology, cobordism categories, Temperley-Lieb algebra and topological quantum field theory. The purpose of this section is to discuss the general idea behind topological quantum field theory, and to illustrate its application to basic quantum mechanics and quantum mechanical formalism. It is useful in this regard to have available the concept of *category*, and we shall begin the section by discussing this far-reaching mathematical concept.

DEFINITION 1. A *category Cat* consists in two related collections:

1. $Obj(Cat)$, the *objects* of *Cat*, and
2. $Morph(Cat)$, the *morphisms* of *Cat*.

satisfying the following axioms:

1. Each morphism f is associated to two objects of *Cat*, the *domain* of f and the *codomain* of f. Letting A denote the domain of f and B denote the codomain of f, it is customary to denote the morphism f by the arrow notation $f : A \longrightarrow B$.
2. Given $f : A \longrightarrow B$ and $g : B \longrightarrow C$ where A, B and C are objects of *Cat*, then there exists an associated morphism $g \circ f : A \longrightarrow C$ called the *composition* of f and g.
3. To each object A of *Cat* there is a unique *identity morphism* $1_A : A \longrightarrow A$ such that $1_A \circ f = f$ for any morphism f with codomain A, and $g \circ 1_A = g$ for any morphism g with domain A.
4. Given three morphisms $f : A \longrightarrow B$, $g : B \longrightarrow C$ and $h : C \longrightarrow D$, then composition is associative. That is
$$(h \circ g) \circ f = h \circ (g \circ f).$$

If Cat_1 and Cat_2 are two categories, then a *functor* $F : Cat_1 \longrightarrow Cat_2$ consists in functions $F_O : Obj(Cat_1) \longrightarrow Obj(Cat_2)$ and $F_M : Morph(Cat_1) \longrightarrow Morph(Cat_2)$ such that identity morphisms and composition of morphisms are preserved under these mappings. That is (writing just F for F_O and F_M),

1. $F(1_A) = 1_{F(A)}$,
2. $F(f : A \longrightarrow B) = F(f) : F(A) \longrightarrow F(B)$,
3. $F(g \circ f) = F(g) \circ F(f)$.

A functor $F : Cat_1 \longrightarrow Cat_2$ is a structure preserving mapping from one category to another. It is often convenient to think of the image of the functor F as an *interpretation* of the first category in terms of the second. We shall use this terminology below and sometimes refer to an interpretation without specifying all the details of the functor that describes it.

The notion of category is a broad mathematical concept, encompassing many fields of mathematics. Thus one has the category of sets where the objects are sets (collections) and the morphisms are mappings between sets. One has the category of topological spaces where the objects are spaces and the morphisms are continuous mappings of topological spaces. One has the category of groups where the objects are groups and the morphisms are homomorphisms of groups. Functors are structure preserving mappings from one category to another. For example, the fundamental group is a functor from the category of topological spaces with base point, to the category of groups. In all the examples mentioned so far, the morphisms in the category are restrictions of mappings in the category of sets, but this is not necessarily the case. For example, any group G can be regarded as a category, $Cat(G)$, with one object $*$. The morphisms from $*$ to itself are the elements of the group and composition is group multiplication. In this example, the object has no internal structure and all the complexity of the category is in the morphisms.

The Artin braid group B_n can be regarded as a category whose single object is an ordered row of points $[n] = \{1, 2, 3, \ldots, n\}$. The morphisms are the braids themselves and composition is the multiplication of the braids. A given ordered row of points is interpreted as the starting or ending row of points at the bottom or the top of the braid. In the case of the braid category, the morphisms have both external and internal structure. Each morphism produces a permutation of the ordered row of points (corresponding to the begiinning and ending points of the individual braid strands), and weaving of the braid is extra structure beyond the object that is its domain and codomain. Finally, for this example, we can take all the braid groups B_n (n a positive integer) under the wing of a single category, $Cat(B)$, whose objects are all ordered rows of points $[n]$, and whose morphisms are of the form $b : [n] \longrightarrow [n]$ where b is a braid in B_n. The reader may wish to have morphisms between objects with different n. We will have this shortly in the Temperley-Lieb category and in the category of tangles.

The *n-Cobordism Category*, *Cob*$[n]$, has as its objects smooth manifolds of dimension n, and as its morphisms, smooth manifolds M^{n+1} of dimension $n+1$ with a partition of the boundary, ∂M^{n+1}, into two collections of n-manifolds that we denote by $L(M^{n+1})$ and $R(M^{n+1})$. We regard M^{n+1} as a morphism from $L(M^{n+1})$ to $R(M^{n+1})$

$$M^{n+1} : L(M^{n+1}) \longrightarrow R(M^{n+1}).$$

As we shall see, these cobordism categories are highly significant for quantum mechanics, and the simplest one, *Cob*[0] is directly related to the Dirac notation of bras and kets and to the Temperley-Lieb algebara. We shall concentrate in this section on these cobordism categories, and their relationships with quantum mechanics.

One can choose to consider either oriented or non-oriented manifolds, and within unoriented manifolds there are those that are orientable and those that are not orientable. In this section we will implicitly discuss only orientable manifolds, but we shall not specify an orientation. In the next section, with the standard definition of topological quantum field theory, the manifolds will be oriented. The definitions of the cobordism categories for oriented manifolds go over mutatis mutandis.

Lets begin with *Cob*[0]. Zero dimensional manifolds are just collections of points. The simplest zero dimensional manifold is a single point p. We take p to be an object of this category and also $*$, where $*$ denotes the empty manifold (i.e. the empty set in the category of manifolds). The object $*$ occurs in $Cob[n]$ for every n, since it is possible that either the left set or the right set of a morphism is empty. A line segment S with boundary points p and q is a morphism from p to q.

$$S : p \longrightarrow q$$

See Figure 35. In this figure we have illustrated the morphism from p to p.

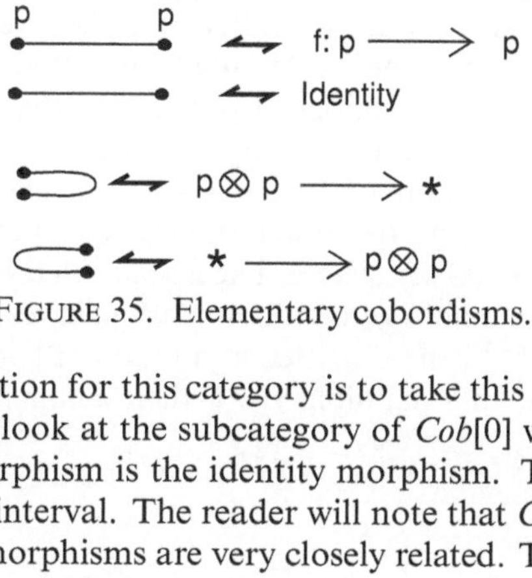

FIGURE 35. Elementary cobordisms.

The simplest convention for this category is to take this morphism to be the identity. Thus if we look at the subcategory of *Cob*[0] whose only object is p, then the only morphism is the identity morphism. Two points occur as the boundary of an interval. The reader will note that *Cob*[0] and the usual arrow notation for morphisms are very closely related. This is a place where notation and mathematical structure share common elements. In general the objects of *Cob*[0] consist in the empty object $*$ and non-empty rows of points, symbolized by

$$p \otimes p \otimes \cdots \otimes p \otimes p.$$

Figure 35 also contains a morphism

$$p \otimes p \longrightarrow *$$

and the morphism

$$* \longrightarrow p \otimes p.$$

The first represents a cobordism of two points to the empty set (via the bounding curved interval). The second represents a cobordism from the empty set to two points.

In Figure 36, we have indicated more morphisms in *Cob*[0], and we have named the morphisms just discussed as

$$|\Omega\rangle : p \otimes p \longrightarrow *,$$
$$\langle\Theta| : * \longrightarrow p \otimes p.$$

The point to notice is that the usual conventions for handling Dirac bra-kets are essentially the same as the compostion rules in this topological category. Thus in Figure 36 we have that

$$\langle\Theta| \circ |\Omega\rangle = \langle\Theta|\Omega\rangle : * \longrightarrow *$$

represents a cobordism from the empty manifold to itself. This cobordism is topologically a circle and, in the Dirac formalism is interpreted as a scalar. In order to interpret the notion of scalar we would have to map the cobordism category to the category of vector spaces and linear mappings. We shall discuss this after describing the similarities with quantum mechanical formalism. Nevertheless, the reader should note that if V is a vector space over the complex numbers C, then a linear mapping from C to C is determined by the image of 1, and hence is characterized by the scalar that is the image of 1. In this sense a mapping $C \longrightarrow C$ can be regarded as a possible image in vector spaces of the abstract structure $\langle\Theta|\Omega\rangle : * \longrightarrow *$. It is therefore assumed that in *Cob*[0] the composition with the morphism $\langle\Theta|\Omega\rangle$ commutes with any other morphism. In that way $\langle\Theta|\Omega\rangle$ behaves like a scalar in the cobordism category. In general, an $n + 1$ manifold without boundary behaves as a scalar in *Cob*[n], and if a manifold M^{n+1} can be written as a union of two submanifolds L^{n+1} and R^{n+1} so that that an n-manifold W^n is their common boundary:

$$M^{n+1} = L^{n+1} \cup R^{n+1}$$

with

$$L^{n+1} \cap R^{n+1} = W^n$$

then, we can write

$$\langle M^{n+1}\rangle = \langle L^{n+1} \cup R^{n+1}\rangle = \langle L^{n+1}|R^{n+1}\rangle,$$

and $\langle M^{n+1}\rangle$ will be a scalar (morphism that commutes with all other morphisms) in the category *Cob*[n].

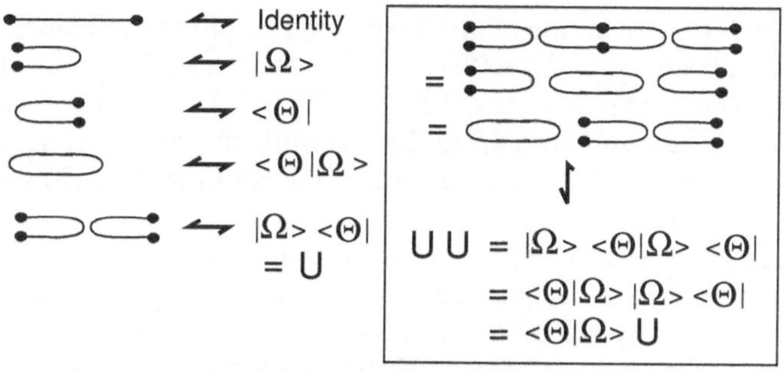

FIGURE 36. Bras, kets and projectors.

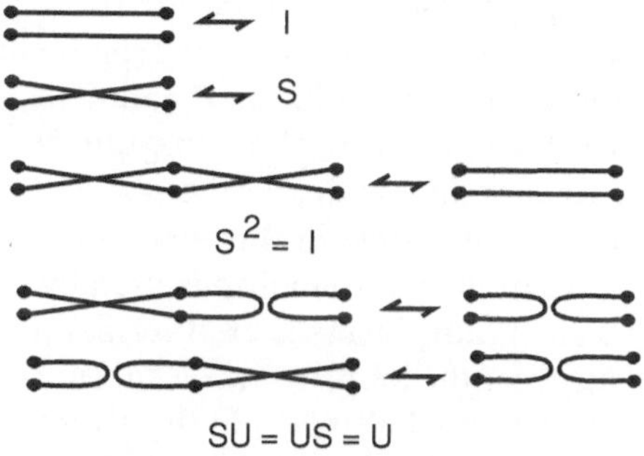

FIGURE 37. Permutations.

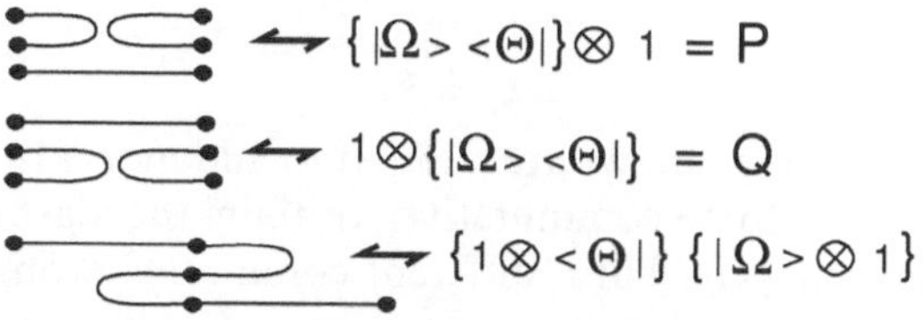

FIGURE 38. Projectors in tensor lines and elementary topology.

Getting back to the contents of Figure 36, note how the zero dimensional cobordism category has structural parallels to the Dirac ket–bra formalism

$$U = |\Omega\rangle\langle\Theta| \quad .$$
$$UU = |\Omega\rangle\langle\Theta|\Omega\rangle\langle\Theta| = \langle\Theta|\Omega\rangle|\Omega\rangle\langle\Theta| = \langle\Theta|\Omega\rangle U.$$

In the cobordism category, the bra–ket and ket–bra formalism is seen as patterns of connection of the one-manifolds that realize the cobordisms.

Now view Figure 37. This Figure illustrates a morphism S in *Cob*[0] that requires two crossed line segments for its planar representation. Thus S can be regarded as a non-trivial permutation, and $S^2 = I$ where I denotes the identity morphisms for a two-point row. From this example, it is clear that

Cob[0] contains the structure of all the syymmetric groups and more. In fact, if we take the subcateogry of *Cob*[0] consisting of all morphisms from $[n]$ to $[n]$ for a fixed positive integer n, then this gives the well-known *Brauer algebra* (see [9]) extending the symmetric group by allowing any connections among the points in the two rows. In this sense, one could call *Cob*[0] the *Brauer category*. We shall return to this point of view later.

In this section, we shall be concentrating on the part of *Cob*[0] that does not involve permutations. This part can be characterized by those morphisms that can be represented by planar diagrams without crossings between any of the line segments (the one-manifolds). We shall call this crossingless subcategory of *Cob*[0] the *Temperley-Lieb Category* and denote it by *CatTL*. In *CatTL* we have the subcategory $TL[n]$ whose only objects are the row of n points and the empty object $*$, and whose morphisms can all be represented by configurations that embed in the plane as in the morphisms P and Q in Figure 38. Note that with the empty object $*$, the morphism whose diagram is a single loop appears in $TL[n]$ and is taken to commute with all other morphisms.

The *Temperley-Lieb Algebra*, $AlgTL[n]$ is generated by the morphisms in $TL[n]$ that go from $[n]$ to itself. Up to multiplication by the loop, the product (composition) of two such morphisms is another flat morphism from $[n]$ to itself. For algebraic purposes the loop $* \longrightarrow *$ is taken to be a scalar algebraic variable δ that commutes with all elements in the algebra. Thus the equation

$$UU = \langle \Theta | \Omega \rangle U.$$

becomes

$$UU = \delta U$$

in the algebra. In the algebra we are allowed to add morphisms formally and this addition is taken to be commutative. Initially the algebra is taken with coefficients in the integers, but a different commutative ring of coefficients can be chosen and the value of the loop may be taken in this ring. For example, for quantum mechanical applications it is natural to work over the complex numbers. The multiplicative structure of $AlgTL[n]$ can be described by generators and relations as follows: Let I_n denote the identity morphism from $[n]$ to $[n]$. Let U_i denote the morphism from $[n]$ to $[n]$ that connects k with k for $k < i$ and $k > i+1$ from one row to the other, and connects i to $i+1$ in each row. Then the algebra $AlgTL[n]$ is generated by $\{I_n, U_1, U_2, \ldots, U_{n-1}\}$ with relations

$$U_i^2 = \delta U_i$$

$$U_i U_{i+1} U_i = U_i$$

$$U_i U_j = U_j U_i \; : \; |i - j| > 1.$$

These relations are illustrated for three strands in Figure 38. We leave the commuting relation for the reader to draw in the case where n is four or greater. For a proof that these are indeed all the relations, see [48].

Figure 38 and Figure 39 indicate how the zero dimensional cobordism category contains structure that goes well beyond the usual Dirac formalism. By tensoring the ket–bra on one side or another by identity morphisms, we obtain the beginnings of the Temperley-Lieb algebra and the Temperley-Lieb category. Thus Figure 39 illustrates the morphisms P and Q obtained by such tensoring, and the relation $PQP = P$ which is the same as $U_1 U_2 U_1 = U_1$

Note the composition at the bottom of the Figure 39. Here we see a composition of the identity tensored with a ket, followed by a bra tensored with the identity. The diagrammatic for this association involves "straightening" the curved structure of the morphism to a straight line. In Figure 40 we have elaborated this situation even further, pointing out that in this category each of the morphisms $\langle\Theta|$ and $|\Omega\rangle$ can be seen, by straightening, as mappings from the generating object to itself. We have denoted these corresponding morphisms by Θ and Ω respectively. In this way there is a correspondence between morphisms $p \otimes p \longrightarrow *$ and morphims $p \longrightarrow p$.

In Figure 40 we have illustrated the generalization of the straightening procedure of Figure 39. In Figure 39 the straightening occurs because the connection structure in the morphism of *Cob*[0] does not depend on the wandering of curves in diagrams for the morphisms in that category. Nevertheless, one can envisage a more complex interpretation of the morphisms where each one-manifold (line segment) has a label, and a multiplicity of morphisms can correspond to a single line segment. This is exactly what we expect in interpretations. For example, we can interpret the line segment $[1] \longrightarrow [1]$ as a mapping from a vector space V to itself. Then $[1] \longrightarrow [1]$ is the diagrammatic abstraction for $V \longrightarrow V$, and there are many instances of linear mappings from V to V.

At the vector space level there is a duality between mappings $V \otimes V \longrightarrow C$ and linear maps $V \longrightarrow V$. Specifically, let

$$\{|0\rangle, \ldots, |m\rangle\}$$

be a basis for V. Then $\Theta : V \longrightarrow V$ is determined by

$$\Theta|i\rangle = \Theta_{ij}\,|j\rangle$$

(where we have used the Einstein summation convention on the repeated index j) corresponds to the bra

$$\langle\Theta| : V \otimes V \longrightarrow C$$

defined by

$$\langle\Theta|ij\rangle = \Theta_{ij}.$$

Given $\langle\Theta| : V \otimes V \longrightarrow C$, we associate $\Theta : V \longrightarrow V$ in this way.

Comparing with the diagrammatic for the category *Cob*[0], we say that $\Theta : V \longrightarrow V$ is obtained by *straightening* the mapping

$$\langle\Theta| : V \otimes V \longrightarrow C.$$

Note that in this interpretation, the bras and kets are defined relative to the tensor product of V with itself and [2] is interpreted as $V \otimes V$. If we interpret [2] as a single vector space W, then the usual formalisms of bras and kets still pass over from the cobordism category.

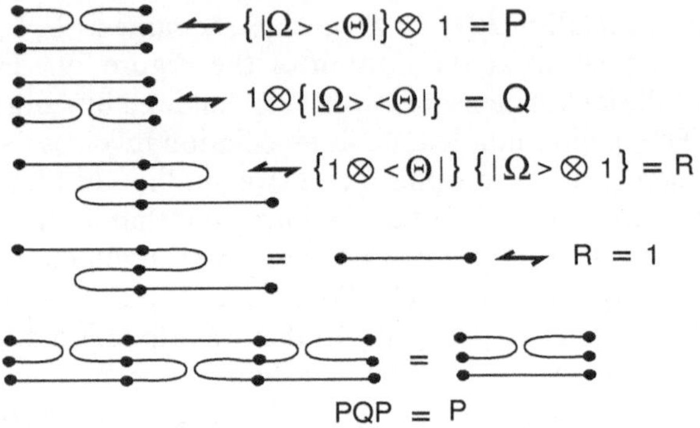

FIGURE 39. The basic Temperley-Lieb relation.

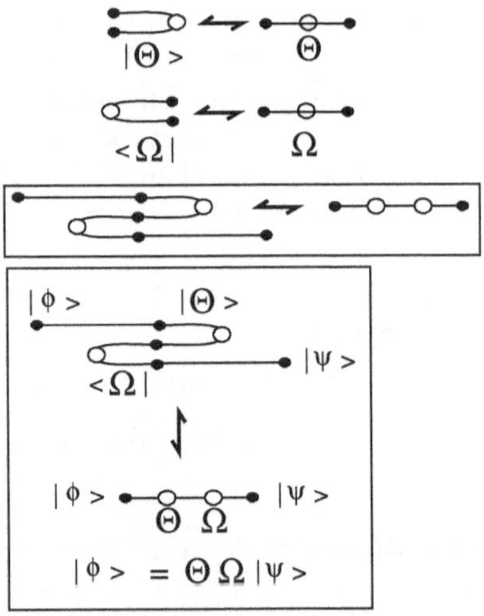

FIGURE 40. The key to teleportation.

Figure 40 illustrates the staightening of $|\Theta\rangle$ and $\langle\Omega|$, and the straightening of a composition of these applied to $|\psi\rangle$, resulting in $|\phi\rangle$. In the left-hand part of the bottom of Figure 40 we illustrate the preparation of the tensor product $|\Theta\rangle \otimes |\psi\rangle$ followed by a successful measurement by $\langle\Omega|$ in the second two tensor factors. The resulting single qubit state, as seen by straightening, is $|\phi\rangle = \Theta \circ \Omega|\psi\rangle$.

From this, we see that it is possible to reversibly, indeed unitarily, transform a state $|\psi\rangle$ via a combination of preparation and measurement just so long as the straightenings of the preparation and measurement (Θ and Ω) are each invertible (unitary). This is the key to teleportation [49, 13, 1]. In the standard teleportation procedure one chooses the preparation Θ to be (up to normalization) the 2 dimensional identity matrix so that $|\theta\rangle = |00\rangle + |11\rangle$. If the successful measurement Ω is also the identity, then the transmitted state $|\phi\rangle$ will be equal to $|\psi\rangle$. In general we will have $|\phi\rangle = \Omega|\psi\rangle$. One can then choose a basis of measurements $|\Omega\rangle$, each corresponding to a unitary transformation Ω so that the recipient of the transmission can rotate the result by the inverse of Ω to reconsitute $|\psi\rangle$ if he is given the requisite information. This is the basic design of the teleportation procedure.

There is much more to say about the category *Cob*[0] and its relationship with quantum mechanics. We will stop here, and invite the reader to explore further. Later in this paper, we shall use these ideas in formulating our representations of the braid group. For now, we point out how things look as we move upward to *Cob*[n] for $n > 0$. In Figure 41 we show typical cobordisms (morphisms) in *Cob*[1] from two circles to one circle and from one circle to two circles. These are often called "pairs of pants". Their composition is a surface of genus one seen as a morphism from two circles to two circles. The bottom of the figure indicates a ket-bra in this dimension in the form of a mapping from one circle to one circle as a composition of a cobordism of a circle to the empty set and a cobordism from the empty set to a circle (circles bounding disks). As we go to higher dimensions the structure of cobordisms becomes more interesting and more complicated. It is remarkable that there is so much structure in the lowest dimensions of these categories.

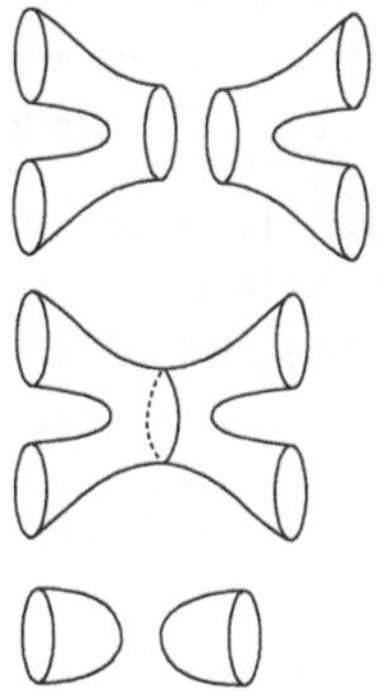

FIGURE 41. Cobordisms of 1-manifolds are surfaces.

§13. Braiding and topological quantum field theory. The purpose of this section is to discuss in a very general way how braiding is related to topological quantum field theory. In the section to follow, we will use the Temperley-Lieb recoupling theory to produce specfic unitary representations of the Artin braid group.

The ideas in the subject of topological quantum field theory (TQFT) are well expressed in the book [6] by Michael Atiyah and the paper [95] by Edward Witten. Here is Atiyah's definition:

DEFINITION 2. A TQFT in dimension d is a functor $Z(\Sigma)$ from the cobordism category $Cob[d]$ to the category $Vect$ of vector spaces and linear mappings which assigns

1. a finite dimensional vector space $Z(\Sigma)$ to each compact, oriented d-dimensional manifold Σ,
2. a vector $Z(Y) \in Z(\Sigma)$ for each compact, oriented $(d+1)$-dimensional manifold Y with boundary Σ.
3. a linear mapping $Z(Y) : Z(\Sigma_1) \longrightarrow Z(\Sigma_2)$ when Y is a $(d+1)$-manifold that is a cobordism between Σ_1 and Σ_2 (whence the boundary of Y is the union of Σ_1 and $-\Sigma_2$.

The functor satisfies the following axioms.

1. $Z(\Sigma^\dagger) = Z(\Sigma)^\dagger$ where $\Sigma^\dagger$ denotes the manifold Σ with the opposite orientation and $Z(\Sigma)^\dagger$ is the dual vector space.
2. $Z(\Sigma_1 \cup \Sigma_2) = Z(\Sigma_1) \otimes Z(\Sigma_2)$ where $\cup$ denotes disjoint union.
3. If Y_1 is a cobordism from Σ_1 to Σ_2, Y_2 is a cobordism from Σ_2 to Σ_3 and Y is the composite cobordism $Y = Y_1 \cup_{\Sigma_2} Y_2$, then
$$Z(Y) = Z(Y_2) \circ Z(Y_1) : Z(\Sigma_1) \longrightarrow Z(\Sigma_2)$$
is the composite of the corresponding linear mappings.
4. $Z(\phi) = C$ (C denotes the complex numbers) for the empty manifold ϕ.
5. With $\Sigma \times I$ (where I denotes the unit interval) denoting the identity cobordism from Σ to Σ, $Z(\Sigma \times I)$ is the identity mapping on $Z(\Sigma)$.

Note that, in this view a TQFT is basically a functor from the cobordism categories defined in the last section to Vector Spaces over the complex numbers. We have already seen that in the lowest dimensional case of cobordisms of zero-dimensional manifolds, this gives rise to a rich structure related to quatum mechanics and quantum information theory. The remarkable fact is that the case of three-dimensions is also related to quantum theory, and to the lower-dimensional versions of the TQFT. This gives a significant way to think about three-manifold invariants in terms of lower dimensional patterns of interaction. Here follows a brief description.

Regard the three-manifold as a union of two handlebodies with boundary an orientable surface S_g of genus g. The surface is divided up into trinions as illustrated in Figure 42. A *trinion* is a surface with boundary that is topologically equivalent to a sphere with three punctures. The trinion constitutes, in itself a cobordism in *Cob*[1] from two circles to a single circle, or from a single circle to two circles, or from three circles to the empty set. The *pattern* of a trinion is a trivalent graphical vertex, as illustrated in Figure 42. In that figure we show the trivalent vertex graphical pattern drawn on the surface of the trinion, forming a graphical pattern for this combordism. It should be clear from this figure that any cobordism in *Cob*[1] can be diagrammed by a trivalent graph, so that the category of trivalent graphs (as morphisms from ordered sets of points to ordered sets of points) has an image in the category of cobordisms of compact one-dimensional manifolds. Given a surface S (possibly with boundary) and a decomposition of that surface into triions, we associate to it a trivalent graph $G(S,t)$ where t denotes the particular trinion decomposition.

In this correspondence, distinct graphs can correspond to topologically identical cobordisms of circles, as illustrated in Figure 44. It turns out that the graphical structure is important, and that it is extraordinarily useful to articulate transformations between the graphs that correspond to the homeomorphisms of the corresponding surfaces. The beginning of this structure is indicated in the bottom part of Figure 44.

In Figure 45 we illustrate another feature of the relationship betweem surfaces and graphs. At the top of the figure we indicate a homeomorphism between a twisted trinion and a standard trinion. The homeomorphism leaves the ends of the trinion (denoted A,B and C) fixed while undoing the internal twist. This can be accomplished as an ambient isotopy of the embeddings in three dimensional space that are indicated by this figure. Below this isotopy we indicate the corresponding graphs. In the graph category there will have to be a transformation between a braided and an unbraided trivalent vertex that corresponds to this homeomorphism.

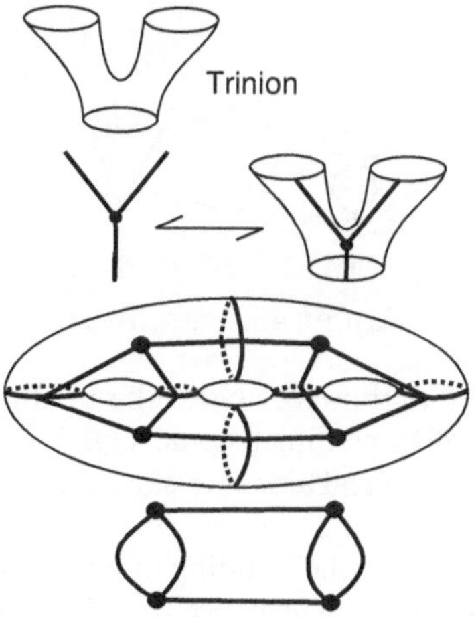

FIGURE 42. Decomposition of a surface into trinions.

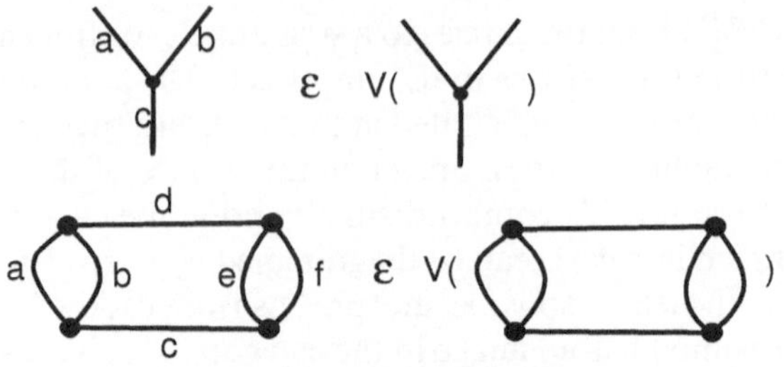

FIGURE 43. Trivalent vectors.

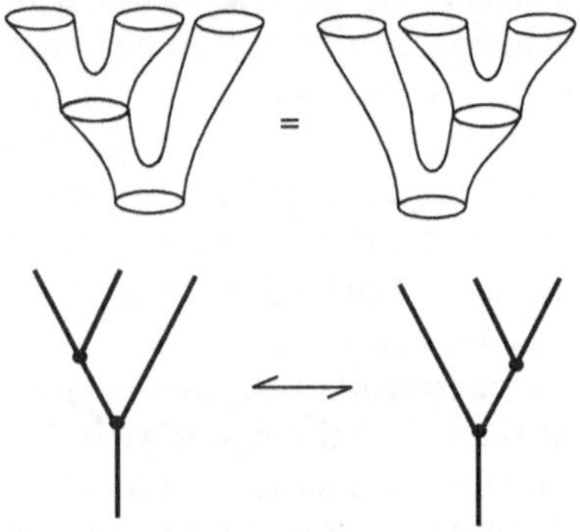

FIGURE 44. Trinion associativity.

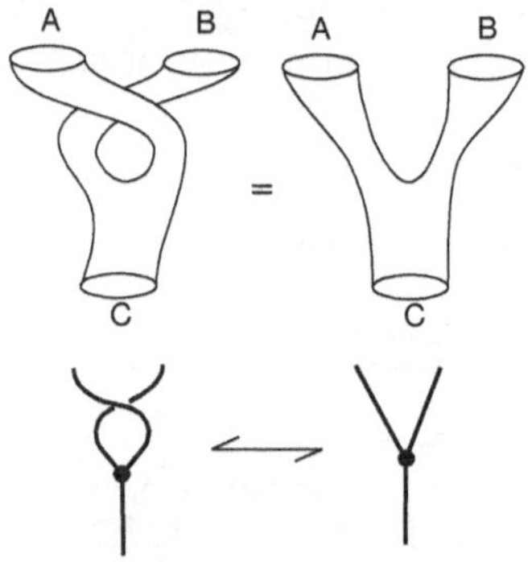

FIGURE 45. Tube twist.

From the point of view that we shall take in this paper, the key to the mathematical structure of three-dimensional TQFT lies in the trivalent graphs, including the braiding of grapical arcs. We can think of these braided graphs as representing idealized Feynman diagrams, with the trivalent vertex as the basic particle interaction vertex, and the braiding of lines representing an interaction resulting from an exchange of particles. In this view one thinks of the particles as moving in a two-dimensional medium, and the diagrams of braiding and trivalent vertex interactions as indications of the temporal events in the system, with time indicated in the direction of the morphisms in the category. Adding such graphs to the category of knots and links is an extension of the *tangle category* where one has already extended braids to allow any embedding of strands and circles that start in n ordered points and end in m ordered points. The tangle category includes the braid category and the Temperley-Lieb category. These are both included in the category of braided trivalent graphs.

Thinking of the basic trivalent vertex as the form of a particle interaction there will be a set of particle states that can label each arc incident to the vertex. In Figure 43 we illustrate the labeling of the trivalent graphs by such particle states. In the next two sections we will see specific rules for labeling such states. Here it suffices to note that there will be some restrictions on these labels, so that a trivalent vertex has a set of possible labelings. Similarly, any trivalent graph will have a set of admissible labelings. These are the possible particle processes that this graph can support. We take the set of admissible labelings of a given graph G as a basis for a vector space $V(G)$ over the complex numbers. This vector space is the space of *processes* associated with the graph G. Given a surface S and a decomposition t of the surface into trinions, we have the associated graph $G(S, t)$ and hence a vector space of processes $V(G(S, t))$. It is desirable to have this vector space independent of the particular decomposition into trinions. If this can be accomplished, then the set of vector spaces and linear mappings associated to the surfaces can consitute a functor from the category of cobordisms of one-manifolds to vector spaces, and hence gives rise to a one-dimensional topological quantum field theory. To this end we need some properties of the particle interactions that will be described below.

A *spin network* is, by definition a lableled trivalent graph in a category of graphs that satisfy the properties outlined in the previous paragraph. We shall detail the requirements below.

The simplest case of this idea is C. N. Yang's original interpretation of the Yang-Baxter equation [98]. Yang articulated a quantum field theory in one dimension of space and one dimension of time in which the R-matrix giving the scattering amplitudes for an interaction of two particles whose (let us say) spins corresponded to the matrix indices so that R_{ab}^{cd} is the amplitude for particles of spin a and spin b to interact and produce particles of spin c and d. Since these interactions are between particles in a line, one takes the convention that the particle with spin a is to the left of the particle with spin b, and the particle with spin c is to the left of the particle with spin d. If one follows the concatenation of such interactions, then there is an underlying permutation that is obtained by following strands from the bottom to the top of the diagram (thinking of time as moving up the page). Yang designed the Yang-Baxter equation for R so that *the amplitudes for a composite process depend only on the underlying permutation corresponding to the process and not on the individual sequences of interactions.*

In taking over the Yang-Baxter equation for topological purposes, we can use the same interpretation, but think of the diagrams with their under- and over-crossings as modeling events in a spacetime with two dimensions of space and one dimension of time. The extra spatial dimension is taken in displacing the woven strands perpendicular to the page, and allows us to use braiding operators R and R^{-1} as scattering matrices. Taking this picture to heart, one can add other particle properties to the idealized theory. In particular one can add fusion and creation vertices where in fusion two particles interact to become a single particle and in creation one particle changes (decays) into two particles. These are the trivalent vertices discussed above. Matrix elements corresponding to trivalent vertices can represent these interactions. See Figure 46.

FIGURE 46. Creation and fusion.

Once one introduces trivalent vertices for fusion and creation, there is the question how these interactions will behave in respect to the braiding operators. There will be a matrix expression for the compositions of braiding and fusion or creation as indicated in Figure 25. Here we will restrict ourselves to showing the diagrammatics with the intent of giving the reader a flavor of these structures. It is natural to assume that braiding intertwines with creation as shown in Figure 49 (similarly with fusion). This intertwining identity is clearly the sort of thing that a topologist will love, since it indicates that the diagrams can be interpreted as embeddings of graphs in three-dimensional space, and it fits with our interpretation of the vertices in terms of trinions. Figure 47 illustrates the Yang-Baxter equation. The intertwining identity is an assumption like the Yang-Baxter equation itself, that simplifies the mathematical structure of the model.

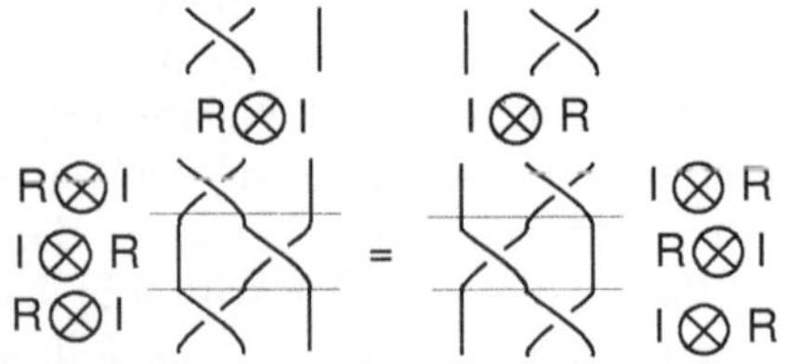

FIGURE 47. Yang-Baxter equation.

It is to be expected that there will be an operator that expresses the recoupling of vertex interactions as shown in Figure 50 and labeled by Q. This corresponds to the associativity at the level of trinion combinations shown in Figure 44. The actual formalism of such an operator will parallel the mathematics of

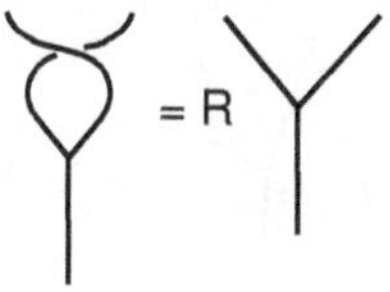

FIGURE 48. Braiding.

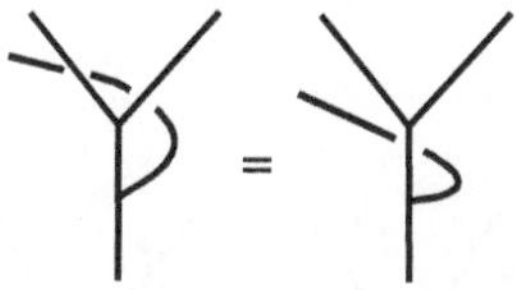

FIGURE 49. Intertwining.

recoupling for angular momentum. See for example [40]. If one just considers the abstract structure of recoupling then one sees that for trees with four branches (each with a single root) there is a cycle of length five as shown in Figure 51. One can start with any pattern of three vertex interactions and go through a sequence of five recouplings that bring one back to the same tree from which one started. *It is a natural simplifying axiom to assume that this composition is the identity mapping.* This axiom is called the *pentagon identity*.

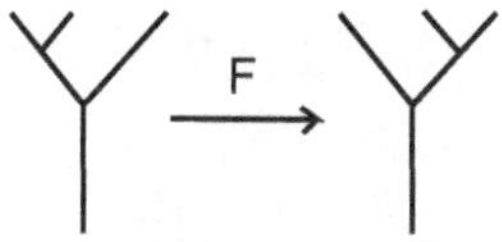

FIGURE 50. Recoupling.

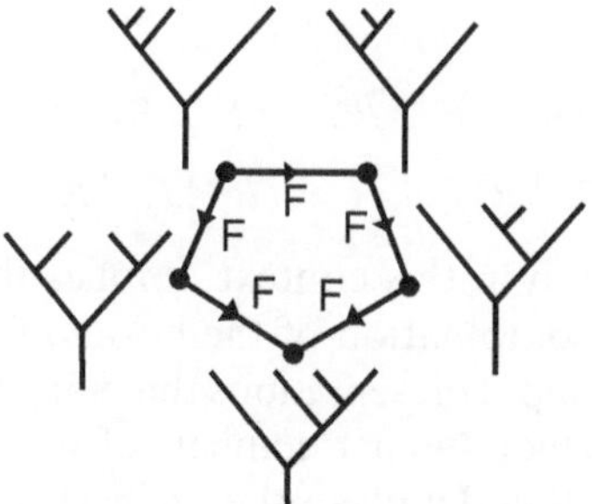

FIGURE 51. Pentagon identity.

Finally there is a hexagonal cycle of interactions between braiding, recoupling and the intertwining identity as shown in Figure 52. One says that the interactions satisfy the *hexagon identity* if this composition is the identity.

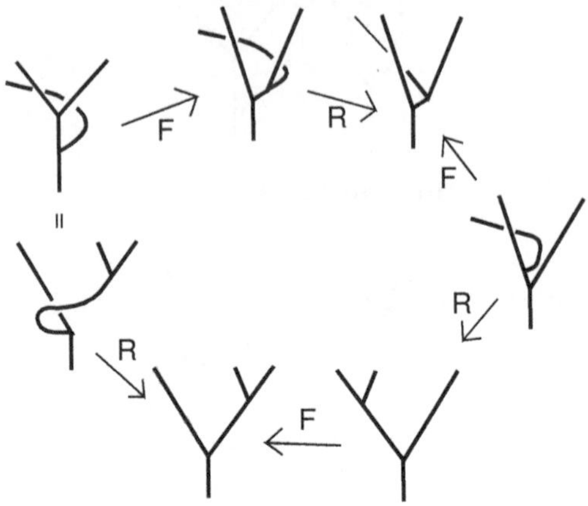

FIGURE 52. Hexagon identity.

Remark 9. It is worth pointing out how these identities are related to the braiding. The hexagon identity tells us that

$$R^{-1}FRF^{-1}RF = I$$

where I is the identity mapping on the process space for trees with three branches. Letting

$$A = R$$

and

$$B = F^{-1}RF,$$

we see that the hexagon identity is equivalent to the statement

$$B = R^{-1}F^{-1}R.$$

Thus

$$\begin{aligned} ABA &= R(R^{-1}F^{-1}R)R = F^{-1}R^2 = (F^{-1}RF)F^{-1}R \\ &= (R^{-1}F^{-1}R)F^{-1}R = (R^{-1}F^{-1}R)R(R^{-1}F^{-1}R) = BAB. \end{aligned}$$

Thus the hexagon relation in this context, implies that A and B satisfy the braiding relation. The combination of the hexagon and pentagon relations ensures that the braid group representations that are generated are well-defined and fit together as we include smaller numbers of strands in larger numbers of strands. We omit the further details of the verification of this statement.

A *graphical three-dimensional topological quantum field theory* is an algebra of interactions that satisfies the Yang-Baxter equation, the intertwining identity, the pentagon identity and the hexagon identity. There is not room in this summary to detail the way that these properties fit into the topology of knots and three-dimensional manifolds, but a sketch is in order. For the case of topological quantum field theory related to the group $SU(2)$ there is a construction based entirely on the combinatorial topology of the bracket polynomial (See Sections 11 to 18 of this paper.). See [39, 40] for more information on this approach.

Now return to Figure 42 where we illustrate trinions, shown in relation to a trivalent vertex, and a surface of genus three that is decomposed into four trinions. It turns out that the vector space $V(S_g) = V(G(S_g, t))$ to a surface with a trinion decomposition as t described above, and defined in terms of the graphical topological quantum field theory, does not depend upon the choice of trinion decomposition. This independence is guaranteed by the braiding, hexagon and pentagon identities. One can then associate a well-defined vector $|M\rangle$ in $V(S_g)$ whenenver M is a three manifold whose boundary is S_g. Furthermore, if a closed three-manifold M^3 is decomposed along a surface S_g into the union of M_- and M_+ where these parts are otherwise disjoint three-manifolds with boundary S_g, then the inner product $I(M) = \langle M_- | M_+ \rangle$ is, up to normalization, an invariant of the three-manifold M_3. With the definition of graphical topological quantum field theory given above, knots and links can be incorporated as well, so that one obtains a source of invariants $I(M^3, K)$ of knots and links in orientable three-manifolds. Here we see the uses of the relationships that occur in the higher dimensional cobordism categories, as described in the previous section.
The invariant $I(M^3, K)$ can be formally compared with the Witten [95] integral

$$Z(M^3, K) = \int DAe^{(ik/4\pi)S(M,A)} W_K(A).$$

It can be shown that up to limits of the heuristics, $Z(M, K)$ and $I(M^3, K)$ are essentially equivalent for appropriate choice of gauge group and corresponding spin networks.

By these graphical reformulations, a three-dimensional *TQFT* is, at base, a highly simplified theory of point particle interactions in $2 + 1$ dimensional spacetime. It can be used to articulate invariants of knots and links and invariants of three manifolds. The reader interested in the $SU(2)$ case of this structure and its implications for invariants of knots and three manifolds can consult [40, 39, 69, 14, 77]. One expects that physical situations involving $2 + 1$ spacetime will be approximated by such an idealized theory. There are also applications to $3 + 1$ quantum gravity[53, 85, 91]. Aspects of the quantum Hall effect may be related to topological quantum field theory [94]. One can study a physics in two dimensional space where the braiding of particles or collective excitations leads to non-trival representations of the Artin braid group. Such particles are called *Anyons*. Such *TQFT* models would describe applicable physics. One can think about applications of anyons to quantum computing along the lines of the topoological models described here.

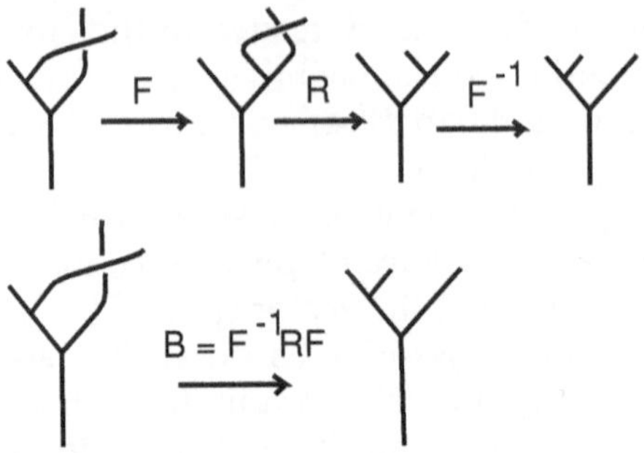

FIGURE 53. A more complex braiding operator.

A key point in the application of *TQFT* to quantum information theory is contained in the structure illustrated in Figure 53. There we show a more complex braiding operator, based on the composition of recoupling with the elementary braiding at a vertex. (This structure is implicit in the Hexagon identity of Figure 52.) The new braiding operator is a source of unitary representations of braid group in situations (which exist mathematically) where the recoupling transformations are themselves unitary. This kind of pattern is utilized in the work of Freedman and collaborators [22, 20, 23, 24, 21] and in the case of classical angular momentum formalism has been dubbed a "spin-network quantum simlator" by Rasetti and collaborators [75, 25]. In the next section we show how certain natural deformations [40] of Penrose spin networks [80] can be used to produce these unitary representations of the Artin braid group and the corresponding models for anyonic topological quantum computation.

§14. Spin networks and Temperley-Lieb recoupling theory. In this section we discuss a combinatorial construction for spin networks that generalizes the original construction of Roger Penrose. The result of this generalization is a structure that satisfies all the properties of a graphical *TQFT* as described in the previous section, and specializes to classical angular momentum recoupling theory in the limit of its basic variable. The construction is based on the properties of the bracket polynomial (as already described in Section 11). A complete description of this theory can be found in the book "Temperley-Lieb Recoupling Theory and Invariants of Three-Manifolds" by Kauffman and Lins [40].

The "q-deformed" spin networks that we construct here are based on the bracket polynomial relation. View Figure 54 and Figure 55.

In Figure 54 we indicate how the basic projector (symmetrizer, Jones-Wenzl projector) is constructed on the basis of the bracket polynomial expansion. In this technology a symmetrizer is a sum of tangles on n strands (for a chosen integer n). The tangles are made by summing over braid lifts of

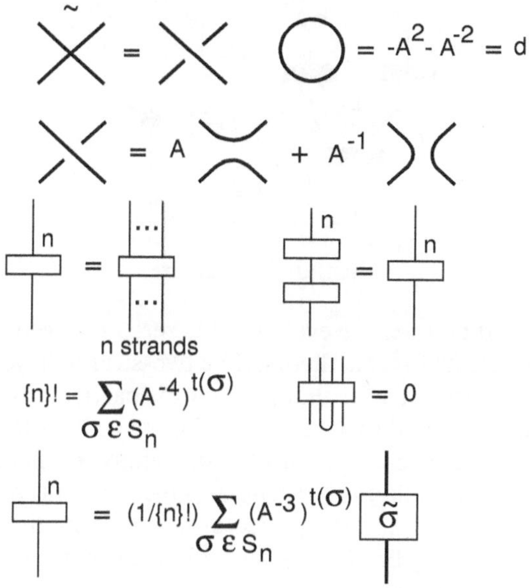

FIGURE 54. Basic projectors.

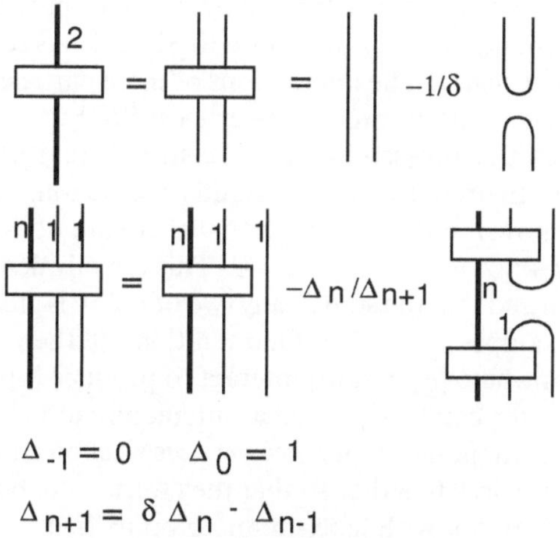

FIGURE 55. Two strand projector.

permutations in the symmetric group on n letters, as indicated in Figure 54. Each elementary braid is then expanded by the bracket polynomial relation as indicated in Figure 54 so that the resulting sum consists of flat tangles without any crossings (these can be viewed as elements in the Temperley-Lieb algebra). The projectors have the property that the concatenation of a projector with itself is just that projector, and if you tie two lines on the top or the bottom

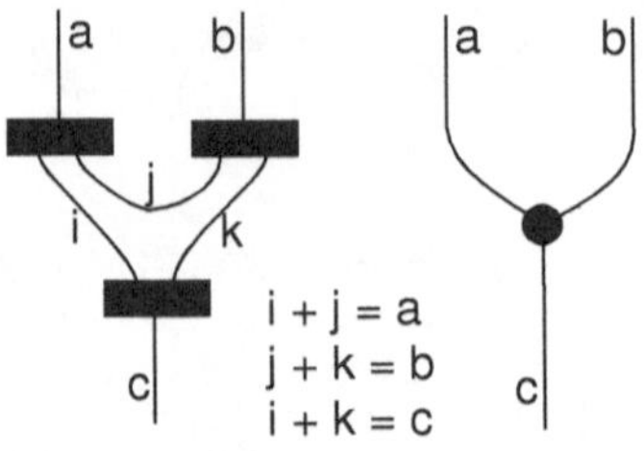

FIGURE 56. Vertex.

of a projector together, then the evaluation is zero. This general definition of projectors is very useful for this theory. The two-strand projector is shown in Figure 55. Here the formula for that projector is particularly simple. It is the sum of two parallel arcs and two turn-around arcs (with coefficient $-1/d$, with $d = -A^2 - A^{-2}$ is the loop value for the bracket polynomial. Figure 55 also shows the recursion formula for the general projector. This recursion formula is due to Jones and Wenzl and the projector in this form, developed as a sum in the Temperley-Lieb algebra (see Section 12 of this paper), is usually known as the *Jones-Wenzl projector*.

The projectors are combinatorial analogs of irreducible representations of a group (the original spin nets were based on $SU(2)$ and these deformed nets are based on the corresponding quantum group to SU(2)). As such the reader can think of them as "particles". The interactions of these particles are governed by how they can be tied together into three-vertices. See Figure 56. In Figure 56 we show how to tie three projectors, of a, b, c strands respectively, together to form a three-vertex. In order to accomplish this interaction, we must share lines between them as shown in that figure so that there are non-negative integers i, j, k so that $a = i + j, b = j + k, c = i + k$. This is equivalent to the condition that $a+b+c$ is even and that the sum of any two of a, b, c is greater than or equal to the third. For example $a + b \geq c$. One can think of the vertex as a possible particle interaction where $[a]$ and $[b]$ interact to produce $[c]$. That is, any two of the legs of the vertex can be regarded as interacting to produce the third leg.

There is a basic orthogonality of three vertices as shown in Figure 57. Here if we tie two three-vertices together so that they form a "bubble" in the middle, then the resulting network with labels a and b on its free ends is a multiple of an a-line (meaning a line with an a-projector on it) or zero (if a is not equal to b). The multiple is compatible with the results of closing the diagram in the equation of Figure 57 so the two free ends are identified with one another. On closure, as shown in the figure, the left hand side of the equation becomes a Theta graph and the right hand side becomes a multiple of a "delta" where Δ_a denotes the bracket polynomial evaluation of the a-strand loop with a projector on it. The $\Theta(a, b, c)$ denotes the bracket evaluation of a theta graph made from three trivalent vertices and labeled with a, b, c on its edges.

There is a recoupling formula in this theory in the form shown in Figure 58. Here there are "6-j symbols", recoupling coefficients that can be expressed, as shown in Figure 58, in terms of tetrahedral graph evaluations and theta graph evaluations. The tetrahedral graph is shown in Figure 59. One derives the formulas for these coefficients directly from the orthogonality relations for the trivalent vertices by closing the left hand side of the recoupling formula and using orthogonality to evaluate the right hand side. This is illustrated in Figure 60. The reader should be advised that there are specific calculational formulas for the theta and tetrahedral nets. These can be found in [40]. Here we are indicating only the relationships and external logic of these objects.

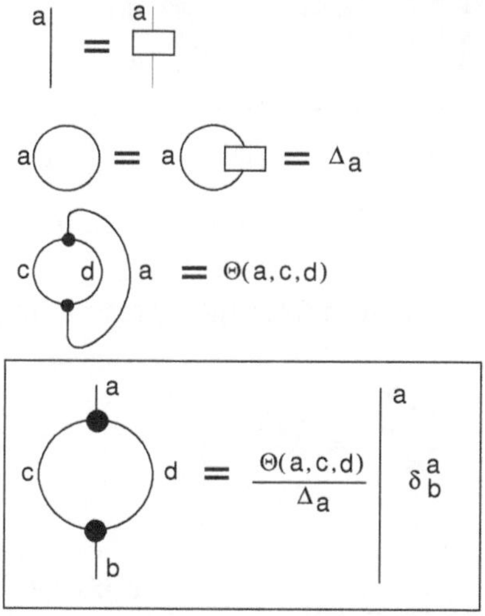

FIGURE 57. Orthogonality of trivalent vertices.

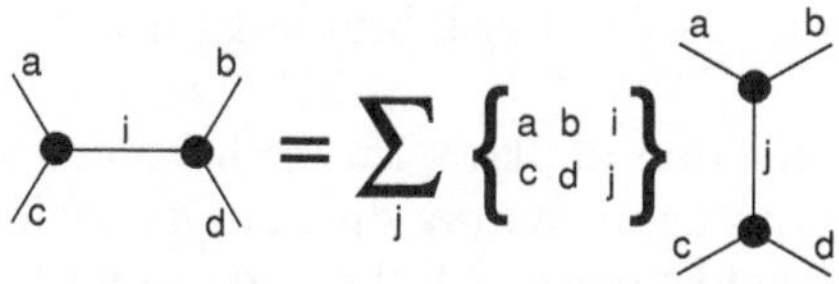

FIGURE 58. Recoupling formula.

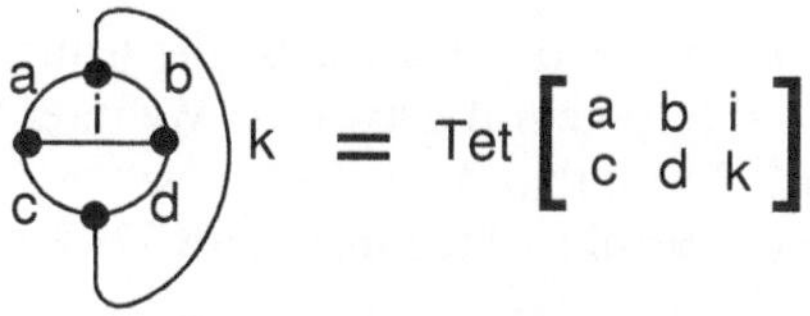

FIGURE 59. Tetrahedron network.

Finally, there is the braiding relation, as illustrated in Figure 36.

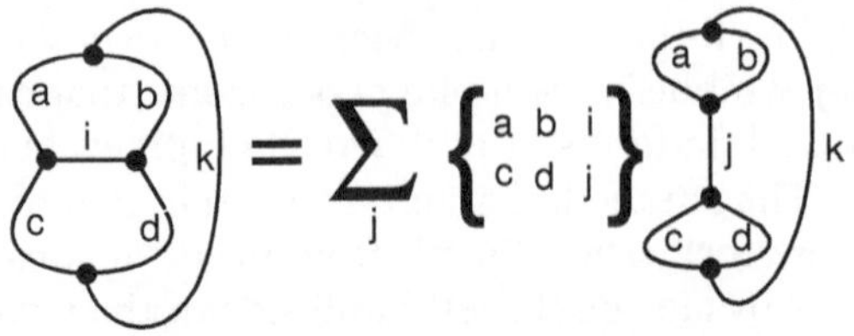

$$= \sum_j \begin{Bmatrix} a & b & i \\ c & d & j \end{Bmatrix} \frac{\Theta(a,b,j)}{\Delta_j} \frac{\Theta(c,d,j)}{\Delta_j} \Delta_j \, \delta_j^k$$

$$= \begin{Bmatrix} a & b & i \\ c & d & k \end{Bmatrix} \frac{\Theta(a,b,k)\,\Theta(c,d,k)}{\Delta_k}$$

$$\boxed{\begin{Bmatrix} a & b & i \\ c & d & k \end{Bmatrix} = \frac{\mathrm{Tet}\begin{bmatrix} a & b & i \\ c & d & k \end{bmatrix} \Delta_k}{\Theta(a,b,k)\,\Theta(c,d,k)}}$$

FIGURE 60. Tetrahedron formula for recoupling coefficients.

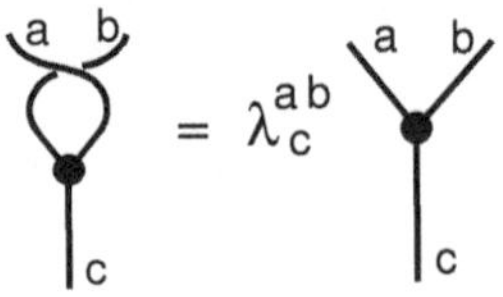

$$\lambda_c^{ab} = (-1)^{(a+b-c)/2} A^{(a'+b'-c')/2}$$

$$x' = x(x+2)$$

FIGURE 61. Local braiding formula.

With the braiding relation in place, this q-deformed spin network theory satisfies the pentagon, hexagon and braiding naturality identities needed for a topological quantum field theory. All these identities follow naturally from the basic underlying topological construction of the bracket polynomial. One can apply the theory to many different situations.

14.1. Evaluations. In this section we discuss the structure of the evaluations for Δ_n and the theta and tetrahedral networks. We refer to [40] for the details behind these formulas. Recall that Δ_n is the bracket evaluation of the closure of the n-strand projector, as illustrated in Figure 57. For the bracket variable A, one finds that

$$\Delta_n = (-1)^n \frac{A^{2n+2} - A^{-2n-2}}{A^2 - A^{-2}}.$$

One sometimes writes the *quantum integer*

$$[n] = (-1)^{n-1}\Delta_{n-1} = \frac{A^{2n} - A^{-2n}}{A^2 - A^{-2}}.$$

If

$$A = e^{i\pi/2r}$$

where r is a positive integer, then

$$\Delta_n = (-1)^n \frac{\sin((n+1)\pi/r)}{\sin(\pi/r)}.$$

Here the corresponding quantum integer is

$$[n] = \frac{\sin(n\pi/r)}{\sin(\pi/r)}.$$

Note that $[n+1]$ is a positive real number for $n = 0, 1, 2, \ldots, r-2$ and that $[r-1] = 0$.

The evaluation of the theta net is expressed in terms of quantum integers by the formula

$$\Theta(a, b, c) = (-1)^{m+n+p} \frac{[m+n+p+1]![n]![m]![p]!}{[m+n]![n+p]![p+m]!}$$

where

$$a = m + p, b = m + n, c = n + p.$$

Note that

$$(a + b + c)/2 = m + n + p.$$

When $A = e^{i\pi/2r}$, the recoupling theory becomes finite with the restriction that only three-vertices (labeled with a, b, c) are *admissible* when $a + b + c \leq 2r - 4$. All the summations in the formulas for recoupling are restricted to admissible triples of this form.

14.2. Symmetry and unitarity. The formula for the recoupling coefficients given in Figure 60 has less symmetry than is actually inherent in the structure of the situation. By multiplying all the vertices by an appropriate factor, we can reconfigure the formulas in this theory so that the revised recoupling transformation is orthogonal, in the sense that its transpose is equal to its inverse. This is a very useful fact. It means that when the resulting matrices are real, then the recoupling transformations are unitary. We shall see particular applications of this viewpoint later in the paper.

Figure 62 illustrates this modification of the three-vertex. Let *Vert*$[a, b, c]$ denote the original 3-vertex of the Temperley-Lieb recoupling theory. Let *ModVert*$[a, b, c]$ denote the modified vertex. Then we have the formula

$$ModVert[a, b, c] = \frac{\sqrt{\sqrt{\Delta_a \Delta_b \Delta_c}}}{\sqrt{\Theta(a, b, c)}}\ Vert[a, b, c].$$

LEMMA 3. *For the bracket evaluation at the root of unity* $A = e^{i\pi/2r}$ *the factor*

$$f(a,b,c) = \frac{\sqrt{\sqrt{\Delta_a\Delta_b\Delta_c}}}{\sqrt{\Theta(a,b,c)}}$$

is real, and can be taken to be a positive real number for (a,b,c) *admissible (i.e.* $a+b+c \leq 2r-4$*).*

PROOF. By the results from the previous subsection,

$$\Theta(a,b,c) = (-1)^{(a+b+c)/2}\hat{\Theta}(a,b,c)$$

where $\hat{\Theta}(a,b,c)$ is positive real, and

$$\Delta_a\Delta_b\Delta_c = (-1)^{(a+b+c)}[a+1][b+1][c+1]$$

where the quantum integers in this formula can be taken to be positive real. It follows from this that

$$f(a,b,c) = \sqrt{\frac{\sqrt{[a+1][b+1][c+1]}}{\hat{\Theta}(a,b,c)}},$$

showing that this factor can be taken to be positive real. ⊣

In Figure 63 we show how this modification of the vertex affects the non-zero term of the orthogonality of trivalent vertices (compare with Figure 57). We refer to this as the "modified bubble identity." The coefficient in the modified bubble identity is

$$\sqrt{\frac{\Delta_b\Delta_c}{\Delta_a}} = (-1)^{(b+c-a)/2}\sqrt{\frac{[b+1][c+1]}{[a+1]}}$$

where (a,b,c) form an admissible triple. In particular $b+c-a$ is even and hence this factor can be taken to be real.

We rewrite the recoupling formula in this new basis and emphasize that the recoupling coefficients can be seen (for fixed external labels a,b,c,d) as a matrix transforming the horizontal "double-Y" basis to a vertically disposed double-Y basis. In Figure 64, Figure 65 and Figure 66 we have shown the form of this transformation,using the matrix notation

$$M\lfloor a,b,c,d\rfloor_{ij}$$

for the modified recoupling coefficients. In Figure 64 we derive an explicit formula for these matrix elements. The proof of this formula follows directly from trivalent–vertex orthogonality (See Figure 57 and Figure 60.), and is given in Figure 64. The result shown in Figure 64 and Figure 65 is the following formula for the recoupling matrix elements.

$$M[a,b,c,d]_{ij} = ModTet\begin{pmatrix} a & b & i \\ c & d & j \end{pmatrix} / \sqrt{\Delta_a\Delta_b\Delta_c\Delta_d}$$

where $\sqrt{\Delta_a\Delta_b\Delta_c\Delta_d}$ is short-hand for the product

$$\sqrt{\frac{\Delta_a\Delta_b}{\Delta_j}}\sqrt{\frac{\Delta_c\Delta_d}{\Delta_j}}\Delta_j$$

$$= (-1)^{(a+b-j)/2}(-1)^{(c+d-j)/2}(-1)^j\sqrt{\frac{[a+1][b+1]}{[j+1]}}\sqrt{\frac{[c+1][d+1]}{[j+1]}}[j+1]$$

$$= (-1)^{(a+b+c+d)/2}\sqrt{[a+1][b+1][c+1][d+1]}$$

In this form, since (a, b, j) and (c, d, j) are admissible triples, we see that this coeffient can be taken to be real, and its value is independent of the choice of i and j. The matrix $M[a, b, c, d]$ is real-valued.
It follows from Figure 58 (turn the diagrams by ninety degrees) that

$$M[a, b, c, d]^{-1} = M[b, d, a, c].$$

In Figure 67 we illustrate the formula

$$M[a, b, c, d]^T = M[b, d, a, c].$$

It follows from this formula that

$$M[a, b, c, d]^T = M[a, b, c, d]^{-1}.$$

Hence $M[a, b, c, d]$ is an orthogonal, real-valued matrix.

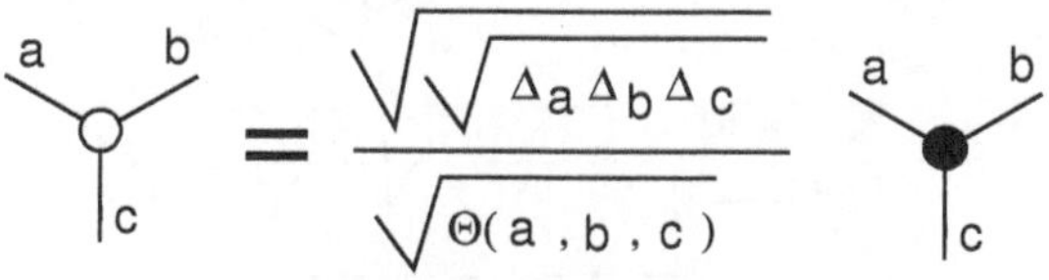

FIGURE 62. Modified three vertex.

THEOREM 5. *In the Temperley-Lieb theory we obtain unitary (in fact real orthogonal) recoupling transformations when the bracket variable A has the form $A = e^{i\pi/2r}$ for r a positive integer. Thus we obtain families of unitary representations of the Artin braid group from the recoupling theory at these roots of unity.*

PROOF. The proof is given the discussion above. ⊣

In Section 16 we shall show explictly how these methods work in the case of the Fibonacci model where $A = e^{3i\pi/5}$.

§15. Fibonacci particles. In this section and the next we detail how the Fibonacci model for anyonic quantum computing [68, 81] can be constructed by using a version of the two-stranded bracket polynomial and a generalization of Penrose spin networks. This is a fragment of the Temperley-Lieb recoupling

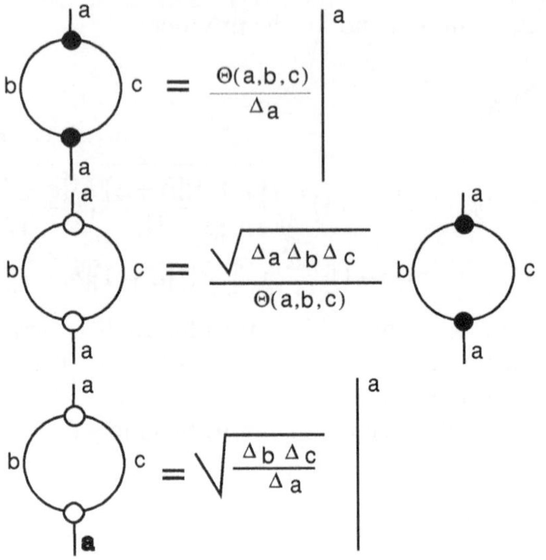

FIGURE 63. Modified bubble identity.

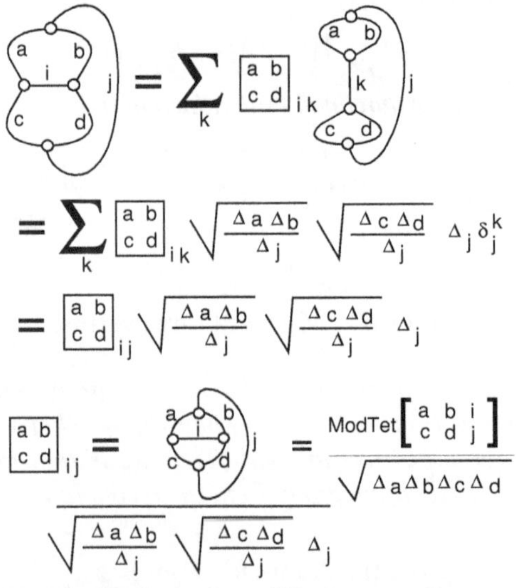

FIGURE 64. Derivation of modified recoupling coefficients.

theory [40]. We already gave in the preceding sections a general discussion of the theory of spin networks and their relationship with quantum computing.

The Fibonacci model is a *TQFT* that is based on a single "particle" with two states that we shall call the *marked state* and the *unmarked state*. The particle in the marked state can interact with itself either to produce a single particle in

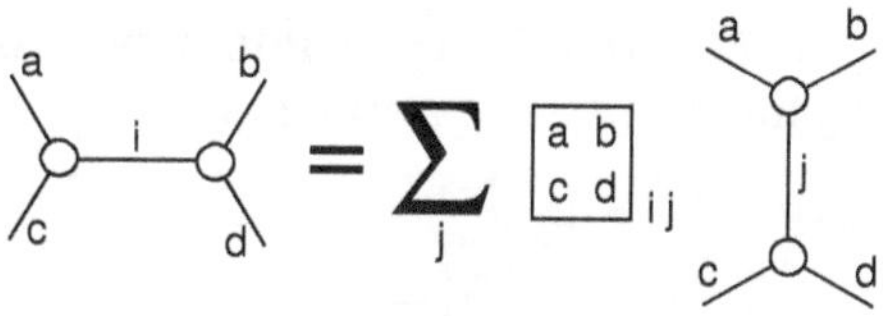

FIGURE 65. Modified recoupling formula.

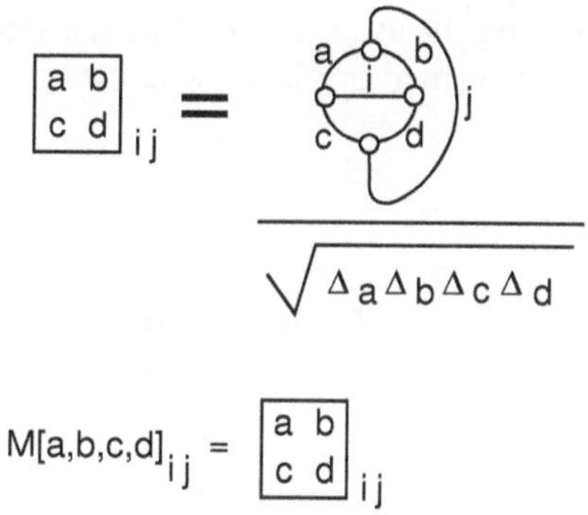

FIGURE 66. Modified recoupling matrix.

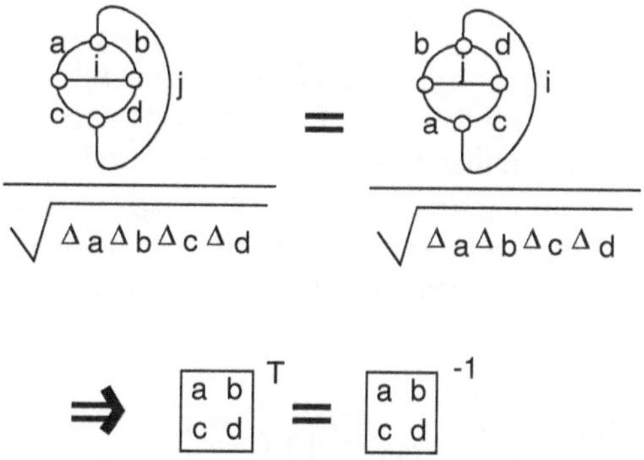

FIGURE 67. Modified matrix transpose.

the marked state, or to produce a single particle in the unmarked state. The particle in the unmarked state has no influence in interactions (an unmarked state interacting with any state S yields that state S). One way to indicate these two interactions symbolically is to use a box,for the marked state and a blank space for the unmarked state. Then one has two modes of interaction of a box with itself:

1. Adjacency: 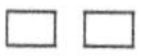
   and
2. Nesting: 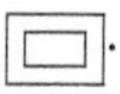.

With this convention we take the adjacency interaction to yield a single box, and the nesting interaction to produce nothing:

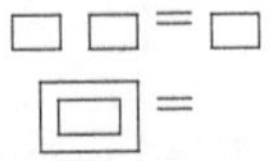

We take the notational opportunity to denote nothing by an asterisk (*). The syntatical rules for operating the asterisk are Thus the asterisk is a stand-in for no mark at all and it can be erased or placed wherever it is convenient to do so. Thus

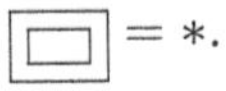

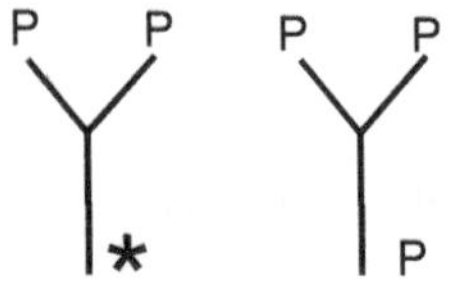

FIGURE 68. Fibonacci particle interaction.

We shall make a recoupling theory based on this particle, but it is worth noting some of its purely combinatorial properties first. The arithmetic of combining boxes (standing for acts of distinction) according to these rules has been studied and formalized in [92] and correlated with Boolean algebra and classical logic. Here *within* and *next to* are ways to refer to the two sides delineated by the given distinction. From this point of view, there are two modes of relationship (adjacency and nesting) that arise at once in the presence of a distinction.

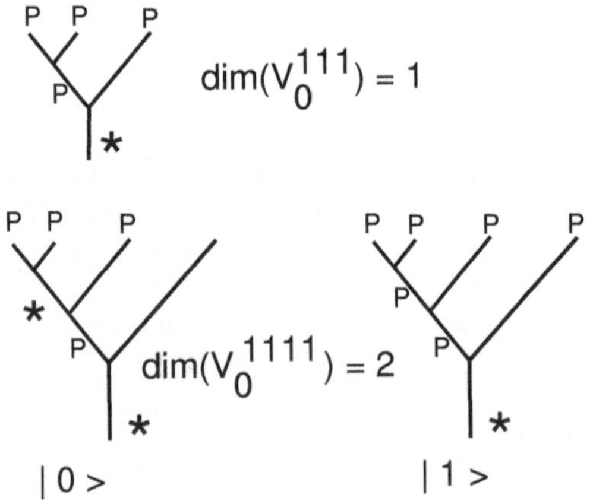

FIGURE 69. Fibonacci trees.

From here on we shall denote the Fibonacii particle by the letter P. Thus the two possible interactions of P with itself are as follows.

1. $P, P \longrightarrow *$
2. $P, P \longrightarrow P$

In Figure 69 we indicate in small tree diagrams the two possible interactions of the particle P with itself. In the first interaction the particle vanishes, producing the asterix. In the second interaction the particle a single copy of P is produced. These are the two basic actions of a single distinction relative to itself, and they constitute our formalism for this very elementary particle.

In Figure 69, we have indicated the different results of particle processes where we begin with a left-associated tree structure with three branches, all marked and then four branches all marked. In each case we demand that the particles interact successively to produce an unmarked particle in the end, at the root of the tree. More generally one can consider a left-associated tree with n upward branches and one root. Let $T(a_1, a_2, \ldots, a_n : b)$ denote such a tree with particle labels $a_1, \ldots, a_n$ on the top and root label b at the bottom of the tree. We consider all possible processes (sequences of particle interactions) that start with the labels at the top of the tree, and end with the labels at the bottom of the tree. Each such sequence is regarded as a basis vector in a complex vector space

$$V_b^{a_1,a_2,\ldots,a_n}$$

associated with the tree. In the case where all the labels are marked at the top and the bottom label is unmarked, we shall denote this tree by

$$V_0^{111\ldots11} = V_0^{(n)}$$

where n denotes the number of upward branches in the tree. We see from Figure 69 that the dimension of $V_0^{(3)}$ is 1, and that

$$dim(V_0^{(4)}) = 2.$$

This means that $V_0^{(4)}$ is a natural candidate in this context for the two-qubit space.

Given the tree $T(1, 1, 1, \ldots, 1 : 0)$ (n marked states at the top, an unmarked state at the bottom), a process basis vector in $V_0^{(n)}$ is in direct correspondence with a string of boxes and asterisks (1's and 0's) of length $n-2$ with no repeated asterisks and ending in a marked state. See Figure 69 for an illustration of the simplest cases. It follows from this that

$$dim(V_0^{(n)}) = f_{n-2}$$

where f_k denotes the k-th Fibonacci number:

$$f_0 = 1, f_1 = 1, f_2 = 2, f_3 = 3, f_4 = 5, f_5 = 8, \ldots$$

where

$$f_{n+2} = f_{n+1} + f_n.$$

The dimension formula for these spaces follows from the fact that there are f_n sequences of length $n-1$ of marked and unmarked states with no repetition of an unmarked state. This fact is illustrated in Figure 70.

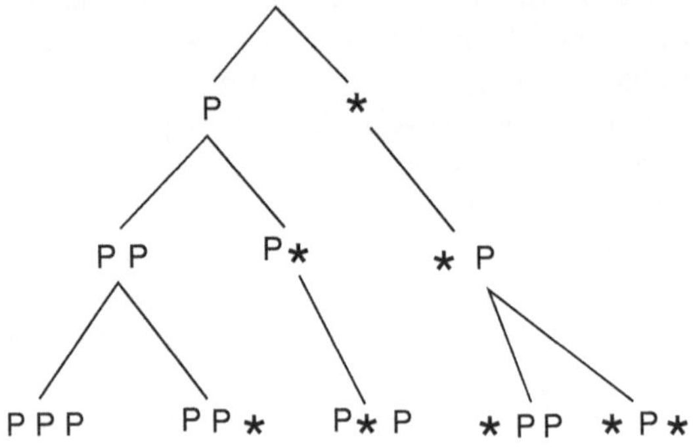

Tree of squences with no occurence of * *

FIGURE 70. Fibonacci sequence.

§16. The Fibonacci recoupling model. We now show how to make a model for recoupling the Fibonacci particle by using the Temperley Lieb recoupling theory and the bracket polynomial. Everything we do in this section will be based on the 2-projector, its properties and evaluations based on the bracket polynomial model for the Jones polynomial. While we have outlined the general recoupling theory based on the bracket polynomial in earlier sections of this paper, the present section is self-contained, using only basic information about the bracket polyonmial, and the essential properties of the 2-projector as shown in Figure 71. In this figure we state the definition of the 2-projector, list its two main properties (the operator is idempotent and a self-attached strand yields a zero evaluation) and give diagrammatic proofs of these properties.

In Figure 72, we show the essence of the Temperley-Lieb recoupling model for the Fibonacci particle. The Fibonaccie particle is, in this mathematical model, identified with the 2-projector itself. As the reader can see from Figure 72, there are two basic interactions of the 2-projector with itself, one giving a 2-projector, the other giving nothing. This is the pattern of self-iteraction of the Fibonacci particle. There is a third possibility, depicted in Figure 72, where two 2-projectors interact to produce a 4-projector. We could remark at the outset, that the 4-projector will be zero if we choose the bracket polynomial variable $A = e^{3\pi/5}$. Rather than start there, we will assume that the 4-projector is forbidden and deduce (below) that the theory has to be at this root of unity.

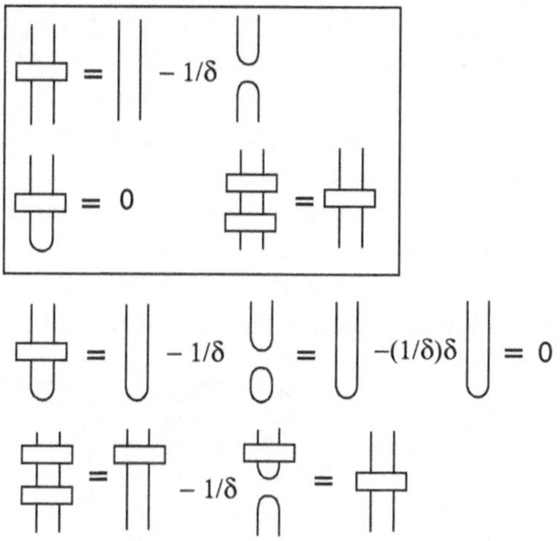

FIGURE 71. The 2-projector.

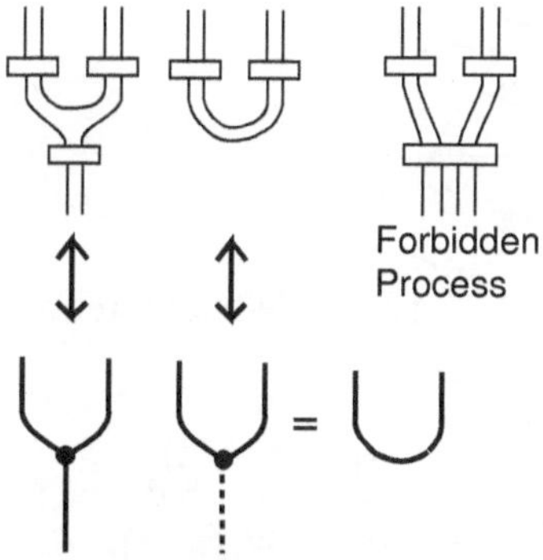

FIGURE 72. Fibonacci particle as 2-projector.

Note that in Figure 72 we have adopted a single strand notation for the particle interactions, with a solid strand corresponding to the marked particle, a dotted strand (or nothing) corresponding to the unmarked particle. A dark vertex indicates either an interaction point, or it may be used to indicate the single strand is shorthand for two ordinary strands. Remember that these are all shorthand expressions for underlying bracket polynomial calculations.

In Figures 73–78 we have provided complete diagrammatic calculations of all of the relevant small nets and evaluations that are useful in the two-strand theory that is being used here. The reader may wish to skip directly to Figure 79 where we determine the form of the recoupling coefficients for this theory. We will discuss the resulting algebra below.

For the reader who does not want to skip the next collection of figures, here is a guided tour. Figure 73 illustrates three basic nets in case of two strands.

These are the theta, delta and tetrahedron nets. In this figure we have shown the decomposition on the theta and delta nets in terms of 2-projectors. The Tetrahedron net will be similarly decomposed in Figure 77 and Figure 78. The theta net is denoted Θ, the delta by Δ, and the tetrahedron by T. In Figure 74 we illustrate how a pedant loop has a zero evaluation. In Figure 75 we use the identity in Figure 74 to show how an interior loop (formed by two trivalent vertices) can be removed and replaced by a factor of Θ/Δ. Note how, in this figure, line two proves that one network is a multiple of the other, while line three determines the value of the multiple by closing both nets.

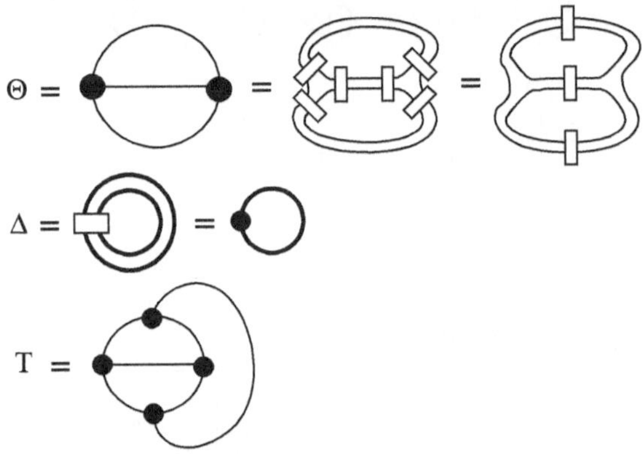

FIGURE 73. Theta, delta and tetrahedron.

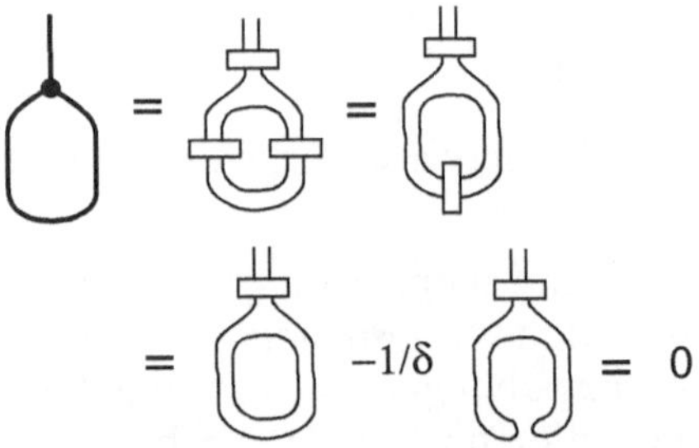

FIGURE 74. LoopEvaluation – 1.

Figure 76 illustrates the explicit calculation of the delta and theta nets. The figure begins with a calculation of the result of closing a single strand of the 2-projector. The result is a single stand multiplied by $(\delta - 1/\delta)$ where $\delta = -A^2 - A^{-2}$, and A is the bracket polynomial parameter. We then find that

$$\Delta = \delta^2 - 1$$

and

$$\Theta = (\delta - 1/\delta)^2\delta - \Delta/\delta = (\delta - 1/\delta)(\delta^2 - 2).$$

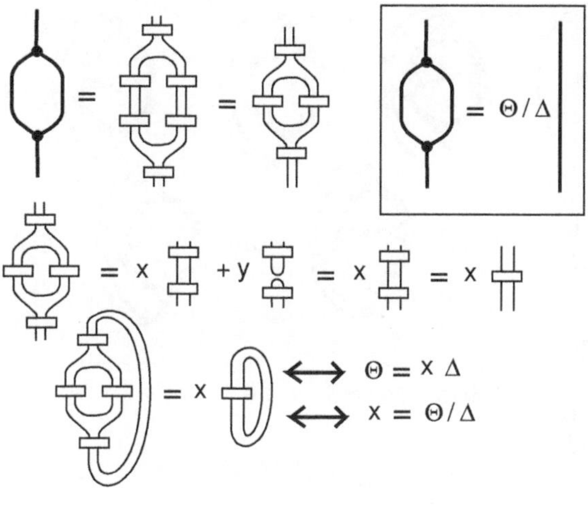

FIGURE 75. LoopEvaluation – 2.

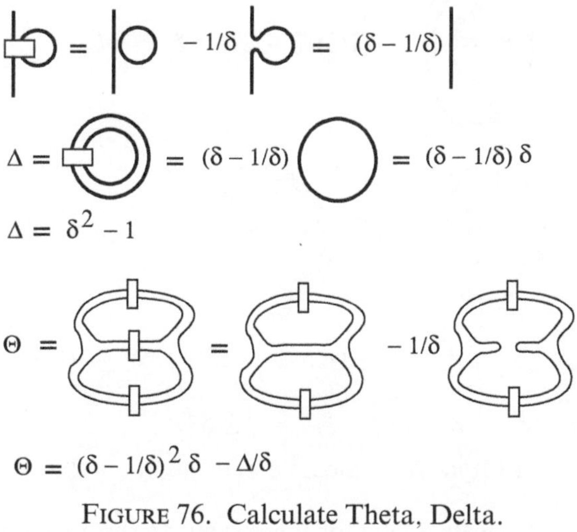

FIGURE 76. Calculate Theta, Delta.

Figure 77 and Figure 78 illustrate the calculation of the value of the tetrahedral network T. The reader should note the first line of Figure 77 where the tetradedral net is translated into a pattern of 2-projectors, and simplified. The rest of these two figures are a diagrammatic calculation, using the expansion formula for the 2-projector. At the end of Figure 78 we obtain the formula for the tetrahedron

$$T = (\delta - 1/\delta)^2(\delta^2 - 2) - 2\Theta/\delta.$$

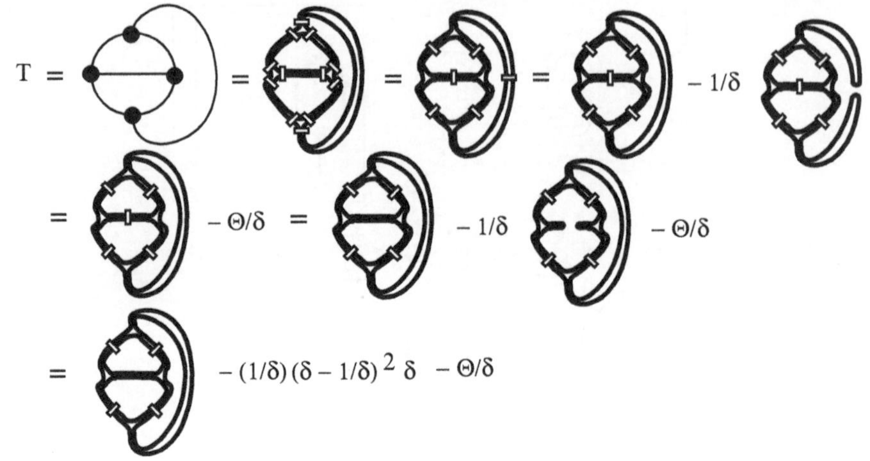

FIGURE 77. Calculate tetrahedron – 1.

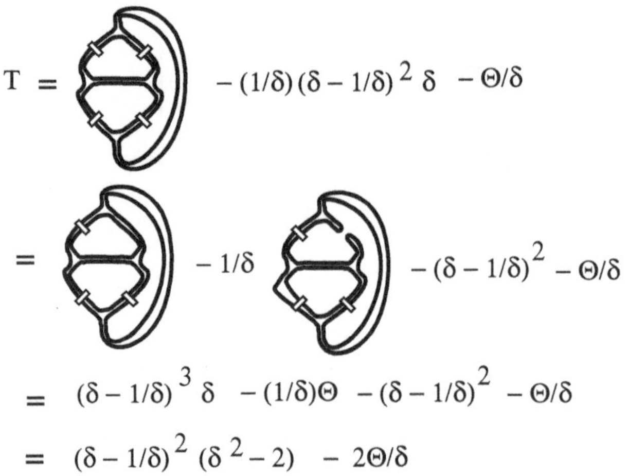

FIGURE 78. Calculate Tetrahedron – 2.

Figure 79 is the key calculation for this model. In this figure we assume that the recoupling formulas involve only 0 and 2 strands, with 0 corresponding to the null particle and 2 corresponding to the 2-projector. ($2+2=4$ is forbidden as in Figure 72.) From this assumption we calculate that the recoupling matrix is given by

$$F = \begin{pmatrix} a & b \\ c & d \end{pmatrix} = \begin{pmatrix} 1/\Delta & \Delta/\Theta \\ \Theta/\Delta^2 & T\Delta/\Theta^2 \end{pmatrix}$$

Figure 80 and Figure 81 work out the exact formulas for the braiding at a three-vertex in this theory. When the 3-vertex has three marked lines, then the

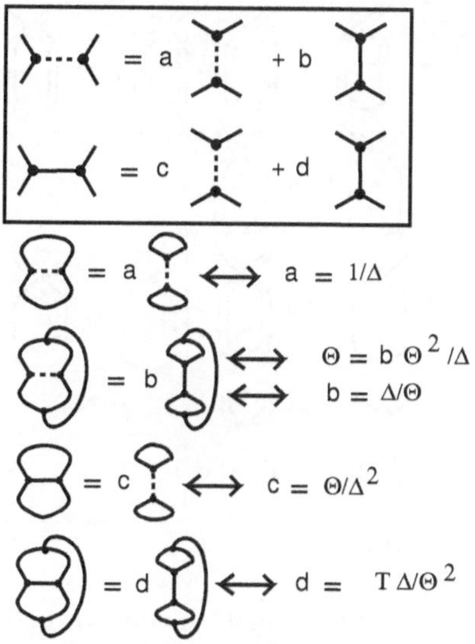

FIGURE 79. Recoupling for 2-projectors.

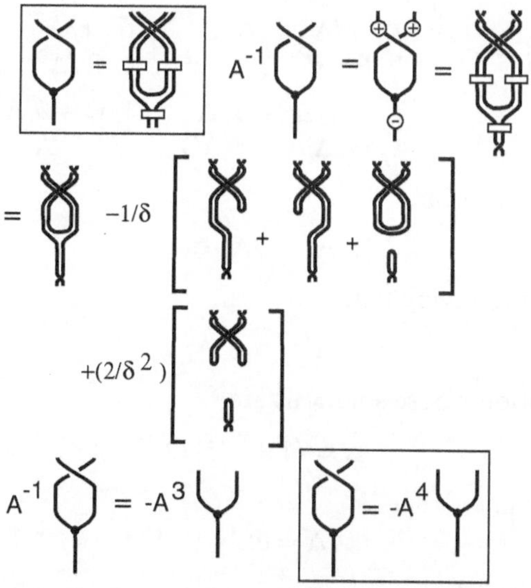

FIGURE 80. Braiding at the three-vertex.

braiding operator is multiplication by $-A^4$, as in Figure 58. When the 3-vertex has two marked lines, then the braiding operator is multiplication by A^8, as shown in Figure 81.

Notice that it follows from the symmetry of the diagrammatic recoupling formulas of Figure 79 that *the square of the recoupling matrix F is equal to the*

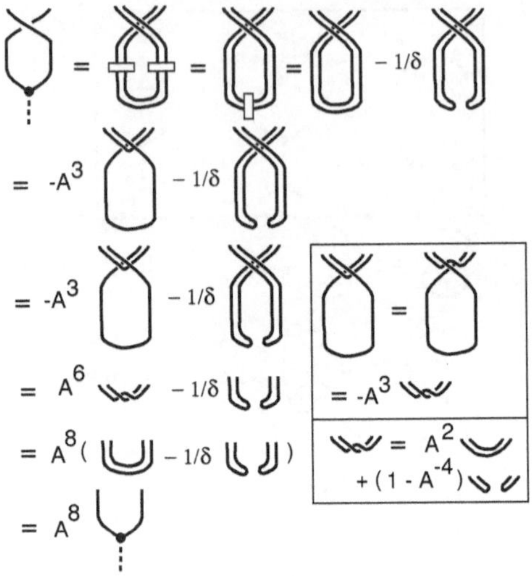

FIGURE 81. Braiding at the null-three-vertex.

identity. That is,

$$\begin{pmatrix} 1 & 0 \\ 0 & 1 \end{pmatrix} = F^2 = \begin{pmatrix} 1/\Delta & \Delta/\Theta \\ \Theta/\Delta^2 & T\Delta/\Theta^2 \end{pmatrix}\begin{pmatrix} 1/\Delta & \Delta/\Theta \\ \Theta/\Delta^2 & T\Delta/\Theta^2 \end{pmatrix}$$
$$= \begin{pmatrix} 1/\Delta^2 + 1/\Delta & 1/\Theta + T\Delta^2/\Theta^3 \\ \Theta/\Delta^3 + T/(\Delta\Theta) & 1/\Delta + \Delta^2 T^2/\Theta^4 \end{pmatrix}.$$

Thus we need the relation

$$1/\Delta + 1/\Delta^2 = 1.$$

This is equivalent to saying that

$$\Delta^2 = 1 + \Delta,$$

a quadratic equation whose solutions are

$$\Delta = (1 \pm \sqrt{5})/2.$$

Furthermore, we know that

$$\Delta = \delta^2 - 1$$

from Figure 76. Hence

$$\Delta^2 = \Delta + 1 = \delta^2.$$

We shall now specialize to the case where

$$\Delta = \delta = (1 + \sqrt{5})/2,$$

leaving the other cases for the exploration of the reader. We then take

$$A = e^{3\pi i/5}$$

so that

$$\delta = -A^2 - A^{-2} = -2\cos(6\pi/5) = (1+\sqrt{5})/2.$$

Note that $\delta - 1/\delta = 1$. Thus

$$\Theta = (\delta - 1/\delta)^2\delta - \Delta/\delta = \delta - 1.$$

and

$$T = (\delta - 1/\delta)^2(\delta^2 - 2) - 2\Theta/\delta = (\delta^2 - 2) - 2(\delta - 1)/\delta$$
$$= (\delta - 1)(\delta - 2)/\delta = 3\delta - 5.$$

Note that

$$T = -\Theta^2/\Delta^2,$$

from which it follows immediately that

$$F^2 = I.$$

This proves that we can satisfy this model when $\Delta = \delta = (1+\sqrt{5})/2$. For this specialization we see that the matrix F becomes

$$F = \begin{pmatrix} 1/\Delta & \Delta/\Theta \\ \Theta/\Delta^2 & T\Delta/\Theta^2 \end{pmatrix} = \begin{pmatrix} 1/\Delta & \Delta/\Theta \\ \Theta/\Delta^2 & (-\Theta^2/\Delta^2)\Delta/\Theta^2 \end{pmatrix} = \begin{pmatrix} 1/\Delta & \Delta/\Theta \\ \Theta/\Delta^2 & -1/\Delta \end{pmatrix}$$

This version of F has square equal to the identity independent of the value of Θ, so long as $\Delta^2 = \Delta + 1$.

The final adjustment. Our last version of F suffers from a lack of symmetry. It is not a symmetric matrix, and hence not unitary. A final adjustment of the model gives this desired symmetry. *Consider the result of replacing each trivalent vertex (with three 2-projector strands) by a multiple by a given quantity* α. Since the Θ has two vertices, it will be multiplied by α^2. Similarly, the tetradhedron T will be multiplied by α^4. The Δ and the δ will be unchanged. Other properties of the model will remain unchanged. The new recoupling matrix, after such an adjustment is made, becomes

$$\begin{pmatrix} 1/\Delta & \Delta/\alpha^2\Theta \\ \alpha^2\Theta/\Delta^2 & -1/\Delta \end{pmatrix}$$

For symmetry we require

$$\Delta/(\alpha^2\Theta) = \alpha^2\Theta/\Delta^2.$$

We take

$$\alpha^2 = \sqrt{\Delta^3}/\Theta.$$

With this choice of α we have

$$\Delta/(\alpha^2\Theta) = \Delta\Theta/(\Theta\sqrt{\Delta^3}) = 1/\sqrt{\Delta}.$$

Hence the new symmetric F is given by the equation

$$F = \begin{pmatrix} 1/\Delta & 1/\sqrt{\Delta} \\ 1/\sqrt{\Delta} & -1/\Delta \end{pmatrix} = \begin{pmatrix} \tau & \sqrt{\tau} \\ \sqrt{\tau} & -\tau \end{pmatrix}$$

where Δ is the golden ratio and $\tau = 1/\Delta$. This gives the Fibonacci model. Using Figure 80 and Figure 81, we have that the local braiding matrix for the model is given by the formula below with $A = e^{3\pi i/5}$.

$$R = \begin{pmatrix} -A^4 & 0 \\ 0 & A^8 \end{pmatrix} = \begin{pmatrix} e^{4\pi i/5} & 0 \\ 0 & -e^{2\pi i/5} \end{pmatrix}.$$

The simplest example of a braid group representation arising from this theory is the representation of the three strand braid group generated by $S_1 = R$ and $S_2 = FRF$ (Remember that $F = F^T = F^{-1}$.). The matrices S_1 and S_2 are both unitary, and they generate a dense subset of the unitary group $U(2)$, supplying the first part of the transformations needed for quantum computing.

§17. Quantum computation of colored Jones polynomials and the Witten-Reshetikhin-Turaev invariant. In this section we make some brief comments on the quantum computation of colored Jones polynomials. This material will be expanded in a subsequent publication.

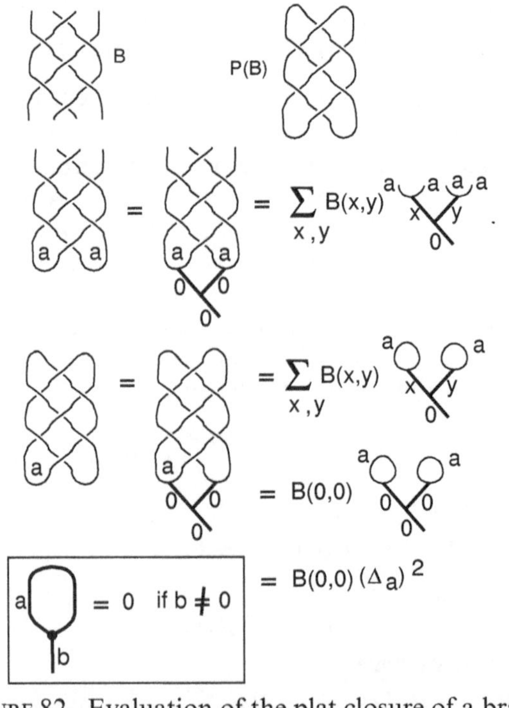

FIGURE 82. Evaluation of the plat closure of a braid.

First, consider Figure 82. In that figure we illustrate the calculation of the evalutation of the *(a)-colored bracket polynomial* for the *plat closure* $P(B)$ of a braid B. The reader can infer the definition of the plat closure from Figure 82.

One takes a braid on an even number of strands and closes the top strands with each other in a row of maxima. Similarly, the bottom strands are closed with a row of minima. It is not hard to see that any knot or link can be represented as the plat closure of some braid. Note that in this figure we indicate the action of the braid group on the process spaces corresponding to the small trees attached below the braids.

The (a)-colored bracket polynonmial of a link L, denoted $< L >_a$, is the evaluation of that link where each single strand has been replaced by a parallel strands and the insertion of Jones-Wenzl projector (as discussed in Section 14). We then see that we can use our discussion of the Temperley-Lieb recoupling theory as in sections 14,15 and 16 to compute the value of the colored bracket polynomial for the plat closure PB. As shown in Figure 82, we regard the braid as acting on a process space $V_0^{a,a,\ldots,a}$ and take the case of the action on the vector v whose process space coordinates are all zero. Then the action of the braid takes the form

$$Bv(0,\ldots,0) = \Sigma_{x_1,\ldots,x_n} B(x_1,\ldots,x_n)v(x_1,\ldots,x_n)$$

where $B(x_1,\ldots,x_n)$ denotes the matrix entries for this recoupling transformation and $v(x_1,\ldots,x_n)$ runs over a basis for the space $V_0^{a,a,\ldots,a}$. Here n is even and equal to the number of braid strands. In the figure we illustrate with $n = 4$. Then, as the figure shows, when we close the top of the braid action to form PB, we cut the sum down to the evaluation of just one term. In the general case we will get

$$< PB >_a = B(0,\ldots,0)\Delta_a^{n/2}.$$

The calculation simplifies to this degree because of the vanishing of loops in the recoupling graphs. The vanishing result is stated in Figure 82, and it is proved in the case $a = 2$ in Figure 74.

The *colored Jones polynomials* are normalized versions of the colored bracket polymomials, differing just by a normalization factor.

In order to consider quantumn computation of the colored bracket or colored Jones polynomials, we therefore can consider quantum computation of the matrix entries $B(0,\ldots,0)$. These matrix entries in the case of the roots of unity $A = e^{i\pi/2r}$ and for the $a = 2$ Fibonacci model with $A = e^{3i\pi/5}$ are parts of the diagonal entries of the unitary transformation that represents the braid group on the process space $V_0^{a,a,\ldots,a}$. *We can obtain these matrix entries by using the Hadamard test as described in section 11.* As a result we get relatively efficient quantum algorithms for the colored Jones polynonmials at these roots of unity, in essentially the same framework as we described in section 11, but for braids of arbitrary size. The computational complexity of these models is essentially the same as the models for the Jones polynomial discussed in [3]. We reserve discussion of these issues to a subsequent publication.

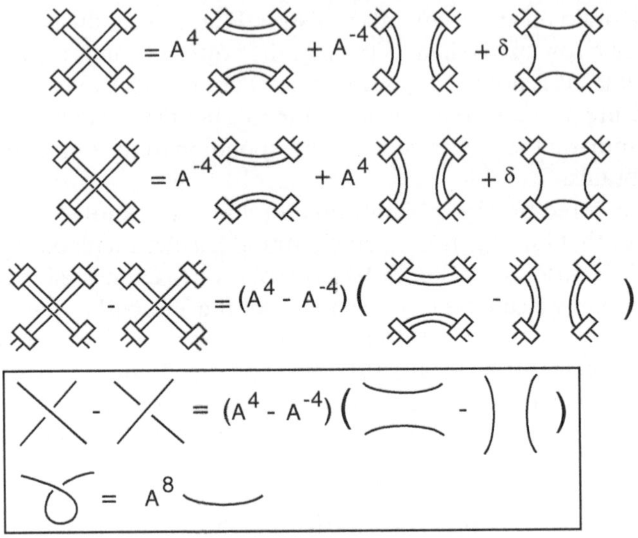

FIGURE 83. Dubrovnik polynomial specialization at two strands.

It is worth remarking here that these algorithms give not only quantum algorithms for computing the colored bracket and Jones polynomials, but also for computing the Witten-Reshetikhin-Turaev (*WRT*) invariants at the above roots of unity. The reason for this is that the *WRT* invariant, in unnormalized form is given as a finite sum of colored bracket polynomials:

$$WRT(L) = \Sigma_{a=0}^{r-2} \Delta_a < L >_a,$$

and so the same computation as shown in Figure 82 applies to the *WRT*. This means that we have, in principle, a quantum algorithm for the computation of the Witten functional integral [95] via this knot-theoretic combinatorial topology. It would be very interesting to understand a more direct approach to such a computation via quantum field theory and functional integration.

Finally, we note that in the case of the Fibonacci model, the (2)-colored bracket polynomial is a special case of the Dubrovnik version of the Kauffman polynomial [38]. See Figure 83 for diagammatics that resolve this fact. The skein relation for the Dubrovnik polynomial is boxed in this figure. Above the box, we show how the double strands with projectors reproduce this relation. This observation means that in the Fibonacci model, the natural underlying knot polynomial is a special evaluation of the Dubrovnik polynomial, and the Fibonacci model can be used to perform quantum computation for the values of this invariant.

§18. A direct construction of the Fibonacci model. In section 10 of this paper, we give elementary constructions for unitary representations of the three strand braid group in $U(2)$. In section 11 we show how to use unitary representations of the three strand brand group to devise a quantum computation for the Jones polynomial. In this section we return to these considerations, and show how to construct the Fibonacci model by elementary means, without using the recoupling theory that we have explained in the previous sections of the paper. This final approach is significant in that it shows an even closer relationship of the Fibonacci model with the Temperley Lieb algebra representation associated with the Jones polynomial.

The constructions in this section are based on the combinatorics of the Fibonacci model. While we do not assume the recoupling theory of the previous sections, we essentially reconstruct its patterns for the particular purposes of the Fibonacci model. Recall that in the Fibonacci model we have a (mathematical) particle P that interacts with itself either to produce P or to produce a neutral particle $*$. If X is any particle then $*$ interacts with X to produce X. Thus $*$ acts as an identity trasformation. These rules of interaction are illustrated in Figure 68, Figure 69, Figure 70 and Figure 84.

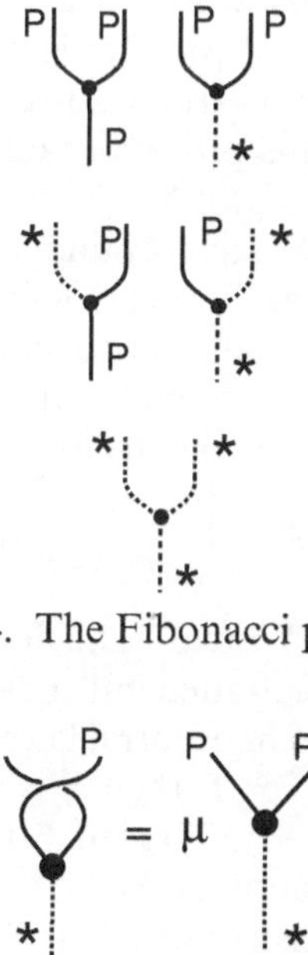

FIGURE 84. The Fibonacci particle P.

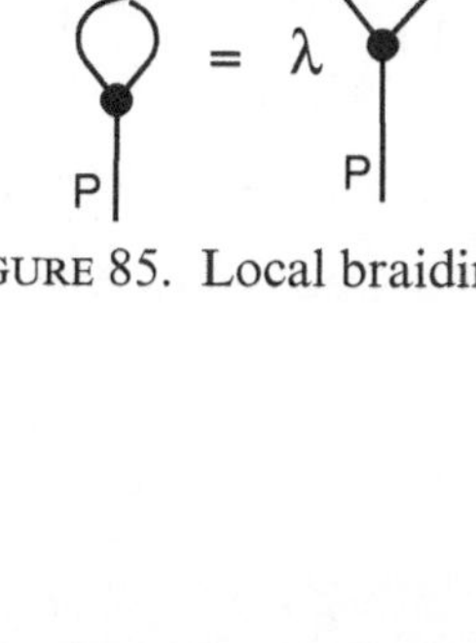

FIGURE 85. Local braiding.

The braiding of two particles is measured in relation to their interaction. In Figure 85 we illustrate braiding of P with itself in relation to the two possible interactions of P with itself. If P interacts to produce $*$, then the braiding gives a phase factor of μ. If P interacts to produce P, then the braiding gives a phase factor of λ. We assume at the outset that μ and λ are unit complex numbers. One should visualize these particles as moving in a plane and the diagrams of interaction are either creations of two particles from one particle, or fusions of two particles to a single particle (depending on the choice of temporal direction). Thus we have a braiding matrix for these "local" particle interactions:

$$R = \begin{pmatrix} \mu & 0 \\ 0 & \lambda \end{pmatrix}$$

written with respect to the basis $\{|*\rangle, |P\rangle\}$ for this space of particle interactions.

We want to make this braiding matrix part of a larger representation of the braid group. In particular, we want a representation of the three-strand braid group on the process space V_3 illustrated in Figure 6. This space starts with three P particles and considers processes associated in the patttern $(PP)P$ with the stipulation that the end product is P. The possible pathways are illustrated in Figure 86. They correspond to $(PP)P \longrightarrow (*)P \longrightarrow P$ and $(PP)P \longrightarrow (P)P \longrightarrow P$. This process space has dimension two and can support a second braiding generator for the second two strands on the top of the tree. In order to articulate the second braiding we change basis to the process space corresponding to $P(PP)$ as shown in Figure 87 and Figure 88. The change of basis is shown in Figure 88 and has matrix F as shown below. We want a unitary representation ρ of three-strand braids so that $\rho(\sigma_1) = R$ and $\rho(\sigma_2) = S = F^{-1}RF$. See Figure 88. We take the form of the matrix F as follows.

$$F = \begin{pmatrix} a & b \\ b & -a \end{pmatrix}$$

where $a^2 + b^2 = 1$ with a and b real. This form of the matrix for the basis change is determined by the requirement that F is symmetric with $F^2 = I$. The symmetry of the change of basis formula essentially demands that $F^2 = I$. If F is real, symmetric and $F^2 = I$, then F is unitary. Since R is unitary we see that $S = FRF$ is also unitary. Thus, if F is constructed in this way then we obtain a unitary representation of B_3.

Now we try to simultaneously construct an F *and* construct a representation of the Temperley-Lieb algebra. We begin by noting that

$$R = \begin{pmatrix} \mu & 0 \\ 0 & \lambda \end{pmatrix} = \begin{pmatrix} \lambda & 0 \\ 0 & \lambda \end{pmatrix} + \begin{pmatrix} \mu - \lambda & 0 \\ 0 & 0 \end{pmatrix} = \begin{pmatrix} \lambda & 0 \\ 0 & \lambda \end{pmatrix} + \lambda^{-1}\begin{pmatrix} \delta & 0 \\ 0 & 0 \end{pmatrix}$$

where $\delta = \lambda(\mu - \lambda)$. Thus $R = \lambda I + \lambda^{-1}U$ where $U = \begin{pmatrix} \delta & 0 \\ 0 & 0 \end{pmatrix}$ so that $U^2 = \delta U$. For the Temperley-Lieb representation, we want $\delta = -\lambda^2 - \lambda^{-2}$.

Hence we need $-\lambda^2 - \lambda^{-2} = \lambda(\mu - \lambda)$, which implies that $\mu = -\lambda^{-3}$. With this restriction on μ, we have the Temperley-Lieb representation and the corresponding unitary braid group representation for 2-strand braids and the 2-strand Temperley-Lieb algebra.

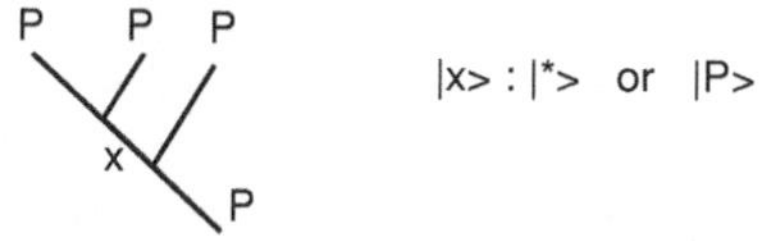

FIGURE 86. Three strands at dimension two.

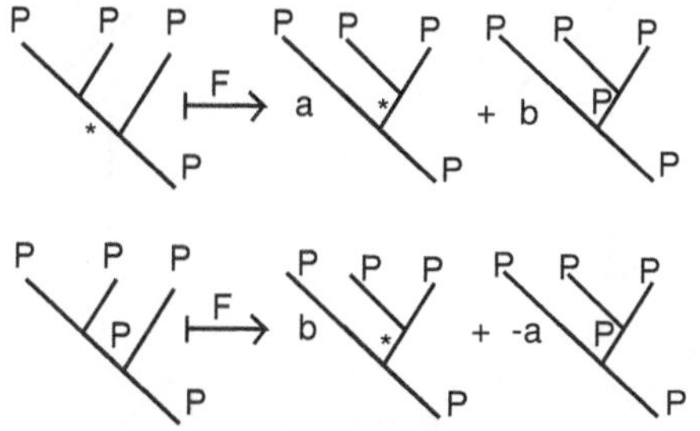

FIGURE 87. Recoupling formula.

Now we can go on to B_3 and TL_3 via $S = FRF = \lambda I + \lambda^{-1} V$ with $V = FUF$. We must examine V^2, UVU and VUV. We find that

$$V^2 = FUFFUF = FU^2F = \delta FUF = \delta V,$$

as desired and

$$V = FUF = \begin{pmatrix} a & b \\ b & -a \end{pmatrix} \begin{pmatrix} \delta & 0 \\ 0 & 0 \end{pmatrix} \begin{pmatrix} a & b \\ b & -a \end{pmatrix} = \delta \begin{pmatrix} a^2 & ab \\ ab & b^2 \end{pmatrix}.$$

Thus $V^2 = V$ and since $V = \delta|v\rangle\langle v|$ and $U = \delta|w\rangle\langle w|$ with $w = (1,0)^T$ and $v = Fw = (a,b)^T$ (T denotes transpose), we see that

$$VUV = \delta^3|v\rangle\langle v|w\rangle\langle w|v\rangle\langle v| = \delta^3 a^2|v\rangle\langle v| = \delta^2 a^2 V.$$

Similarly $UVU = \delta^2 a^2 U$. Thus, we need $\delta^2 a^2 = 1$ and so we shall take $a = \delta^{-1}$. With this choice, we have a representation of the Temperley-Lieb algebra TL_3 so that $\sigma_1 = AI + A^{-1}U$ and $\sigma_2 = AI + A^{-1}V$ gives a unitary representation of the braid group when $A = \lambda = e^{i\theta}$ and $b = \sqrt{1 - \delta^{-2}}$ is real. This last reality condition is equivalent to the inequality

$$cos^2(2\theta) \geq \frac{1}{4},$$

which is satisfied for infinitely many values of θ in the ranges

$$[0, \pi/6] \cup [\pi/3, 2\pi/3] \cup [5\pi/6, 7\pi/6] \cup [4\pi/3, 5\pi/3].$$

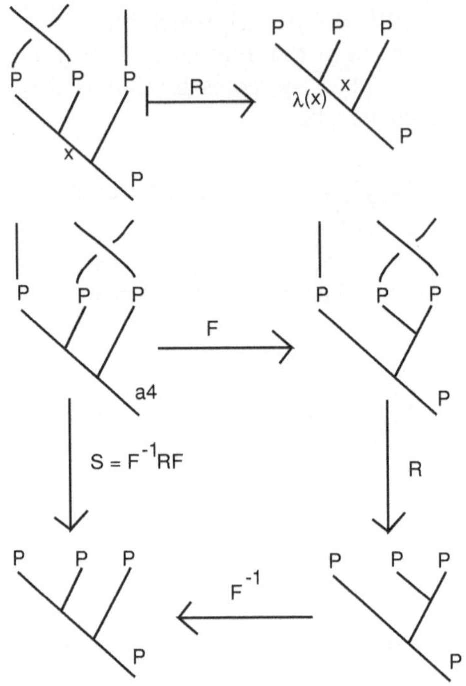

FIGURE 88. Change of basis.

With these choices we have

$$F = \begin{pmatrix} a & b \\ b & -a \end{pmatrix} = \begin{pmatrix} 1/\delta & \sqrt{1-\delta^{-2}} \\ \sqrt{1-\delta^{-2}} & -1/\delta \end{pmatrix}$$

real and unitary, and for the Temperley-Lieb algebra,

$$U = \begin{pmatrix} \delta & 0 \\ 0 & 0 \end{pmatrix}, V = \delta \begin{pmatrix} a^2 & ab \\ ab & b^2 \end{pmatrix} = \begin{pmatrix} a & b \\ b & \delta b^2 \end{pmatrix}.$$

Now examine Figure 89. Here we illustrate the action of the braiding and the Temperley-Lieb Algebra on the first Fibonacci process space with basis $\{|*\rangle, |P\rangle\}$. Here we have $\sigma_1 = R, \sigma_2 = FRF$ and $U_1 = U, U_2 = V$ as described above. Thus we have a representation of the braid group on three strands and a representation of the Temperley-Lieb algebra on three strands with no further restrictions on δ.

So far, we have arrived at exactly the 3-strand braid representations that we used in our papers [44, 56] giving a quantum algorithm for the Jones polynomial for three-strand braids. In this paper we are working in the context of the Fibonacci process spaces and so we wish to see how to make a representation of the Temperley-Lieb algebra to this model as a whole, not restricting ourselves to only three strands. The generic case to consider is the

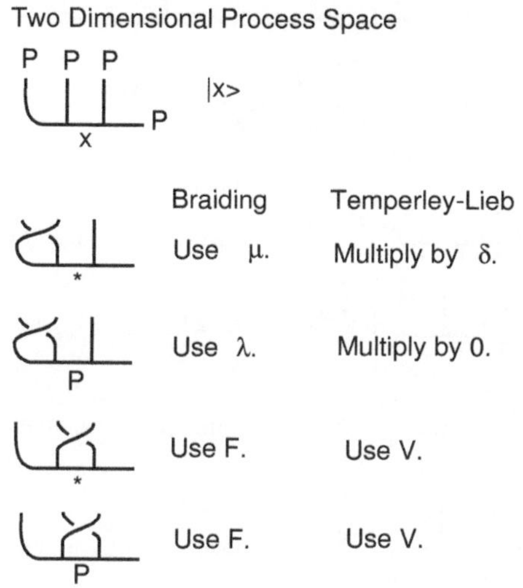

FIGURE 89. Algebra for a two dimensional process space.

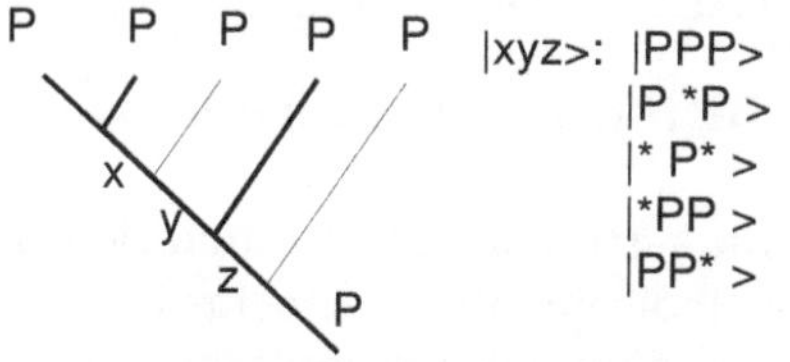

FIGURE 90. A five dimensional process space.

action of the Temperley-Lieb algebra on process spaces of higher dimension as shown in Figure 90 and Figure 91. In Figure 91 we have illustrated the triplets from the previous figure as part of a possibly larger tree and have drawn the strings horizontally rather than diagonally. In this figure we have listed the effects of braiding the vertical strands 3 and 4. We see from this figure that the action of the Temperley-Lieb algebra must be as follows:

$$U_3|P * P\rangle = a|P * P\rangle + b|PPP\rangle,$$

$$U_3|PPP\rangle = b|P * P\rangle + \delta b^2|PPP\rangle,$$

$$U_3| * P*\rangle = \delta| * P*\rangle,$$

$$U_3| * PP\rangle = 0,$$

$$U_3|PP*\rangle = 0.$$

Here we have denoted this action as U_3 because it connotes the action on the third and fourth vertical strands in the sequences shown in Figure 91. Note that in a larger sequence we can recognize U_j by examining the triplet surrounding the $j-1$-th element in the sequence, just as the pattern above is

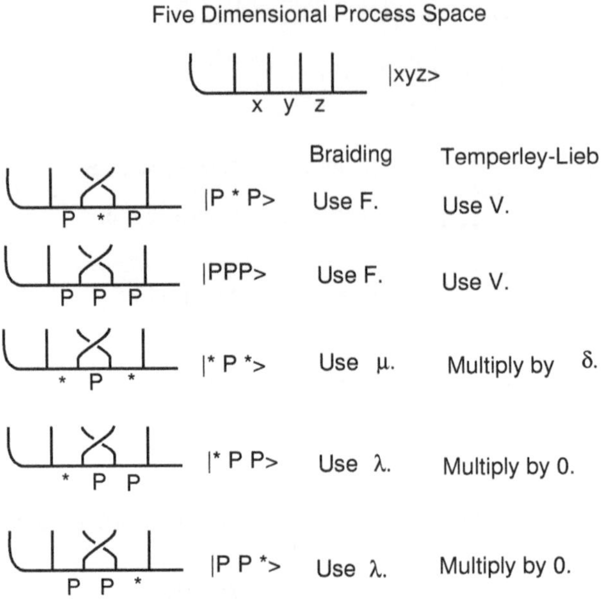

FIGURE 91. Algebra for a five dimensional process space.

governed by the elements surrounding the second element in the sequence. For simplicity, we have only indicated three elements in the sequences above. Note that in a sequence for the Fibonacci process there are never two consecutive appearances of the neutral element $*$.

We shall refer to a sequence of $*$ and P as a *Fibonacci sequence* if it contains no consecutive appearances of $*$. Thus $|PP*P*P*P\rangle$ is a Fibonacci sequence. In working with this representation of the braid group and Temperley-Lieb algebra, it is convenient to assume that the ends of the sequence are flanked by P as in Figure 90 and Figure 91 for sequences of length 3. It is convenient to leave out the flanking P's when notating the sequence.

Using these formulas we can determine conditions on δ such that this is a representation of the Temperley-Lieb algebra for all Fibonacci sequences. Consider the following calculation:

$$\begin{aligned} U_4U_3U_4|PPPP\rangle &= U_3U_2(b|PP*P\rangle + \delta b^2|PPPP\rangle) \\ &= U_4(bU_3|PP*P\rangle + \delta b^2 U_3|PPPP\rangle) \\ &= U_4(0 + \delta b^2(b|P*PP\rangle + \delta b^2|PPPP\rangle)) \\ &= \delta b^2(bU_4|P*PP\rangle + \delta b^2 U_4|PPPP\rangle)) \\ &= \delta^2 b^4 U_4|PPPP\rangle. \end{aligned}$$

Thus we see that in order for $U_4U_3U_4 = U_4$, we need that $\delta^2 b^4 = 1$.

It is easy to see that $\delta^2 b^4 = 1$ is the only remaining condition needed to make sure that the action of the Temperley-Lieb algebra extends to all Fibonacci Model sequences.

Note that $\delta^2 b^4 = \delta^2(1 - \delta^{-2})^2 = (\delta - 1/\delta)^2$. Thus we require that

$$\delta - 1/\delta = \pm 1.$$

When $\delta - 1/\delta = 1$, we have the solutions $\delta = \frac{1 \pm \sqrt{5}}{2}$. However, for the reality of F we require that $1 - \delta^{-2} \geq 0$, ruling out the choice $\delta = \frac{1-\sqrt{5}}{2}$. When $\delta - 1/\delta = -1$, we have the solutions $\delta = \frac{-1 \pm \sqrt{5}}{2}$. This leaves only $\delta = \pm\phi$ where $\phi = \frac{1+\sqrt{5}}{2}$ (the Golden Ratio) as possible values for δ that satisfy the reality condition for F. Thus, up to a sign we have arrived at the well-known value of $\delta = \phi$ (the Fibonacci model) as essentially the only way to have an extension of this form of the representation of the Temperley-Lieb algebra for n strands. Let's state this positively as a Theorem.

THEOREM 6 (Fibonacci Theorem). *Let V_{n+2} be the complex vector space with basis $\{|x_1 x_2 \ldots x_n\rangle\}$ where each x_i equals either P or $*$ and there do not occur two consecutive appearances of $*$ in the sequence $\{x_1, \ldots x_n\}$. We refer to this basis for V_n as the set of Fibonacci sequences of length n. Then the dimension of V_n is equal to f_{n+1} where f_n is the n-th Fibonacci number: $f_0 = f_1 = 1$ and $f_{n+1} = f_n + f_{n-1}$. Let $\delta = \pm\phi$ where $\phi = \frac{1+\sqrt{5}}{2}$. Let $a = 1/\delta$ and $b = \sqrt{1 - a^2}$. Then the Temperley-Lieb algebra on $n + 2$ strands with loop value δ acts on V_n via the formulas given below. First we give the left-end actions.*

$$\begin{aligned}
U_1| * x_2 x_3 \ldots x_n\rangle &= \delta| * x_2 x_3 \ldots x_n\rangle,\\
U_1|P x_2 x_3 \ldots x_n\rangle &= 0,\\
U_2| * P x_3 \ldots x_n\rangle &= a| * P x_3 \ldots x_n\rangle + b|PP x_3 \ldots x_n\rangle,\\
U_2|P * x_3 \ldots x_n\rangle &= 0,\\
U_2|PP x_3 \ldots x_n\rangle &= b| * P x_3 \ldots x_n\rangle + \delta b^2|PP x_3 \ldots x_n\rangle.
\end{aligned}$$

Then we give the general action for the middle strands.

$$\begin{aligned}
U_i|x_1 \ldots x_{i-3} P * P x_{i+1} \ldots x_n\rangle &= a|x_1 \ldots x_{i-3} P * P x_{i+1} \ldots x_n\rangle\\
&\quad + b|x_1 \ldots x_{i-3} PPP x_{i+1} \ldots x_n\rangle,\\
U_i|x_1 \ldots x_{i-3} PPP x_{i+1} \ldots x_n\rangle &= b|x_1 \ldots x_{i-3} P * P x_{i+1} \ldots x_n\rangle\\
&\quad + \delta b^2|x_1 \ldots x_{i-3} PPP x_{i+1} \ldots x_n\rangle,\\
U_i|x_1 \ldots x_{i-3} * P * x_{i+1} \ldots x_n\rangle &= \delta|x_1 \ldots x_{i-3} * P * x_{i+1} \ldots x_n\rangle,\\
U_i|x_1 \ldots x_{i-3} * PP x_{i+1} \ldots x_n\rangle &= 0,\\
U_i|x_1 \ldots x_{i-3} PP * x_{i+1} \ldots x_n\rangle &= 0.
\end{aligned}$$

Finally, we give the right-end action.

$$U_{n+1}|x_1 \ldots x_{n-2} * P\rangle = 0,$$
$$U_{n+1}|x_1 \ldots x_{n-2}P*\rangle = 0,$$
$$U_{n+1}|x_1 \ldots x_{n-2}PP\rangle = b|x_1 \ldots x_{n-2}P*\rangle + \delta b^2|x_1 \ldots x_{n-2}PP\rangle.$$

Remark 10. Note that the left and right end Temperley-Lieb actions depend on the same basic pattern as the middle action. The Fibonacci sequences $|x_1 x_2 \ldots x_n\rangle$ should be regarded as flanked left and right by P's just as in the special cases discussed prior to the proof of the Fibonacci Theorem.

COROLLARY 1. *With the hypotheses of Theorem 2, we have a unitary representation of the Artin Braid group B_{n+2} to TL_{n+2}, $\rho : B_{n+2} \longrightarrow TL_{n+2}$ given by the formulas*

$$\rho(\sigma_i) = AI + A^{-1}U_i,$$
$$\rho(\sigma_i^{-1}) = A^{-1}I + AU_i,$$

where $A = e^{3\pi i/5}$ where the U_i connote the representation of the Temperley-Lieb algebra on the space V_{n+2} of Fibonacci sequences as described in the Theorem above.

Remark 11. The Theorem and Corollary give the original parameters of the Fibonacci model and shows that this model admits a unitary representation of the braid group via a Jones representation of the Temperley-Lieb algebra.

In the original Fibonacci model [60], there is a basic non-trivial recoupling matrix F.

$$F = \begin{pmatrix} 1/\delta & 1/\sqrt{\delta} \\ 1/\sqrt{\delta} & -1/\delta \end{pmatrix} = \begin{pmatrix} \tau & \sqrt{\tau} \\ \sqrt{\tau} & -\tau \end{pmatrix}$$

where $\delta = \frac{1+\sqrt{5}}{2}$ is the golden ratio and $\tau = 1/\delta$. The local braiding matrix is given by the formula below with $A = e^{3\pi i/5}$.

$$R = \begin{pmatrix} A^8 & 0 \\ 0 & -A^4 \end{pmatrix} = \begin{pmatrix} e^{4\pi i/5} & 0 \\ 0 & -e^{2\pi i/5} \end{pmatrix}.$$

This is exactly what we get from our method by using $\delta = \frac{1+\sqrt{5}}{2}$ and $A = e^{3\pi i/5}$. Just as we have explained earlier in this paper, the simplest example of a braid group representation arising from this theory is the representation of the three strand braid group generated by $\sigma_1 = R$ and $\sigma_2 = FRF$ (Remember that $F = F^T = F^{-1}$.). The matrices σ_1 and σ_2 are both unitary, and they generate a dense subset of $U(2)$, supplying the local unitary transformations needed for quantum computing. The full braid group representation on the Fibonacci sequences is computationally universal for quantum computation. In our earlier paper [60] and in the previous sections of the present work, we gave a construction for the Fibonacci model based on Temperely-Lieb recoupling theory. In this section, we have reconstructed the Fibonacci model on the more elementary grounds of the representation of the Temperley-Lieb algebra summarized in the statement of the Fibonacci Theorem and its Corollary.

REFERENCES

[1] S. Abramsky and B. Coecke, *Categorical quantum mechanics*, ***Handbook of Quantum Logic and Quantum Structures: Quantum Logic***, Elsevier/North-Holland, Amsterdam, 2009, pp. 261–323.

[2] D. Aharonov and I. Arad, *The BQP-hardness of approximating the Jones polynomial*, ***quant-ph/0605181***.

[3] D. Aharonov, V. F. R. Jones, and Z. Landau, *A polynomial quantum algorithm for approximating the Jones polynomial*, ***STOC'06: Proceedings of the 38th Annual ACM Symposium on Theory of Computing*** (New York), ACM, 2006, quant-ph/0511096, pp. 427–436.

[4] Y. Akutsu and M. Wadati, *Knot invariants and critical statistical systems*, ***J. Phys. Soc. Japan***, vol. 56 (1987), pp. 839–842.

[5] P. K. Aravind, *Borromean entanglement of the GHZ state*, ***Potentiality, Entanglement and Passion-at-a-Distance*** (R. S. Cohen et al., editors), Kluwer, 1997, pp. 53–59.

[6] M. F. Atiyah, ***The Geometry and Physics of Knots***, Cambridge University Press, 1990.

[7] R. J. Baxter, ***Exactly Solved Models in Statistical Mechanics***, Acad. Press, 1982.

[8] C. W. J. Beenakker, *Search for majorana fermions in superconductors*, ***arXiv: 1112.1950***.

[9] G. Benkart, *Commuting actions – a tale of two groups*, ***Lie algebras and their representations (Seoul 1995)***, Contemp. Math. Series, vol. 194, American Mathematical Society, 1996, pp. 1–46.

[10] N. E. Bonesteel, L. Hormozi, G. Zikos, and S. H. Simon, *Braid topologies for quantum computation*, ***Physical Review Letters***, vol. 95 (2005), pp. 140503, 4 pp., quant-ph/0505665.

[11] J. L. Brylinski and R. Brylinski, *Universal quantum gates*, ***Mathematics of Quantum Computation*** (R. Brylinski and G. Chen, editors), Chapman & Hall/CRC Press, Boca Raton, Florida, 2002.

[12] G. Chen, L. Kauffman, and S. Lomonaco (editors), ***Mathematics in Quantum Computation and Quantum Technology***, Chapman & Hall/CRC, 2007.

[13] B. Coecke, *The logic of entanglement*, ***quant-phy/0402014***.

[14] L. Crane, *2-d physics and 3-d topology*, ***Communications in Mathematical Physics***, vol. 135 (1991), pp. 615–640.

[15] P. A. M. Dirac, ***Principles of Quantum Mechanics***, Oxford University Press, 1958.

[16] H. A. Dye, *Unitary solutions to the yang-baxter equation in dimension four*, ***Quantum Inf. Process***, vol. 2 (2002/3), pp. 117–151, arXiv:quant-ph/0211050.

[17] C. Ernst and D. W. Sumners, *A calculus for rational tangles: Applications to DNA recombination*, ***Mathematical Proceedings of the Cambridge Philosophical Society***, vol. 108 (1990), pp. 489–515.

[18] D. Finkelstein, ***Quantum Relativity: A Synthesis of the Ideas of Einstein and Heisenberg***, Springer-Verlag, 1996.

[19] E. Fradkin and P. Fendley, *Realizing non-abelian statistics in time-reversal invariant systems*, ***Theory Seminar, Physics Department, UIUC***, 4/25/2005.

[20] M. Freedman, *Topological views on computational complexity*, ***Documenta Mathematica, Extra Volume***, ICM, 1998, pp. 453–464.

[21] ———, *Quantum computation and the localization of modular functors*, ***Foundations of Computational Mathematics***, vol. 1 (2001), pp. 183–204, quant-ph/0003128.

[22] ———, *A magnetic model with a possible chern-simons phase*, ***Communications in Mathematical Physics***, vol. 234 (2003), pp. 129–183, With an appendix by F. Goodman and H. Wenzl; quant-ph/0110060v1 9 Oct 2001.

[23] M. Freedman, M. Larsen, and Z. Wang, *A modular functor which is universal for quantum computation*, ***Communications in Mathematical Physics***, vol. 227 (2002), pp. 605–622, quant-ph/0001108v2, 1 Feb 2000.

[24] M. H. Freedman, A. Kitaev, and Z. Wang, *Simulation of topological field theories by quantum computers*, ***Communications in Mathematical Physics***, vol. 227 (2002), pp. 587–603, quant-ph/0001071.

[25] S. Garnerone, A. Marzuoli, and M. Rasetti, *Quantum automata, braid group and link polynomials*, ***quant-ph/0601169***.

[26] L. S. Georgiev, *Topological quantum computation with the universal R matrix for Ising anyons*, ***Lie Theory and Its Applications in Physics VII*** (Sofia) (H. D. Doebner and V. K. Dobrev, editors), Heron Press, 2008, pp. 256–265.

[27] V. G.Turaev, *The Yang-Baxter equations and invariants of links*, ***Inventiones mathematicae***, vol. 92, Fasc. 3, pp. 527–553, LOMI preprint E-3-87, Steklov Institute, Leningrad, USSR.

[28] D. A. Ivanov, *Non-abelian statistics of half-quantum vortices in p-wave superconductors*, ***Physical Review Letters***, vol. 86 (2001), p. 268.

[29] V. F. R. Jones, *Hecke algebra representations of braid groups and link polynomials*, ***Annals of Mathematics***, vol. 126 (1987), pp. 335–338.

[30] ———, *On knot invariants related to some statistical mechanics models*, ***Pacific J. Math.***, vol. 137 (1989), pp. 311–334.

[31] L. H. Kauffman, *Reflexivity and foundations of physics*, ***Search for Fundamental Theory, The VIIth Intenational Symposium Honoring French Mathematical Physicist Jean-Pierre Vigier, Imperial College, London, UK, 12-14 July 2010*** (Melville, N.Y.) (R. Amaroso, P. Rowlands, and S. Jeffers, editors), AIP, American Institute of Physics Pub., pp. 48–89.

[32] ———, *Space and time in computation, topology and discrete physics*, ***Proceedings of the Workshop on Physics and Computation, PhysComp '94, Nov. 1994, Dallas, Texas***, IEEE Computer Society Press, pp. 44–53.

[33] ———, *Sign and space*, ***Religious Experience and Scientific Paradigms. Proceedings of the 1982 IASWR Conference, Stony Brook*** (New York), Institute of Advanced Study of World Religions, 1985, pp. 118–164.

[34] ———, *Self-reference and recursive forms*, ***Journal of Social and Biological Structures***, vol. 10 (1987), pp. 53–72.

[35] ———, *State models and the Jones polynomial*, ***Topology***, vol. 26 (1987), pp. 395–407.

[36] ———, *New invariants in the theory of knots*, ***Amer. Math. Monthly***, vol. 95 (1988), pp. 195–242.

[37] ———, *Statistical mechanics and the Jones polynomial*, ***AMS Contemp. Math. Series***, vol. 78 (1989), pp. 263–297.

[38] ———, *An invariant of regular isotopy*, ***Trans. Amer. Math. Soc.***, vol. 318 (1990), pp. 417–471.

[39] ———, ***Knots and Physics***, World Scientific Publishers, 1991, Second Edition (1993), Third Edition (2002), Fourth Edition (2012).

[40] ———, ***Temperley-Lieb Recoupling Theory and Invariants of Three-Manifolds***, Annals Studies, vol. 114, Princeton University Press, 1994.

[41] ———, *Knot logic*, ***Knots and Applications*** (L. H. Kauffman, editor), World Scientific Pub., 1995, pp. 1–110.

[42] L. H. Kauffman (editor), ***Knots and Applications***, World Scientific Pub., 1995.

[43] L. H. Kauffman (editor), ***The Interface of Knots and Physics***, AMS PSAPM, vol. 51, American Mathematical Society, Providence, RI, 1996.

[44] ———, *Quantum computing and the Jones polynomial*, ***Quantum Computation and Information*** (S. Lomonaco Jr., editor), AMS CONM/305, American Mathematical Society, 2002, math.QA/0105255, pp. 101–137.

[45] ———, *Time imaginary value, paradox sign and space*, ***Computing Anticipatory Systems, CASYS, Fifth International Conference, Liege, Belgium (2001)*** (D. Dubois, editor), AIP Conference Proceedings, vol. 627, 2002.

[46] ———, *Non-commutative worlds*, ***New Journal of Physics***, vol. 6 (2004), pp. 173.1–173.47, (Short version in "Spin, Proceedings of ANPA 25", K. Bowden (ed.), Pub. May 2004).

[47] ———, *Eigenform*, ***Kybernetes, The International Journal of Systems and Cybernetics***, vol. 34 (2005), pp. 129–150.

[48] ———, *Knot diagrammatics*, ***Handbook of Knot Theory*** (Menasco and Thistlethwaite, editors), Elsevier B. V., Amsterdam, 2005, math.GN/0410329, pp. 233–318.

[49] ———, *Teleportation topology*, ***Opt. Spectrosc.***, vol. 9 (2005), pp. 227–232, quant-ph/0407224, (in the Proceedings of the 2004 Byelorus Conference on Quantum Optics).

[50] ———, *Glafka-*2004: *non-commutative worlds*, ***International Journal of Theoretical Physics***, vol. 45 (2006), pp. 1443–1470.

[51] ———, *Reflexivity and eigenform: The shape of process*, ***Kybernetes***, vol. 4(3) (2009).

[52] ———, *Eigenforms, discrete processes and quantum processes*, ***Journal of Physics: Conference Series. EmerQuM11: Emergent Quantum Mechanics 2011 (Heinz von Foerster Congress), Vienna Austria 11-13 November 2011*** (G. Grossing, editor), vol. 361, IOP Publishing, 2012, p. 012034.

[53] L. H. Kauffman and T. Liko, *Knot theory and a physical state of quantum gravity*, ***Classical and Quantum Gravity***, vol. 23 (2006), p. R63, hep-th/0505069.

[54] L. H. Kauffman and S. J. Lomonaco Jr., *Entanglement criteria: Quantum and topological*, ***Quantum Information and Computation, Spie Proceedings, 21-22 April, 2003, Orlando, FL*** (E. Donkor, A. R. Pirich, and H. E. Brandt, editors), vol. 5105, pp. 51–58.

[55] ———, ***Topological Quantum Information Theory***, (Book in preparation).

[56] ———, *Quantum entanglement and topological entanglement*, ***New Journal of Physics***, vol. 4 (2002), pp. 73.1–73.18.

[57] ———, *Braiding operators are universal quantum gates*, ***New Journal of Physics***, vol. 6 (2004), pp. 1–39.

[58] ———, *Quantum knots*, ***Quantum Information and Computation II, Proceedings of Spie, 12 -14 April 2004*** (E. Donkor, A. R. Pirich, and H. E. Brandt, editors), 2004, pp. 268–284.

[59] ———, *q-deformed spin networks, knot polynomials and anyonic topological quantum computation*, ***J. Knot Theory Ramifications***, vol. 16 (2007), pp. 267–332.

[60] ———, *Spin networks and quantum computation*, ***Lie Theory and Its Applications in Physics VII***, Heron Press, Sofia, 2008, pp. 225–239.

[61] ———, *The Fibonacci Model and the Temperley-Lieb Algebra*, ***International J. Modern Phys. B***, vol. 22 (2008), pp. 5065–5080.

[62] ———, *Topological quantum information theory*, ***Proceedings of the AMS Short Course in Quantum Computation and Quantum Information held in Washington, D.C. January 3-4, 2009*** (S. J. Lomonaco Jr., editor), Proceedings of Symposia in Applied Mathematics, vol. 68, American Mathematical Society, 2010, pp. 103–176.

[63] ———, *Quantizing knots groups and graphs*, ***Quantum Information and Computation IX, Spie Proceedings, April 2011*** (H. E. Brandt, E. Donkor, and A. R. Pirich, editors), Proceedings of Spie, vol. 8057, SPIE, 2011, pp. 80570T–1–80570T–15.

[64] L. H. Kauffman and P. Noyes, *Discrete physics and the derivation of electromagnetism from the formalism of quantum mechanics*, ***Proceedings of the Royal Society of London A***, vol. 452 (1996), pp. 81–95.

[65] ———, *Discrete physics and the dirac equation*, ***Physics Letters A***, vol. 218 (1996), pp. 139–146.

[66] L. H. Kauffman and D. E. Radford, *Invariants of 3-manifolds derived from finite dimensional Hopf algebras*, ***Journal of Knot Theory and its Ramifications***, vol. 4 (1995), pp. 131–162.

[67] L. H. Kauffman and F. G. Varela, *Form dynamics*, ***Journal of Social and Biological Structures***, (1980), pp. 171–206.

[68] A. Kitaev, *Anyons in an exactly solved model and beyond*, ***Annals of Physics***, vol. 321 (2006), pp. 2–111, arXiv.cond-mat/0506438 v1 17 June 2005.

[69] T. Kohno, ***Conformal Field Theory and Topology***, AMS Translations of Mathematical Monographs, vol. 210, American Mathematical Society, 1998.

[70] M. Leijnse and K. Flensberg, *Introduction to topological superconductivity and Majorana ferminons*, ***Semiconductor Science and Technology***, vol. 27 (2012), p. 124003, arXiv:1206.1736.

[71] S. J. Lomonaco and L. H. Kauffman, *Quantum knots and mosaics*, ***Journal of Quantum Information Processing***, vol. 7 (2008), pp. 85–115, http://arxiv.org/abs/0805.0339.

[72] ———, *Quantum knots and lattices, or a blueprint for quantum systems that do rope tricks*, ***Quantum Information Science and its Contributions to Mathematics*** (Providence, RI), Proc. Sympos. Appl. Math., vol. 68, Amer. Math. Soc., 2010, pp. 209–276.

[73] ———, *Quantizing braids and other mathematical structures: the general quantization procedure*, ***Quantum Information and Computation IX, Spie Proceedings, April 2011*** (H. E. Brandt, E. Donkor, and A. R. Pirich, editors), Proceedings of Spie, vol. 8057, SPIE, 2011, pp. 805702–1–805702–14.

[74] E. Majorana, *A symmetric theory of electrons and positrons*, ***I Nuovo Cimento***, vol. 14 (1937), pp. 171–184.

[75] A. Marzuoli and M. Rasetti, *Spin network quantum simulator*, ***Physics Letters A***, vol. 306 (2002), pp. 79–87.

[76] G. Moore and N. Read, *Noabelions in the fractional quantum Hall effect*, ***Nuclear Physics B***, vol. 360 (1991), pp. 362–396.

[77] G. Moore and N. Seiberg, *Classical and quantum conformal field theory*, ***Communications in Mathematical Physics***, vol. 123 (1989), pp. 177–254.

[78] V. Mourik, K. Zuo, S. M. Frolov, S. R. Plissard, E. P. A. M. Bakkers, and L. P. Kouwenhuven, *Signatures of Majorana fermions in hybred superconductor-semiconductor devices*, ***arXiv: 1204.2792***.

[79] M. A. Nielsen and I. L. Chuang, ***Quantum Computation and Quantum Information***, Cambridge University Press, 2000.

[80] R. Penrose, *Angular momentum: An approach to combinatorial spacetime*, ***Quantum Theory and Beyond*** (T. Bastin, editor), Cambridge University Press, 1969.

[81] J. Preskill, *Topological computing for beginners*, (*slide presentation*), Technical report, Caltech, http://www.iqi.caltech.edu/~preskill/ph219.

[82] N. Y. Reshetikhin and V. Turaev, *Ribbon graphs and their invariants derived from quantum groups*, ***Communications in Mathematical Physics***, vol. 127 (1990), pp. 1–26.

[83] ———, *Invariants of three manifolds via link polynomials and quantum groups*, ***Inventiones mathematicae***, vol. 103 (1991), pp. 547–597.

[84] M. Roetteles, (Private conversation, fall 2003).

[85] C. Rovelli and L. Smolin, *Spin networks and quantum gravity*, ***Physical Review D***, vol. 52 (1995), pp. 5743–5759.

[86] P. Rowlands, ***Zero to Infinity: The Foundations of Physics***, Series on Knots and Everything, vol. 41, World Scientific Publishing Company, 2007.

[87] B. Schmeikal, ***Primordial Space: Point-free Space and Logic Case***, Nova Publishers, 2012.

[88] B. Schumacher, ***Communication, Correlation, and Complementarity***, Ph.D. thesis, University of Texas at Austin, 1990.

[89] P. W. Shor and S. P. Jordan, *Estimating Jones polynomials is a complete problem for one clean qubit*, ***Quantum Information & Computation***, vol. 8 (2008), pp. 681–714.

[90] S. H. Simon, N. E. Bonesteel, M. H. Freedman, N. Petrovic, and L. Hormozi, *Topological quantum computing with only one mobile quasiparticle*, ***Physical Review Letters***, vol. 96 (2006), pp. 070503, 4 pp., quant-ph/0509175.

[91] L. Smolin, *Link polynomials and critical points of the chern-simons path integrals*, ***Modern Physics Letters A***, vol. 4 (1989), pp. 1091–1112.

[92] G. Spencer-Brown, ***Laws of Form***, George Allen and Unwin Ltd., London, 1969.

[93] V. G. Turaev and O. Viro, *State sum invariants of 3-manifolds and quantum 6j symbols*, ***Topology***, vol. 31 (1992), pp. 865–902.

[94] F. Wilczek, ***Fractional Statistics and Anyon Superconductivity***, World Scientific Publishing Company, 1990.

[95] E. Witten, *Quantum field Theory and the Jones Polynomial*, ***Communications in Mathematical Physics***, vol. 121 (1989), pp. 351–399.

[96] L. Wittgenstein, ***Tractatus Logico-Philosophicus***, New York: Harcourt, Brace and Company, Inc., London: Kegan Paul, Trench, Trubner and Co. Ltd., 1922.

[97] P. Wocjan and J. Yard, *The Jones polynomial: quantum algorithms and applications in quantum complexity theory*, ***quant-ph/0603069***.

[98] C. N. Yang, *Some exact results for the many-body problem in one dimension with repulsive delta-function interaction*, ***Physical Review Letters***, vol. 19 (1967), p. 1312.

[99] Y. Zhang, L. H. Kauffman, and M. L. Ge, *Yang-Baxterizations, universal quantum gates and Hamiltonians*, ***Quantum Information Processing***, vol. 4 (2005), pp. 159–197.

DEPARTMENT OF MATHEMATICS, STATISTICS AND COMPUTER SCIENCE (M/C 249)
UNIVERSITY OF ILLINOIS AT CHICAGO
851 SOUTH MORGAN STREET
CHICAGO, ILLINOIS 60607-7045
E-mail: kauffman@uic.edu

INDEX